S Baker

# PRACTICAL NAVIGATION FOR SECOND MATES

Including chartwork to cover the
practical navigation and
chartwork papers for
D.O.T. certificates
Class V, Class IV, and Class III

# Practical Navigation for Second Mates

Including chartwork to cover the
practical navigation and
chartwork papers for
D.O.T. certificates
Class V, Class IV, and Class III

BY

A. FROST, B.Sc., MASTER MARINER, M.R.I.N.

GLASGOW
BROWN, SON & FERGUSON LTD. NAUTICAL PUBLISHERS
4-10 DARNLEY STREET

Copyright in all countries signatory to the Berne Convention
All rights reserved

| | |
|---|---|
| *First Edition* | 1955 |
| *Second Edition* | 1969 |
| *Third Edition* | 1974 |
| *Fourth Edition* | 1977 |
| *Fifth Edition* | 1981 |
| *Reprinted* | 1985 |
| *Reprinted* | 1991 |
| *Reprinted* | 1994 |
| *Reprinted* | 1996 |

ISBN 0 85174 397 8
ISBN 0 85174 300 5 (4th edition)
ISBN 0 85174 221 4 (3rd edition)

©1996—BROWN, SON & FERGUSON, LTD., GLASGOW, G41 2SD
*Printed and Made in Great Britain*

## FOREWORD TO THE FIFTH EDITION

In this revision 'Practical Navigation for Second Mates' has been extended to include chartwork and tidal calculations. The intention in doing this was to provide a text for candidates preparing for the chartwork and practical navigation of the Department of Trade Class V and Class IV certificates. Worked examples of all problems and calculations encountered in these papers together with ample exercises for self examination, are included, and specimen papers similar to Class V 'Chartwork and Practical Navigation', Class IV 'Chartwork', and Class IV 'Practical Navigation' are provided.

All practical navigation problems and tidal calculations used may be worked with the extracts from the Nautical Almanac, and the extracts from the Admiralty Tide Tables included. Chartwork exercises however, inevitably require the use of a chart. Such exercises have been set on Admiralty charts 1179, 5050 or 5051, as these particular charts are published as inexpensive instructional charts, which, although not suitable for navigation, are full size and authentic and entirely adequate for practice purposes.

CARDIFF JUNE 1980                                                             A. FROST


## ACKNOWLEDGEMENTS

Extracts from the 1980 Nautical Almanac are published by permission of the Controller of Her Majesty's Stationery Office.

Extracts from the Admiralty Tide Tables 1980, Avonmouth, Devonport and Milford Haven, are published with permission of the Hydrographer of the Navy.

# CONTENTS

## Introduction

### Section 1

CHAPTER 1—THE MEASUREMENT OF POSITION AND DISTANCE ON THE EARTH'S SURFACE
Latitude and longitude. Difference of latitude, difference of longitude. The sea mile. The nautical mile.     1

CHAPTER 2—THE MEASUREMENT OF DIRECTION. THE EFFECT OF WIND AND TIDE.
The three figure notation. The gyro compass. The magnetic compass. Variation and deviation. Compass error. The three norths. Relative bearing. The deviation table and its use. The effect of tide. Course made good. Counteracting the tide. To reach a position in a required time while counteracting a tide. To find the set and drift between two observed positions.
To find the course to steer to pick up a point of land at a required angle on the bow.
The effect of wind. Leeway. Counteracting the leeway.     8

CHAPTER 3—POSITION LINES
The position line defined. Great circle bearings. Mercator bearings. The half convergency correction.
The position circle. Vertical sextant angles. Horizontal sextant angles. Rising and dipping distances:—Ranges of lights.
The transit bearing. Danger angles.
The hyperbolic position line.     33

CHAPTER 4—THE SAILINGS.
Parallel sailing. Plane sailing. The middle latitude. Mercator sailing.     50

CHAPTER 5—THE TRAVERSE TABLE. THE TRANSFERRED POSITION LINE.
The traverse table and its use. Running up the D.R. position.
Transferring the position line. The running fix.
The running fix with tide.
Transferring the position circle.
Transferring the position line by traverse table.
Doubling the angle on the bow.     65

# CONTENTS

CHAPTER 6—TIDES AND TIDAL STREAMS.
Tidal streams and currents. Tidal stream information on Admiralty charts. Tidal stream atlases.
Tides. Prediction of tidal times and heights.
To find the times and heights of high and low waters at a standard port. To find the height of tide at times between high and low water at a standard port. To find the time at which there will be a given depth of water at a standard port.
Secondary ports. To find the times and heights of high and low water at a secondary port. To find the height of tide at times between high and low water at a secondary port. To find the time when there will be a given depth of water at a secondary port. Drying heights. Heights of terrestrial objects. 84

CHAPTER 7—THE CELESTIAL SPHERE AND THE NAUTICAL ALMANAC.
The celestial sphere. Definition of position on the celestial sphere. The geographical position. Definition of a celestial bodies G.P. on the terrestrial sphere.
The Nautical Almanac. To extract the declination and GHA. To find the times of meridian passage. To find the times of sunrise and sunset, and moonrise and moonset. 101

CHAPTER 8—COMPASS ERROR BY ASTRONOMICAL OBSERVATION.
The azimuth problem and the amplitude problem.
Specimen papers for Class V chartwork and practical navigation, and for Class IV chartwork. 122

**Section 2**

CHAPTER 9—POSITION CIRCLES AND POSITION LINES.
The Marcq St. Hilaire method.
Plotting of position lines. To plot the position from two simultaneous position lines by Marcq St. Hilaire. To plot the position from two position lines by Marcq St. Hilaire with a run between.
The Longitude by Chronometer method. To plot the position from two simultaneous position lines by longitude by chronometer.
To plot the position from two position lines by longitude by chronometer with a run between.
Position line by meridian altitude observation. Plotting the noon position from a forenoon observation and meridian observation.
Plotting the position by celestial observation combined with a terrestrial observation. 141

## CONTENTS

CHAPTER 10—CORRECTION OF ALTITUDES.
The individual corrections. Correction of altitudes by individual corrections. Total correction tables. Correction of altitudes by total correction. ... 157

CHAPTER 11—LATITUDE BY MERIDIAN ALTITUDE.
Calculation of the latitude by observation of a body on the observer's meridian. To find the altitude to set on a sextant to observe the meridian passage.
Lower meridian passage. Calculation of the latitude by observation of a body on the observer's lower meridian. ... 170

CHAPTER 12—CALCULATION OF POSITION LINES BY OBSERVATION OF BODIES OUT OF THE MERIDIAN.
The PZX triangle. The Marcq St. Hilaire method. Calculation of intercept and bearing. The Longitude by Chronometer method. Noon position by forenoon observation and meridian observation. Use of the 'c' correction to find the noon longitude. ... 182

CHAPTER 13—THE EX MERIDIAN PROBLEM.
Calculation of position lines by observation of bodies near the meridian. Ex meridian tables. ... 202

CHAPTER 14—THE POLE STAR PROBLEM.
Calculation of the latitude by observation of the pole star. ... 211

CHAPTER 15—GREAT CIRCLE SAILING.
Calculation of courses and distance by great circle sailing. Calculation of courses and distance by composite great circle sailing. ... 215

Specimen papers in practical navigation for Class IV. ... 227
Answers to exercises. ... 232
Extracts from the Nautical Almanac 1980. ... 245
Extracts from the Admiralty Tide Tables 1980. ... 276
Index ... 279

# SECTION 1

The work contained in this section is required for the 'Chartwork and Practical Navigation' paper for Department of Trade Class V certificate.

# CHAPTER 1

# THE MEASUREMENT OF POSITION AND DISTANCE ON THE EARTH'S SURFACE

## Definitions of terms used in the measurement of position
*Great Circle.* A circle drawn on the surface of a sphere, whose plane passes through the centre of the sphere. It is the largest circle that can be drawn on a sphere, and given any two points on the sphere, there is only one great circle that can be drawn through those points. The shortest distance between those two points will lie along the shorter arc of that great circle. The exception is if the two points are at opposite ends of the same diameter, and in this case an infinite number of great circles can be drawn passing through the two points.
*Small Circle.* Any circle drawn on the surface of a sphere, whose plane does not pass through the centre of the sphere.

## Geographical poles of the earth
Those points at which the axis of the earth's rotation cuts the earth's surface.

## The measurement of position
To define a position on any surface we require two axes of reference, usually at right angles to each other. The definition of a point is obtained by stating the distance of the point from each of these axes. Thus in the construction of a mathematical graph, we draw an $x$-axis and a $y$-axis at right angles to each other and the co-ordinates of any position on the graph give us the distance of that position from the $x$-axis and from the $y$-axis. So defined, the position is unambiguous. On the earth's surface we use two great circles as the axes of reference, and we use angular distances instead of linear distances.

The great circles used are:
*The Equator.* A great circle on the earth's surface, the plane of which is at right angles to the axis of rotation of the earth. Every point on the equator is at an angular distance of 90° from each pole.
*The Prime Meridian.* This is a semi great circle on the earth's surface which runs between the two geographical poles, and passes through an arbitrary point in Greenwich. Any semi great circle which runs between the poles is called a meridian. All meridians

cut the equator at their mid point at right angles, and all meridians intersect at the poles.

The position of any point on the earth is defined as an angular distance from these two circles, the two co-ordinates being called Latitude and Longitude.

*Parallel of Latitude*. A small circle on the earth's surface, the plane of which is parallel to that of the equator.

*The Latitude* of any point is the arc of any meridian contained between the equator and the parallel of latitude through the point. Thus all positions on the same parallel of latitude have the same latitude. The latitude of the equator is 0° and that of each pole is 90°. Latitude is named North or South of the equator.

*The Longitude* of any point is the lesser arc of the equator contained between the prime meridian and the meridian which passes through the point. It is measured from 0° to 180° on either side of the prime meridian and named East or West.

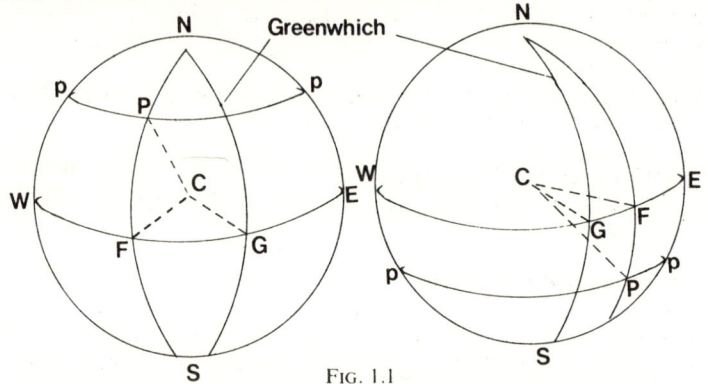

FIG. 1.1

Latitude North, Longitude West                    Latitude South, Longitude East

NGS   = Prime Meridian                      NPFS = Meridian through P
WFGE = Equator                              Angle PCF or arc PF = Latitude of P
pp    = Parallel of latitude through P      Angle FCG or arc FG = Longitude of P

When sailing between any two positions on the earth's surface, a knowledge of the necessary change in latitude and change in longitude between the two positions is essential.

The Difference of Latitude (d. lat.) between any two positions is the arc of a meridian which is contained between the two parallels of latitude through the positions. From figure 1.2, if the two positions are on the same side of the equator (latitudes same name), then the d. lat. will be the numerical difference between the two latitudes. If they are on opposite sides of the equator (latitudes different names), then the d. lat. will be the sum of the two latitudes.

# POSITION AND DISTANCE

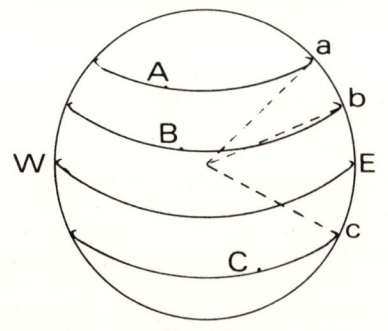

Fig. 1.2

WE = Equator
Aa, Bb, Cc = Parallels of latitude through A, B and C
Then D. lat. between A and B = Lat A − Lat B = arc ab
D. lat. between A and C = Lat A + Lat C = arc ac
D. lat. between B and C = Lat B + Lat C = arc bc

D. lat. is named according to the direction travelled. North or South.

The Difference of Longitude (d. long.) between any two positions is the lesser arc of the equator contained between the two meridians which pass through the positions. If the longitudes of the points lie on the same side of the prime meridian as each other then the d. long. will be the numerical difference between the longitudes (longitudes same name). If they lie on opposite sides of the prime meridian the d. long. will be the sum of the two longitudes (longitudes opposite name). If, however, the d. long. found thus is greater than 180°, as the d. long. is the LESSER arc of the equator between the two positions, then the d. long. is found by subtracting this sum from 360° (see examples). The d. long. is named according to the direction travelled East or West.

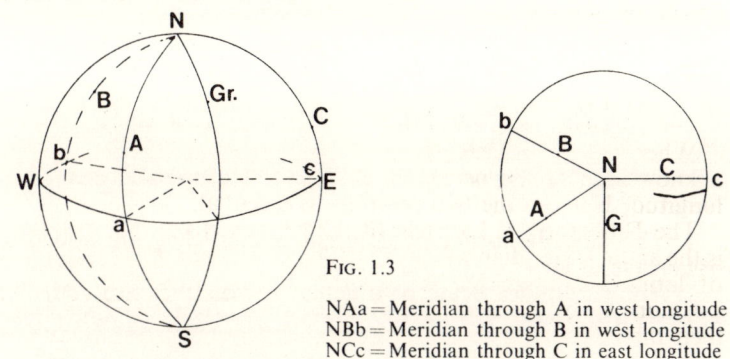

Fig. 1.3

NAa = Meridian through A in west longitude
NBb = Meridian through B in west longitude
NCc = Meridian through C in east longitude

Then D. long. between A and B = Long. A − Long. B = arc ba
D. long. between A and C = Long. A + Long. C = arc ac
D. long. between B and C = 360° − (Long. B + Long. C) = arc bc

# 4    PRACTICAL NAVIGATION FOR SECOND MATES

**Note**

D. lats. and d. longs. are usually required in minutes of arc. They are therefore expressed thus in the following examples.

**Examples**

1. Find the d. lat. and d. long. between lat. 25° 46′ N., long. 15° 28′ W. and lat. 52° 56′ N., long. 39° 47′ W.

```
A lat.   = 25° 46′ N.       A long.  = 15° 28′ W.
B lat.   = 52° 56′ N.       B long.  = 39° 47′ W.
         ―――――――――                   ―――――――――
d. lat.  = 27° 10′ N.       d. long. = 24° 19′ W.
           60                          60
         ―――――――――                   ―――――――――
         = 1630′ N.                  = 1459′ W.
         ═════════                   ═════════
```

The degrees are multiplied by 60 to change them into minutes and the odd minutes are added on.

2. Find the d. lat. and d. long. between lat. 44° 25′ N., long. 75° 46′ W., and lat. 36° 19′ S., long. 09° 26′ W.

```
A lat.   = 44° 25′ N.       A long.  = 75° 46′ W.
B lat.   = 36° 19′ S.       B long.  = 09° 26′ W.
         ―――――――――                   ―――――――――
d. lat.  = 80° 44′ S.       d. long. = 66° 20′ E.
           60                          60
         ―――――――――                   ―――――――――
         = 4844′ S.                  = 3980′ E.
         ═════════                   ═════════
```

3. Required the d. lat. and d. long. made good by a vessel which sails from position A 22° 10′ S., 09° 15′ W., to a position B, 15° 30′ N., 29° 30′ E.

```
A lat.   = 22° 10′ S.       long.    = 09° 15′ W.
B lat.   = 15° 30′ N.       long.    = 29° 30′ E.
         ―――――――――                   ―――――――――
d. lat.  = 37° 40′ N.       d. long. = 38° 45′ E.
         = 2260′ N.                  = 2325′ E.
         ═════════                   ═════════
```

**Notes**

1. The latitudes being of different names, they are added to obtain the d. lat.
2. The longitudes being of different names, they are added to obtain the d. long.

4. A vessel steams from position P 18° 40′ S., 136° 40·6′ W., to position Q 31° 15·2′ S., 126° 35·8′ E. Find the d. lat. and the d. long.

```
P  lat. = 18° 40·0′ S.      long.    = 136° 40·6′ W.
Q  lat. = 31° 15·2′ S.      long.    = 126° 35·8′ E.

   d. lat. = 12° 35·2′ S.   d. long. = 263° 16·4′
          = 755·2′ S.                  360°

                                     =  96° 43·6′ W.
                                     = 5803·6′ W.
```

## Notes

1. The vessel is steaming from a West longitude across the 180° meridian to a position in East longitude, and is therefore proceeding in a westerly direction.

2. The d. long. could have been obtained by adding the differences of each longitude from 180°.

**5.** A vessel steams on a course which lies between North and East, and makes a d. lat. of 925·8′ N., and a d. long. of 1392·6′ E. The initial position was 25° 20·7′ N., 46° 45·2′ W. Find the D.R. position.

```
    lat.    = 25° 20·7′ N.     long.    = 46° 45·2′ W.
    d. lat. = 15° 25·8′ N.     d. long. = 23° 12·6′ E.

    D.R. lat. = 40° 46·5′ N.   long.    = 23° 32·6′ W.
```

## Note

The d. lat., having the same name as the latitude, is added to it, while the d. long. being of opposite name to the longitude, is subtracted from it.

## EXERCISE 1A

### Find the d. lat. and d. long between the following positions

|    |   | Latitude   | Longitude   |   | Latitude   | Longitude   |
|----|---|------------|-------------|---|------------|-------------|
| 1. | P | 40° 10′ N. | 9° 25′ W.   | Q | 47° 15′ N. | 21° 14′ W.  |
| 2. | A | 35° 15′ N. | 22° 12′ W.  | B | 50° 25′ N. | 11° 37′ W.  |
| 3. | X | 10° 12′ N. | 5° 03′ E.   | Y | 5° 18′ S.  | 7° 18′ W.   |
| 4. | L | 20° 40′ S. | 170° 09′ E. | M | 13° 06′ N. | 178° 51′ E. |
| 5. | A | 30° 03′ N. | 152° 43′ W. | B | 42° 24′ N. | 174° 01′ W. |
| 6. | F | 11° 31′ N. | 178° 00′ E. | K | 5° 14′ S.  | 177° 00′ W. |
| 7. | A | 8° 42′ S.  | 162° 41′ W. | Z | 7° 53′ N.  | 135° 27′ E. |
| 8. | B | 15° 20′ S. | 130° 35′ E. | K | 33° 10′ N. | 155° 40′ W. |
| 9. | V | 52° 10′ S. | 171° 08′ E. | W | 27° 02′ S. | 34° 02′ E.  |
| 10.| L | 60° 40′ S. | 151° 23′ W. | M | 10° 57′ S. | 92° 47′ W.  |

## EXERCISE 1B

1. The initial longitude is 4° 30′ W. and the d. long. is 104′ E. Find the final longitude.
2. Initial lat.=20° 50′ S., long.=178° 49′ E., d. lat.=33° 14′ N., d. long.=15° 37′ E. Find the final position.
3. Initial lat.=39° 40′ N., long.=9° 21′ W., d. lat.=3° 57′ N., d. long.=27° 07′ E. Find the final position.
4. Final position lat.=30° 10·6′ S., long.=4° 40·3′ E., d. lat. was 72° 18·8′ S., and d. long. was 38° 54·7′ E. What was the initial position?
5. A ship steered a course between N. and E. making a d. lat. of 38° 55·5′ and a d. long. of 20° 41·8′. If the final position was lat. 21° 10·4′ N., long. 168° 18·7′ W., what was the initial position?

**The measurement of distance**

The measurements and calculations required to find position on the earth's surface are in units of angular measure. Position on the celestial sphere and on the earth's surface is defined in the same units of angular measure. It is convenient therefore, at sea, to use as a unit of linear distance, the length of one minute of arc of a great circle on the surface of the earth. The great circles used are the terrestrial meridians so that the latitude scale of a navigator's chart becomes his scale of distance, one minute of latitude being equal to one mile.

The exact length of the mile however varies due to the fact that the earth is not a true sphere but an oblate spheroid. The earth is flattened at the poles and bulges at the equator due to the forces of its own rotation. A meridian and its opposite meridian form therefore an approximate ellipse rather than a circle.

The nautical mile or 'sea mile' is defined as the length of a meridian which subtends an angle of one minute at the centre of curvature of that part of the meridian being considered. Figure 1.4 shows that because of the flattening at the poles, the radius of curvature of the polar regions is greater than that at the equatorial regions. The linear distance represented by an arc of one minute is therefore greater at the poles than at the equator.

The length of a sea mile at the equator is approximately 1842·9 metres.

The length of a sea mile at the poles is approximately 1861·7 metres.

This variation is of little or no significance in practical navigation at sea and a standard length close to the mean value is adopted. The International Nautical Mile is an adopted value of 1852 metres.

In practical navigation any unit of d'lat is taken as a nautical mile,

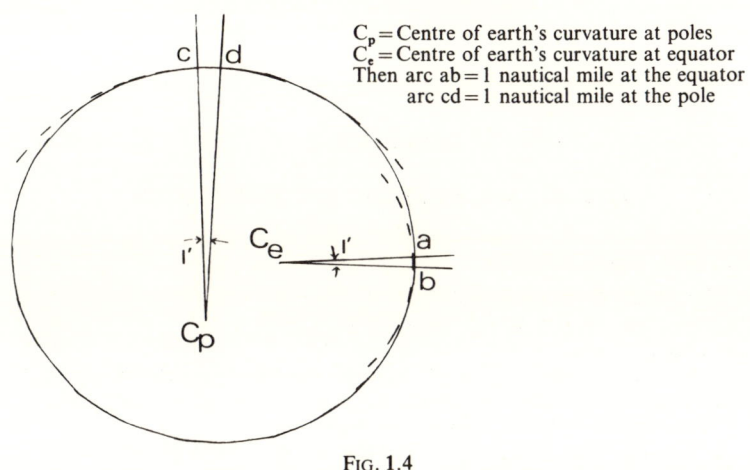

$C_p$ = Centre of earth's curvature at poles
$C_e$ = Centre of earth's curvature at equator
Then arc ab = 1 nautical mile at the equator
arc cd = 1 nautical mile at the pole

Fig. 1.4

so that the difference of latitude between two places on the same meridian is, when expressed in minutes of arc, equal to the distance between them in nautical miles. Units of d'lat and distance are consistent units and may be used as such in any navigational formula.

The unit of speed at sea is the nautical mile per hour. This unit is called a knot.

The equator is the one great circle on the earth's surface which is actually a true circle. One minute of arc of the equator is therefore of constant length, about 1855·3 metres. This unit is called a geographical mile.

# CHAPTER 2

# THE MEASUREMENT OF DIRECTION

### The three figure notation

The observer is imagined to be at the centre of his compass and the direction of the north geographical pole is taken to be 000°. The observer's horizon is divided into 360°, and any direction from the observer is expressed as a number of degrees measured clockwise from the direction of north.

Three figure notation is used to express:
1. Course. The direction of movement of the observer.
2. Bearing. The direction of an object from the observer.

Any instrument designed for the measurement of direction is called a compass. To measure direction correctly the zero mark of the compass must point towards the zero of direction, i.e. the direction of the north pole.

This is not always the case. If it is not then the direction that the compass zero points in must be ascertained in order to apply the necessary correction.

### The Gyro Compass

Gyroscopic compasses are liable to small variable errors, which should never exceed one or two degrees.

If the zero mark, or north point of the compass card points to the left (to the West) of true North then all indications of direction taken from the card will be greater than the true value.

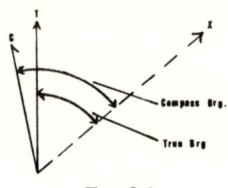

Fig. 2.1

In this case the gyro is said to be reading high, and any compass error will be negative to the compass reading to obtain the true reading.

If the north point of the compass card is pointing to the right, or East, of the true North direction, then all readings taken from the compass card will be less than the true value.

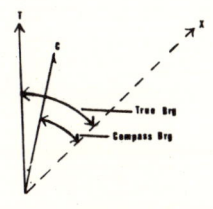

FIG. 2.2

In this case the gyro is said to be reading low, and any compass error will be additive to compass reading to obtain true reading.

Methods of calculating the value of the error will be explained in a later chapter. (See 'The Azimuth Problem'.)

## The Magnetic Compass

*Variation*

The magnetic poles of the earth are not coincident with the geographical poles. The north point of the compass therefore will not point towards the true direction of North.

The direction in which the compass needle aligns itself can be thought of as the magnetic meridian. The angle between the true meridian and the magnetic meridian is called the VARIATION, and this angle varies with position on the earth's surface. It is named WEST if the compass needle points to the left of true North, and EAST if the compass needle points to the right of true North.

*Deviation*

The compass needle will only align itself with the magnetic meridian if it is free from all other influences except the magnetic field of the earth.

This is rarely so, particularly on a ship which is constructed of steel. The magnetism induced in the steel by the earth's magnetic field causes the compass needle to deviate from the magnetic meridian, by an amount which is called the deviation. This will vary for any particular vessel for a number of reasons such as course, angle of heel, position on the earth's surface. Deviation is named WEST if the compass needle points to the left of the magnetic meridian, and EAST if the compass needle points to the right of the magnetic meridian. The direction of the magnetic meridian is called Magnetic North. The direction indicated by the compass needle is called Compass North.

## Compass Error

The actual error of the compass at any time will be the combination of the variation and the deviation. If they are of the same name then the error will be the sum of the two and it will be named as they are. If they are of different names the compass error will be the difference between the two and will be named as the greater of the two.

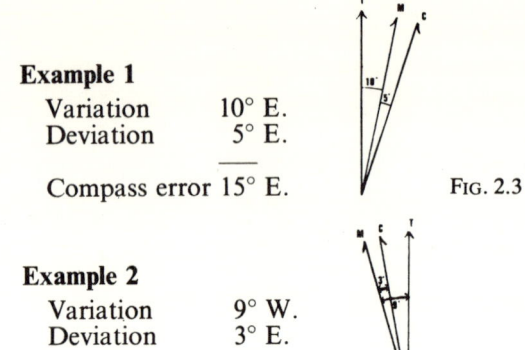

### Example 1
  Variation       10° E.
  Deviation        5° E.
  —————————————
  Compass error  15° E.

FIG. 2.3

### Example 2
  Variation        9° W.
  Deviation        3° E.
  —————————————
  Compass error   6° W.

FIG. 2.4

## EXERCISE 2A

**Find the compass error given**

1. Dev. 15° W., Var. 30° E.
2. Dev. 14° E., Var. 5° E.
3. Dev. 3° W., Var. 30° W.
4. Dev. 5° W., Var. 25° W.
5. Dev. 6° W., Var. 20° E.
6. Dev. 10° W., Var. 5° W.
7. Dev. 21° W., Var. 4° E.
8. Dev. 8° E., Var. 8° W.
9. Dev. 5° W., Var. 50° W.
10. Dev. 3° E., Var. 35° E.

We have defined three directions which we can call north.

*True North.* The direction of the north geographical pole.
*Magnetic North.* The direction of the magnetic meridian at any place.
*Compass North.* The direction indicated by the north point of the compass.

The difference between True North and Magnetic North is the variation.

The difference between Magnetic North and Compass North is the deviation.

The difference between True North and Compass North is the compass error.

# THE MEASUREMENT OF DIRECTION

Any course or bearing can be denoted using any of these three directions of north.

*True Course or Bearing.* The angle at the observer between the direction of True North and the direction being measured, measured clockwise from North.

*Magnetic Course or Bearing.* The angle at the observer between the direction of magnetic meridian and the direction being measured, measured clockwise from North.

*Compass Course or Bearing.* The angle at the observer between the direction of compass north and the direction being measured, measured clockwise from North.

The angle indicated by the compass is the compass course or bearing and this must be corrected to true course or bearing, before use.

If the compass error is west the compass course or bearing will be greater than the true course or bearing.

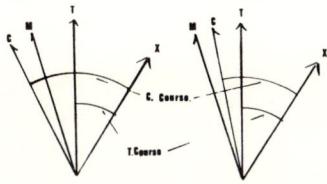

Fig. 2.5

If the compass error is east the compass course or bearing will be less than the true course or bearing.

Fig. 2.6

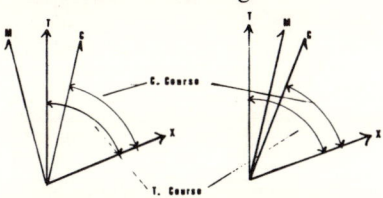

From which may be deduced the mnemonic:
Error WEST, compass BEST
Error EAST, compass LEAST

**Note**

**Deviation is dependent upon course, or ship's head. For any particular ship's head the deviation will be the same for ALL BEARINGS.**

**Example 1**

A vessel is steering 240° by compass. Deviation for the ship's head is 10° E. Variation for the place is 20° W. Find the true course.

12    PRACTICAL NAVIGATION FOR SECOND MATES

| Compass course | 240° | *or* | Variation | 20° W. |
| Deviation | 10° E. | | Deviation | 10° E. |

| Magnetic course | 250° | | Compass error | 10° W. |
| Variation | 20° W. | | Compass course | 240° |

| True course | 230° | | True course | 230° |

FIG. 2.7

### Example 2

Find the compass course to steer to make good a True Course of 130° if the variation is 20° W., and the deviation is 10° E.

| True course | 130° | *or* | Variation | 20° W. |
| Variation | 20° W. | | Deviation | 10° E. |

| Magnetic course | 150° | | Compass error | 10° W |
| Deviation | 10° E. | | True course | 130° |

| Compass course | 140° | | Compass course | 140° |

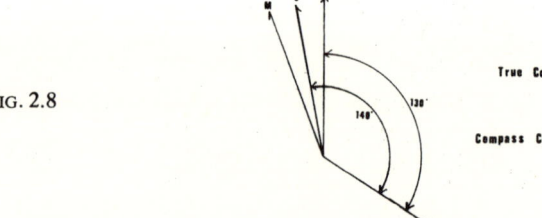

FIG. 2.8

Variation for any particular place is found either from the centre of the compass rose on the Admiralty chart of the area, or from the Admiralty variation charts.

Deviation is obtained from a deviation card compiled for a particular compass by the compass adjuster, or by direct observation as explained in a later chapter.

# THE MEASUREMENT OF DIRECTION

## EXERCISE 2B

**Find the true course**

|     | Course   | Dev.   | Var.   |
| --- | -------- | ------ | ------ |
| 1.  | 226° C.  | 3° W.  | 16° W. |
| 2.  | 010° C.  | 1° W.  | 18° W. |
| 3.  | 358° C.  | 2° E.  | 15° W. |
| 4.  | 267° C.  | 4° W.  | 20° E. |
| 5.  | 034° C.  | 3° E.  | 15° W. |
| 6.  | 332° C.  | 4° W.  | 10° W. |
| 7.  | 116° C.  | 2° W.  | 8° W.  |
| 8.  | 218° C.  | 3° W.  | 11° W. |
| 9.  | 084° C.  | 5° W.  | 17° E. |
| 10. | 178° C.  | 6° E.  | 11° E. |

## EXERCISE 2C

**Find the compass course**

|     | Course   | Dev.   | Var.   |
| --- | -------- | ------ | ------ |
| 1.  | 222° T.  | 4° E.  | 15° E. |
| 2.  | 356° T.  | 5° W.  | 20° W. |
| 3.  | 172° T.  | 3° E.  | 18° W. |
| 4.  | 200° T.  | 2° E.  | 1° W.  |
| 5.  | 005° T.  | 1° E.  | 5° E.  |
| 6.  | 086° T.  | 1° W.  | Nil    |
| 7.  | 106° T.  | 2° W.  | 10° W. |
| 8.  | 173° T.  | 3° E.  | 8° W.  |
| 9.  | 306° T.  | 2° W.  | 11° W. |
| 10. | 185° T.  | 3° W.  | 10° W. |

**Given the error and the variation to find the deviation**

If, when the error and variation are given, it is desired to find the deviation, then the variation must be subtracted from the error as the error is the sum of the two. The variation may be subtracted by reversing its name. The deviation is then named according to the greater.

The following examples indicate the method:

# 14 PRACTICAL NAVIGATION FOR SECOND MATES

**Examples**

| | | | | | |
|---|---|---|---|---|---|
| error | 20° W. | error | 6° E. | error | 0° W. |
| var. | 15° W. (E.) | var. | 20° E. (W.) | var. | 5° E. (W.) |
| dev. | 5° W. | dev. | 14° W. | dev. | 5° W. |

| | | | |
|---|---|---|---|
| error | 10° E. | error | 20° E. |
| var. | 15° W. (E.) | var. | 6° E. (W.) |
| dev. | 25° E. | dev. | 14° E. |

## EXERCISE 2D

**Find the deviation given**

1. error 3° E., var. 21° W.
2. error 15° W., var. 24° W.
3. error 37° E., var. 34° E.
4. error 11° W., var. 7° W.
5. error 23° E., var. 25° E.
6. error 34° W., var. 39° W.
7. error 2° W., var. 12° W.
8. error 7° E., var. 9° W.
9. error 24° W., var. 30° W.
10. error Nil    var. 5° E.

**Given the true bearing and the compass bearing of a body, also the variation, to find the deviation**

Remember that if the error is East, it is added to a compass direction to obtain the true direction; it will be noted that the latter must be numerically greater than the former. Therefore, if the error is to be found, the rule is:

> True greater than Compass—Error is East
> Compass greater than True—Error is West

**Example 1**

The sun bore 120° T. and 110° C., find the compass error, and if the variation was 10° W., find the deviation.

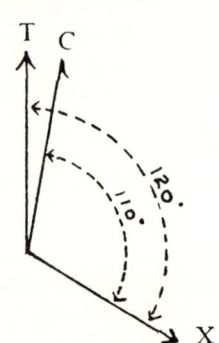

☉ bearing = 110° C.
  bearing = 120° T.

Error = 10° E.
Var.  = 10° W.

Dev.  = 20° E.

*Note.* Compass Least—
        Error East

FIG. 2.9

## THE MEASUREMENT OF DIRECTION

**Example 2**

The sun's true amplitude is W. 10° 20′ S. and the observed amplitude W. 20° N. Find the compass error, and if the variation is 25° W., find the deviation.

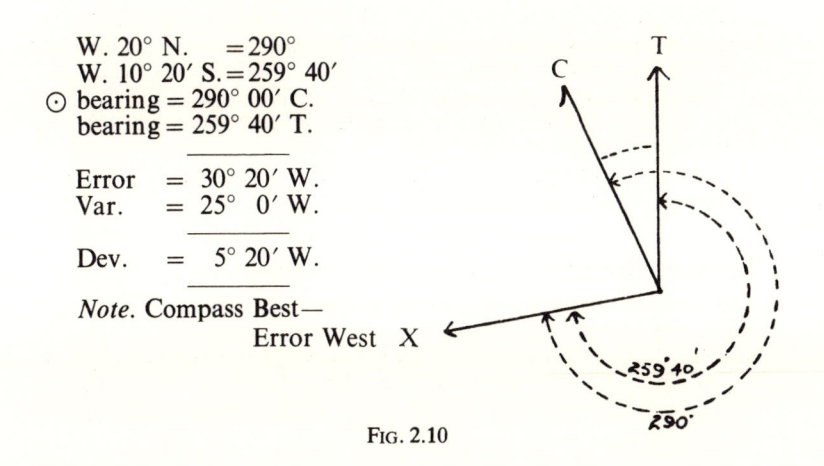

W. 20° N.   = 290°
W. 10° 20′ S. = 259° 40′
☉ bearing = 290° 00′ C.
  bearing = 259° 40′ T.

Error = 30° 20′ W.
Var.  = 25°  0′ W.

Dev.  =  5° 20′ W.

*Note.* Compass Best—
       Error West   X

Fig. 2.10

### EXERCISE 2E

**Find the deviation**

|     | Compass bearing | True bearing | Variation |
|-----|-----------------|--------------|-----------|
| 1.  | 050° C.         | 060° T.      | 12° E.    |
| 2.  | 010° C.         | 005° T.      | 11° W.    |
| 3.  | 075° C.         | 060° T.      | 19° W.    |
| 4.  | 140° C.         | 115° T.      | 24° W.    |
| 5.  | 242° C.         | 248° T.      | 13° E.    |
| 6.  | 201° C.         | 201° T.      | 8° E.     |
| 7.  | 309° C.         | 322° T.      | 8° E.     |
| 8.  | 037° C.         | 022° T.      | 12° W.    |
| 9.  | 341° C.         | 320° T.      | 23° W.    |
| 10. | 289° C.         | 310° T.      | 33° E.    |
| 11. | 260° C.         | 294° T.      | 49° E.    |
| 12. | 134° C.         | 120° T.      | 21° W.    |

16    PRACTICAL NAVIGATION FOR SECOND MATES

|  |  |  |  |
|---|---|---|---|
| 13. | 163° C. | 200° T. | 62° E. |
| 14. | 219° C. | 175° T. | 40° W. |
| 15. | 278° C. | 262° T. | 11° W. |

## Relative bearing

Bearings measured by pelorus, which is a 'dummy' compass card whose zero mark is aligned with the vessel's fore and aft line, are said to be relative bearings, that is, relative to the ship's head.

The relative bearing may be defined as the angle at the observer measured clockwise from the direction of the ship's head, to the direction of the point observed.

A relative bearing is also obtained from measurements by radio direction finder. (see Chapter 3).

## To convert a relative bearing to a true bearing

In order to obtain a true bearing from a relative measurement, the vessel's true heading at the time of the observation must be applied. It is not sufficient to apply the ship's course steered, as at the moment of observation the vessel may be one or more degrees off course. The heading should be observed at the instant of the observation.

True bearing=True ship's head+Relative bearing.
(–360° if necessary)

## Example 1

A relative bearing of 105°, (object 105° on the starboard bow), was observed on a ship whose heading at that time was 085° T. Find the true bearing.

| Relative bearing | 105° |
|---|---|
| True ship's head | 085°T. |
| True bearing | 190°T. |

## Example 2

A relative bearing of 248° was observed from a vessel whose true heading at the time was 176°T. Find the true bearing.

| Relative bearing | 248° |
|---|---|
| True ship's head | 176 T. |
| True bearing | 424° |
|  | –360° |
|  | =064°T. |

# THE MEASUREMENT OF DIRECTION

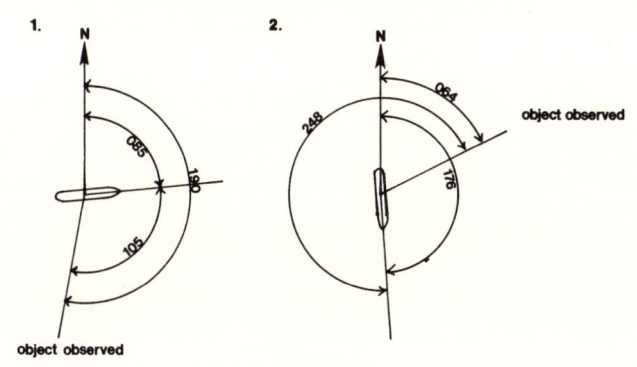

Fig. 2.11

## EXERCISE 2F

**Fill in the blanks**

|    | Compass Course or bearing | Dev. | Magnetic Course or bearing | Var. | True Course or bearing |
|----|---------|------|---------|-------|---------|
| 1. | 050° C. | —    | 056° M. | —     | 036° T. |
| 2. | —       | 3° E.| 220° M. | —     | 225° T. |
| 3. | —       | 4° W.| 280° M. | 18° W.| —       |
| 4. | 003° C. | —    | 358° M. | —     | 013° T. |
| 5. | —       | 4° W.| 241° M. | 11° W.| —       |
| 6. | 169° C. | 3° E.| —       | —     | 184° T. |
| 7. | —       | 2° E.| —       | 20° E.| 008° T. |
| 8. | 286° C. | 6° W.| —       | 5° W. | —       |
| 9. | 088° C. | —    | 091° M. | —     | 066° T. |
| 10.| —       | 4° E.| 205° M. | 30° W.| —       |
| 11.| 332° C. | —    | 332° M. | —     | 014° T. |
| 12.| 180° C. | —    | 178° M. | —     | 178° T. |

## The deviation table

A deviation table and deviation curve is compiled, after correction of the compass, by direct observation of the residual errors and deviations. In subsequent use of the deviation table it should be remembered that it was compiled for a particular condition of the ship with respect to moveable structures, condition of loading, draft, etc., and may not be accurate for other conditions. It should be used only when the deviation is not available by direct observation.

A deviation table is provided with the chartwork paper in Department of Trade examinations on which a deviation is given against compass heading at intervals of 10°. Deviations should be extracted using the compass heading as argument, interpolating between tabulated values. A sample deviation card is given which is used in examples and exercises to follow. The values included in the deviation column are rather large to be acceptable in practice but these values are used the better to illustrate the principles of interpolation required in examinations.

**To extract a deviation given the compass heading of the vessel**

Enter the deviation table with the given compass heading and extract the deviation, interpolating between tabulated values.

Example

A vessel is steering 113° C. Find the true course if the variation is 8° W.

From deviation table;

|  | compass course | deviation |  |
|---|---|---|---|
|  | 110° C. | 15° W. |  |
| difference 3° | 113° C. |  | difference 3° |
|  | 120° C. | 12° W. |  |

$$\text{deviation for } 113° \text{ C} = 15° - \frac{3 \times 3}{10}$$
$$= 15° \quad -1° \text{ almost}$$
$$= 14° \text{ W.}$$

| Compass course | 113° C. |
|---|---|
| deviation | 14° W. |
| Magnetic course | 099° M. |
| variation | 8° W. |
| True course | 091° T. |

**To find the deviation from a deviation table given the true course and the variation**

Again the argument used in the deviation table must be the compass course. As yet however this is not known so that the following procedure should be adopted.

1. Apply the variation to the true course to obtain the magnetic course.

# THE MEASUREMENT OF DIRECTION

2. Extract from the deviation table the two values of compass course and of deviation which when combined will give two values of magnetic heading which 'straddle' the ship's magnetic course found in (1).

3. Interpolate between the two values of deviation according to the value of the ship's magnetic course as compared with the two values, either side of this magnetic course. See example.

4. Apply the deviation obtained to the magnetic course to give the compass course. By entering the deviation table with this compass course obtained, a value of deviation should be extracted which is the same as that used in (3).

## Example

A vessel requires to make good a true course of 213° T. Using the deviation table provided find the compass course to steer if the variation is 9½° W.

(1)  True course         213° T.
     variation           9½° W.
     Magnetic course     222½° M.

(2) From deviation table;

| Compass course | Deviation | | Mag. Co. | |
|---|---|---|---|---|
| 210° C. | 6° E. | = | 216° M. | } diff. 6½° |
|  | diff. 3° |  | 222½° M. |  |
| 220° C. | 9° E. | = | 229° M. |  |

diff. 13°

(3) Interpolating between 216 and 229 magnetic for a magnetic course of 222½°, gives;

$$\text{deviation} = 6° \text{ E.} + \frac{3 \times 6\frac{1}{2}}{13}$$

$$= 6° \;\; +1.5° \;\; = 7.5° \text{ E.}$$

(4)  Magnetic course    222½° M.
     deviation          7½° E.
     compass course     215° C.

## EXERCISE 2G

In the following cases, given the true course and the variation, using the specimen deviation table provided, find the compass course to steer.

1. True course 100°  variation 6° W.
2. True course 024°  variation 9° W.
3. True course 352°  variation 2° E.
4. True course 262°  variation 5° E.
5. True course 148°  variation 12° W.

## Specimen deviation table

| Ship's head by compass | Deviation | Ship's head by compass | Deviation |
|---|---|---|---|
| 000° | 2° W. | 180° | 1° W. |
| 010° | 4° W. | 190° | 2° E. |
| 020° | 6° W. | 200° | 4° E. |
| 030° | 7° W. | 210° | 6° E. |
| 040° | 8° W. | 220° | 9° E. |
| 050° | 10° W. | 230° | 11° E. |
| 060° | 13° W. | 240° | 13° E. |
| 070° | 15° W. | 250° | 15° E. |
| 080° | 16° W. | 260° | 17° E. |
| 090° | 19° W. | 270° | 20° E. |
| 100° | 17° W. | 280° | 18° E. |
| 110° | 15° W. | 290° | 15° E. |
| 120° | 12° W. | 300° | 12° E. |
| 130° | 10° W. | 310° | 9° E. |
| 140° | 8° W. | 320° | 7° E. |
| 150° | 6° W. | 330° | 5° E. |
| 160° | 4° W. | 340° | 3° E. |
| 170° | 3° W. | 350° | nil |

## Wind and tide

The direction in which a vessel progresses may differ from that in which the vessel is heading, due to the effects of wind and tide.

The vessel may be assumed to move through the water in the direction in which it is steered, but a tidal stream, which is a horizontal movement of water due to differences in tidal height at different geographical positions, will carry the ship with it, and the resultant ship's motion will be that of the ship through the water and that of the water itself relative to the sea bed. It is necessary therefore to differentiate between the course steered and the course made good.

### Course steered

This is the heading indicated by the lubber line of the compass, this being the direction in which the vessel is heading. Note that this is the course to which all relative bearings must be applied in order to convert to true bearings, irrespective of the direction in which the

# THE MEASUREMENT OF DIRECTION

vessel is moving. Lines drawn on a chart to represent a course steered should be marked with a single arrow.

*Course made good*

This is the true direction of the ship's movement relative to the sea bottom. It may be found by the vectorial addition of the velocity of the ship through the water, and the velocity of the tidal stream. Lines drawn on a chart to represent a course made good should be marked with a double arrow.

*Rate of the tide*

This is the speed of the tide in nautical miles per hour or knots.

*Drift of the tide*

This is the distance moved by the tide in nautical miles, in a specified time interval.

$$\text{Thus rate of tide} = \frac{\text{drift of tide}}{\text{time interval}}$$

**To find the course and speed made good given the ship's course steered and speed and the set and rate of the tidal stream**

**Example**

Find the course and speed made good by a vessel steering 035° T. at 12 knots through a tidal stream setting 110° T. at 2.5 knots.

*Procedure* (refer to figure 2.12)

1. Lay off from a departure position the course steered (035°), and mark off a distance along this line equal to the vessel's speed. (It is often convenient to use an interval of one hour, but if appropriate an interval of half an hour or any other convenient interval may be used. The distance to be marked off along the course line will then be the distance steamed in that interval.) Mark this line with a single arrow.

2. From the position reached in (1), lay off the direction of the tidal stream or current, and mark off a distance equal to the rate of the tide. (Or if the chosen interval is not one hour then mark off a distance equal to the drift of the tide in the chosen interval.)

3. Join the position reached in (2), to the original departure position to represent the track along which the vessel will progress, i.e. the course made good. The length of this line will give the distance made good in the interval and hence the vessel's speed.

**Counteracting the tide**

The course steered may be adjusted for the effect of the tide in

22  PRACTICAL NAVIGATION FOR SECOND MATES

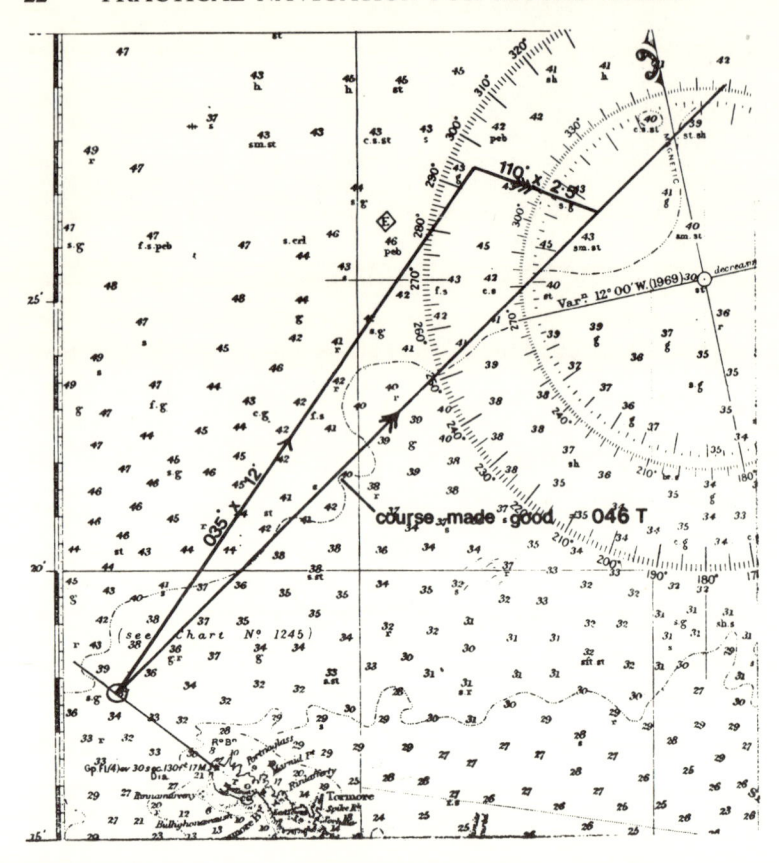

Fig. 2.12

order that the vessel progresses in a required direction. To counteract a tide the vessel must adjust its course up into the tide so that the tide will carry the vessel back down onto its required course line again.

**To find the course to steer to counteract a given current, to make good a required course**

### Example

Find the course to steer to make good a course of 035° T, when steaming through a current setting 110° T at 2·5 knots, if the ship's speed is 12 knots.

## THE MEASUREMENT OF DIRECTION

*Procedure* (refer to figure 2.13)

1. Lay off the course which it is required to make good from the ship's departure position to the destination. Mark this line with double arrow.
2. From the departure position lay off the direction of the tide or current and mark off the drift of the tide for any convenient chosen interval. (One hour is usually convenient.) Mark this line with a treble arrow.
3. From the position at the end of the tide found in (2), with compasses, describe an arc of radius the distance steamed by the ship in the chosen interval (the ship's speed if the interval used is one hour), to cut the course to be made good.

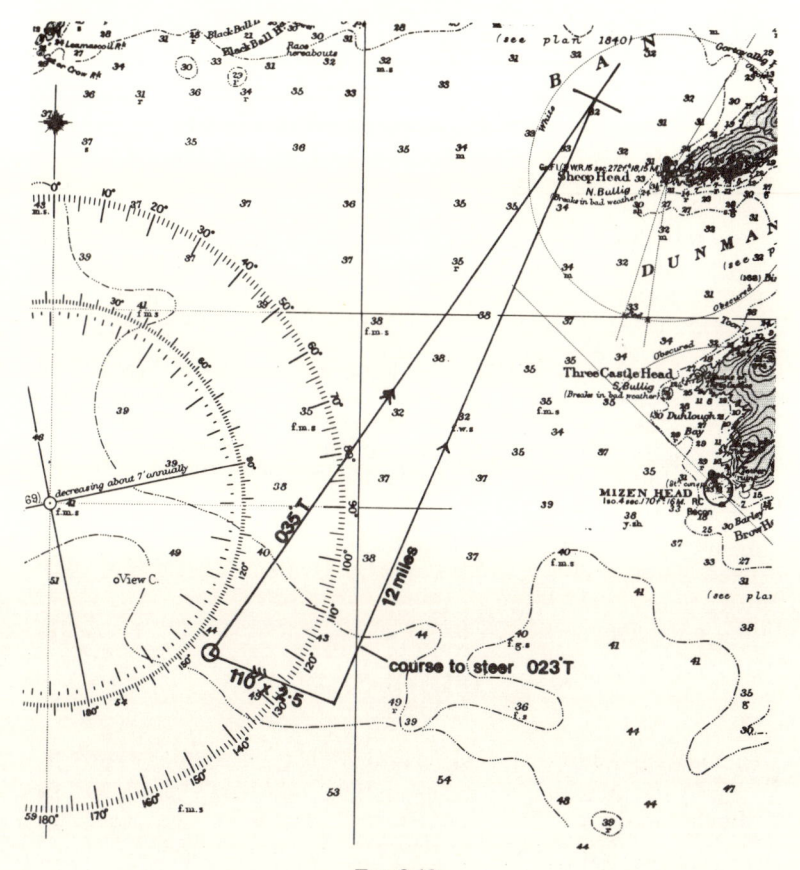

FIG. 2.13

24    PRACTICAL NAVIGATION FOR SECOND MATES

4. Join the position where the arc cuts the course to be made good with the end of the tide. The direction of this line will give the course to be steered. Mark this line with a single arrow.

5. Measure the speed which will be made good. This will be given by the length of the line which represents the course to be made good, from the departure position to the position where the arc described cuts this line.

**Note**

Although the vessel will progress along the course to be made good, the ship's head will be that of the course steered, so that any relative bearing will be with respect to the course steered. Particular note should be taken of this when finding the position where a point of land will be abeam. The ship will be on the course line made good but the beam bearing will be at right angles to the course steered, or ship's head.

**To find the distance and time at which the vessel will pass a point of land when abeam**

**Example**

At 1000 hrs Lizard Point Lt. bore 000° T by 2·5 miles. Find the course to steer to make good a course of 050° T in order to counteract a current setting 280° T at 2 knots. Ship's speed 10 knots. Find the distance off Black Head when abeam and the time when abeam.

Figure 2.14 shows the construction to counteract the current. The course to steer is 059° T, and therefore the beam bearing of Black Head will be 329° T. This will not be at right angles to the course made good, that is the track along which the vessel progresses. The beam distance will therefore not be the least distance.

From figure 2.14;

distance when abeam = 1·5 miles.
speed made good = 8·4 knots.
distance to beam bearing = 5 miles
time when abeam = $1000 + \dfrac{5}{8 \cdot 4}$ hrs.
= 1000 + 0·6
= 1036 hrs.

**To reach a position at a required time while counteracting a current**

To arrive at a position at a given time the vessel's speed must be adjusted so that;

$$\text{speed} = \frac{\text{distance to steam to required position}}{\text{required time interval}}$$

# THE MEASUREMENT OF DIRECTION

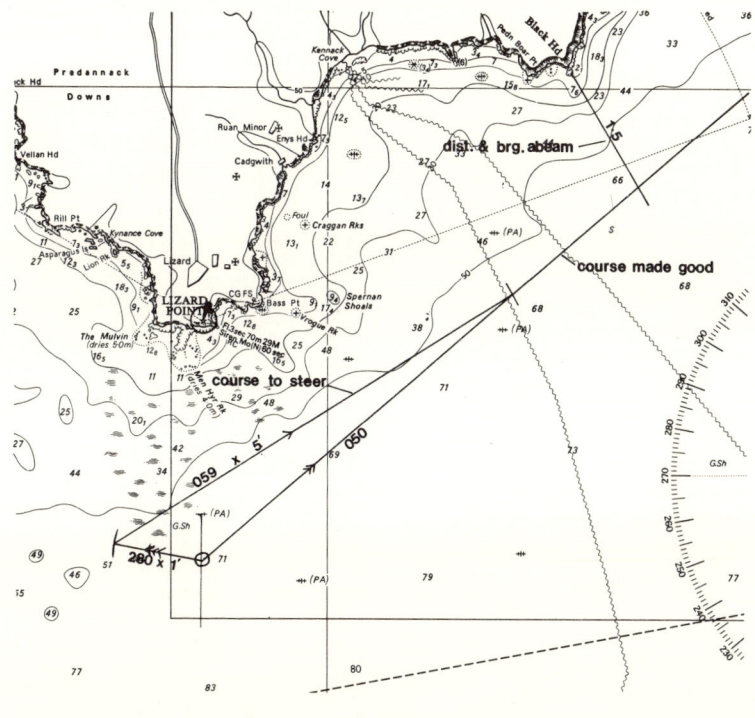

Fig. 2.14

This will give the speed which must be made good over the ground. If there is a current or tidal stream, the vessel's log speed or speed through the water may be more or less than the speed made good. As the speed through the water will be determined by the engine revolutions, it is this speed which must be found.

## Example

A vessel observes Fastnet Rock to bear 340° T. by 7 miles. She wishes to arrive off Cork, a distance of 48 miles, in 6 hours. A tidal stream is estimated to set 285° T. at an average rate of 1·3 knots over the next 6 hours. Find the course to steer to make good a required course of 075° T., and the speed necessary to cover the 48 miles in six hours.

*Procedure* (refer to figure 2.15)

1. Having laid off the required course of 075° T. and measured

# 26 PRACTICAL NAVIGATION FOR SECOND MATES

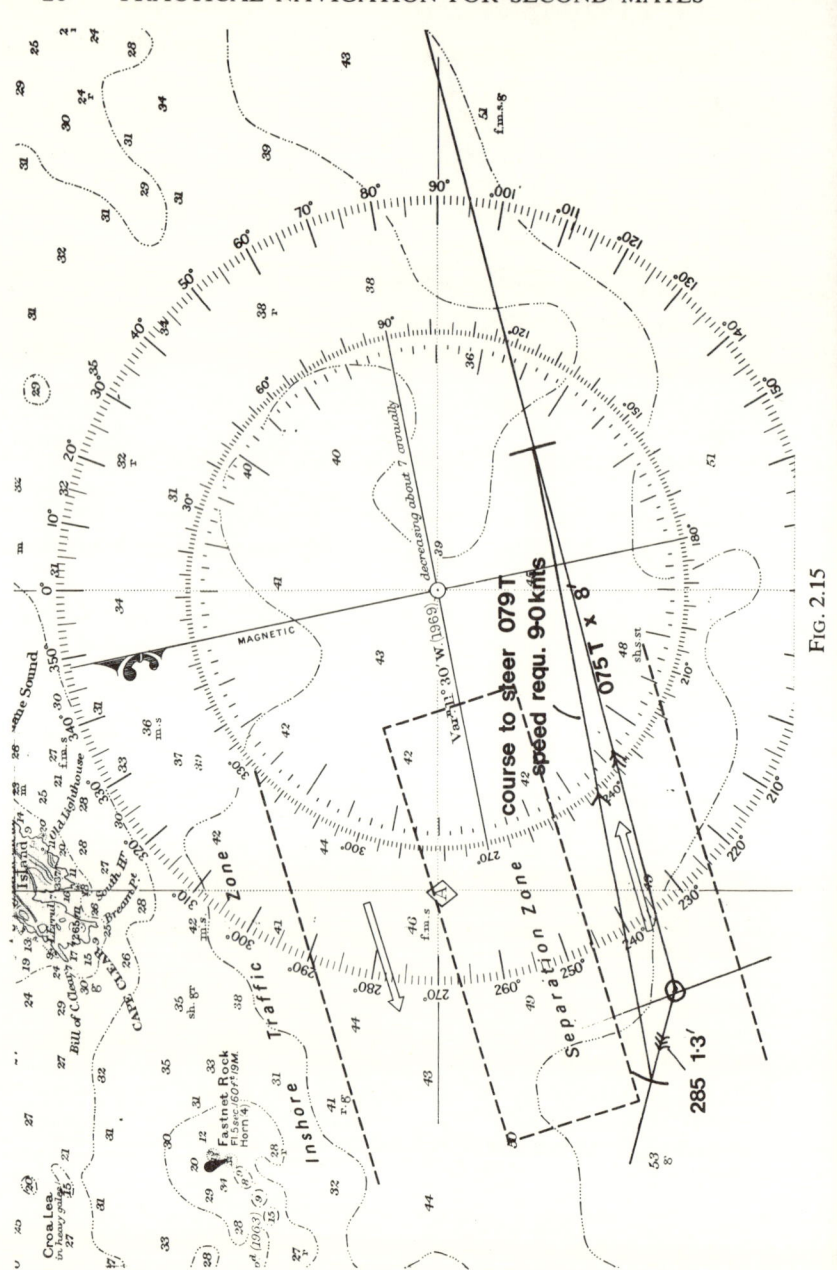

FIG. 2.15

the distance to go to be 48 miles, the speed to make good can be calculated to be

$$\frac{48}{6} = 8 \text{ knots.}$$

2. From the departure position lay off the current in the direction given (285° T.) and mark off the rate (1·3 miles).
3. From the departure position mark off the speed to make good, along the course to be made good.
4. Join the end of the tide found in (2), to the end of the speed made good found in (3). The direction of this line will give the course to steer, and its length will give the speed to make good through the water, that is the speed to use when determining the engine revolutions to order.

**To find the set and drift of the tide between two observed positions**

The set and drift, or rate, of the tidal stream or current may be found if two observed positions are available, and the courses and distances steamed between the two observations are known. The difference between the D.R. position at the time of the second observation, and the actual position as observed, is due to the tide.

**Example**

At 0800 a point of land is observed to bear 120° T. by 5 miles. At 0830 the same point of land was observed to bear 220° T. by 6·5 miles. Find the set and rate of the current, if the course steered in the interval was 078° T. speed 20 knots.

*Procedure* (refer to figure 2.16)

1. Lay off the two observed positions and label with their respective times.
2. From the first observed position lay off the course steered, and the distance steamed in the interval between the observations, to find the D.R. position for the time of the second observation. Mark this line with a single arrow.
3. Join the D.R. position found in (2) to the second observed position and mark the line with a treble arrow. Measure the direction to give the set of the current and the length to give the drift.

**To find the course to steer to pick up a point of land at a required angle on the bow at a given distance.**

This may be done by solving the right angled triangle PAB in figure 2.17 to find the beam distance PB. The angle PAB is the required angle on the bow and the side PA is the required distance.

28    PRACTICAL NAVIGATION FOR SECOND MATES

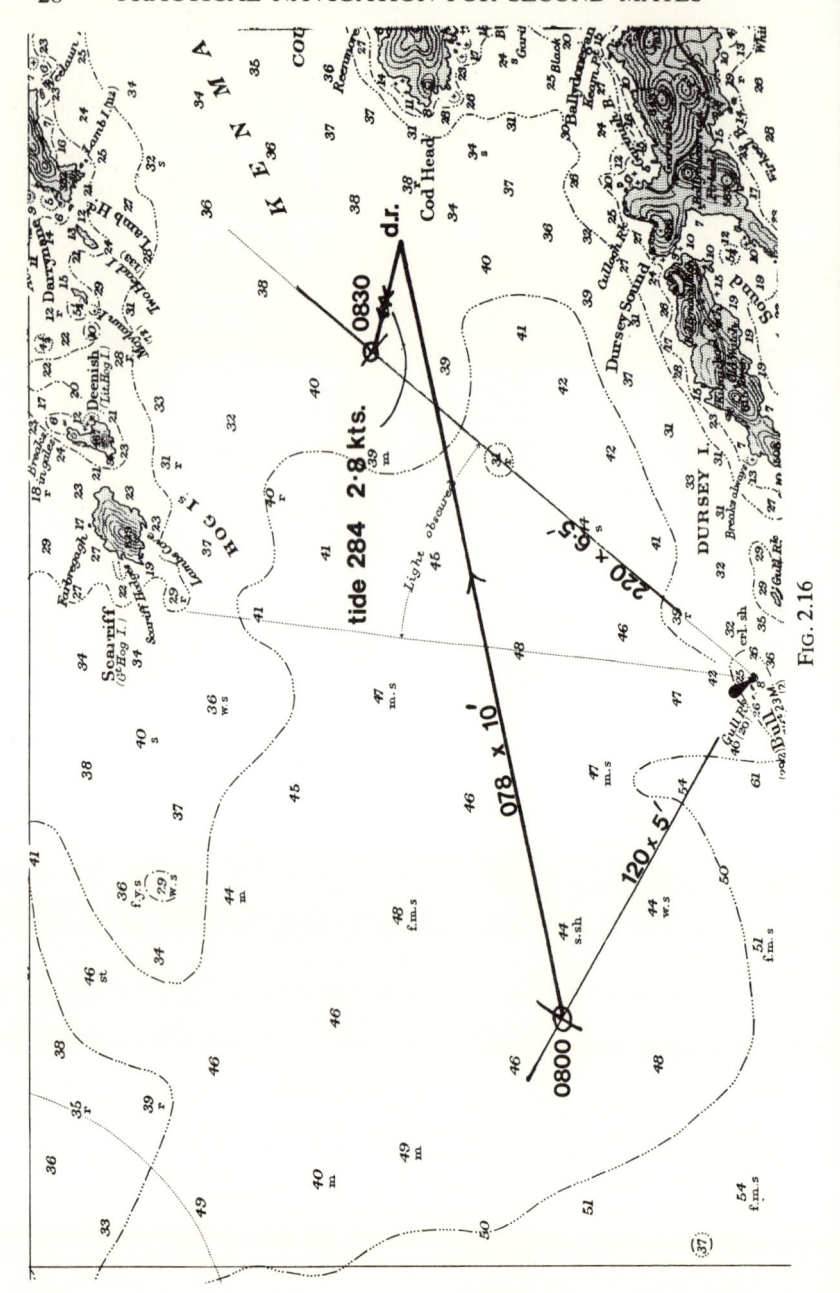

Fig. 2.16

# THE MEASUREMENT OF DIRECTION

When the beam distance is found the required course can be drawn tangential to a circle of that radius centred upon the point of land. The solution of the triangle may be done by traverse table.

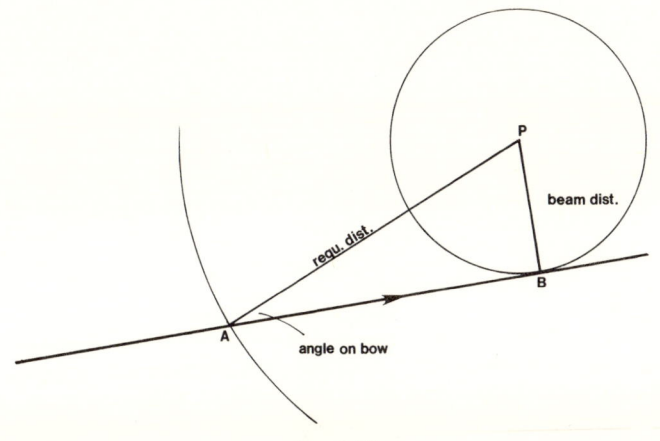

FIG. 2.17

*Procedure*

1. Calculate the beam distance. If traverse tables are used, the tables are entered with the angle on the bow as the course angle, and the required distance off as the distance (hypotenuse). The beam distance is extracted from the departure (opposite) column.
2. Draw a circle centred upon the point of land, of radius the beam distance found in (1).
3. Draw the course line to steer from the departure position, tangential to the beam distance circle.
4. Draw an arc of radius the given distance off, and centred upon the point of land, to cut the course line drawn. The intersection will give the position at which the point of land will be at the given distance and at the required angle on the bow.

## Effect of the wind. Leeway

A vessel may be deviated from her course steered, or from her course made good due to the effect of the tide, by the wind. The change in a vessel's course angle due to the effect of a wind is called the leeway. The course made good due to the wind may be found by applying the leeway to the course steered in the direction in which the wind is causing the vessel to drift, that is down wind. The effect of the wind may be counteracted by applying the leeway to the course steered up into the wind.

# PRACTICAL NAVIGATION FOR SECOND MATES

To find the effect of a tide and a wind, the leeway is applied to the courses steered before laying them on the chart.

To allow for a tide and for the effect of leeway, the leeway is applied to the course to be steered, after allowing for the tide.

## Example 1

Given a vessel's course is 135° T., wind S.W., leeway 5°, find the track.

Course = 135° T.
Leeway = 5°      (wind on the starboard side, subtract)
─────
Track = 130° T.
─────

## Example 2 (refer to figure 2.18)

Find the course to steer to counteract a current setting 085° T at 1·5 knots, and a S.W.'ly wind causing a leeway of 4°, in order to make good a course of 120° T. Ship's speed 10 knots.

### EXERCISE 2H

**Find the track**

|     | Course    | Dev.   | Var.    | Leeway | Wind   |
|-----|-----------|--------|---------|--------|--------|
| 1.  | 055° C.   | 3° E.  | 13° W.  | 4°     | N.N.W. |
| 2.  | 140° C.   | 4° W.  | 10° W.  | 5°     | S.W.   |
| 3.  | 246° C.   | 2° E.  | 15° E.  | 4°     | N.W.   |
| 4.  | 330° C.   | 3° W.  | 8° W.   | 3°     | S.W.   |
| 5.  | 104° C.   | 6° E.  | 12° W.  | 7°     | N.E.   |
| 6.  | 084° C.   | 2° W.  | 20° E.  | 5°     | North  |
| 7.  | 354° C.   | 5° W.  | 18° E.  | 6°     | West   |
| 8.  | 190° C.   | Nil    | 22° W.  | 10°    | E.S.E. |
| 9.  | 240° C.   | 3° E.  | 5° E.   | 8°     | W.N.W. |
| 10. | 280° C.   | 1° W.  | 25° W.  | 4°     | N.N.W. |

### EXERCISE 2I

1. Find the course to steer to make good a course of 160° T. on a vessel of speed 18 knots, steaming through a current setting 215° T. at 3 knots. What will be the speed made good in the direction 160° T.?

2. Find the course and speed made good if a vessel steams 305° T. at 12 knots through a current setting 243° T. at 2·5 knots.

# THE MEASUREMENT OF DIRECTION

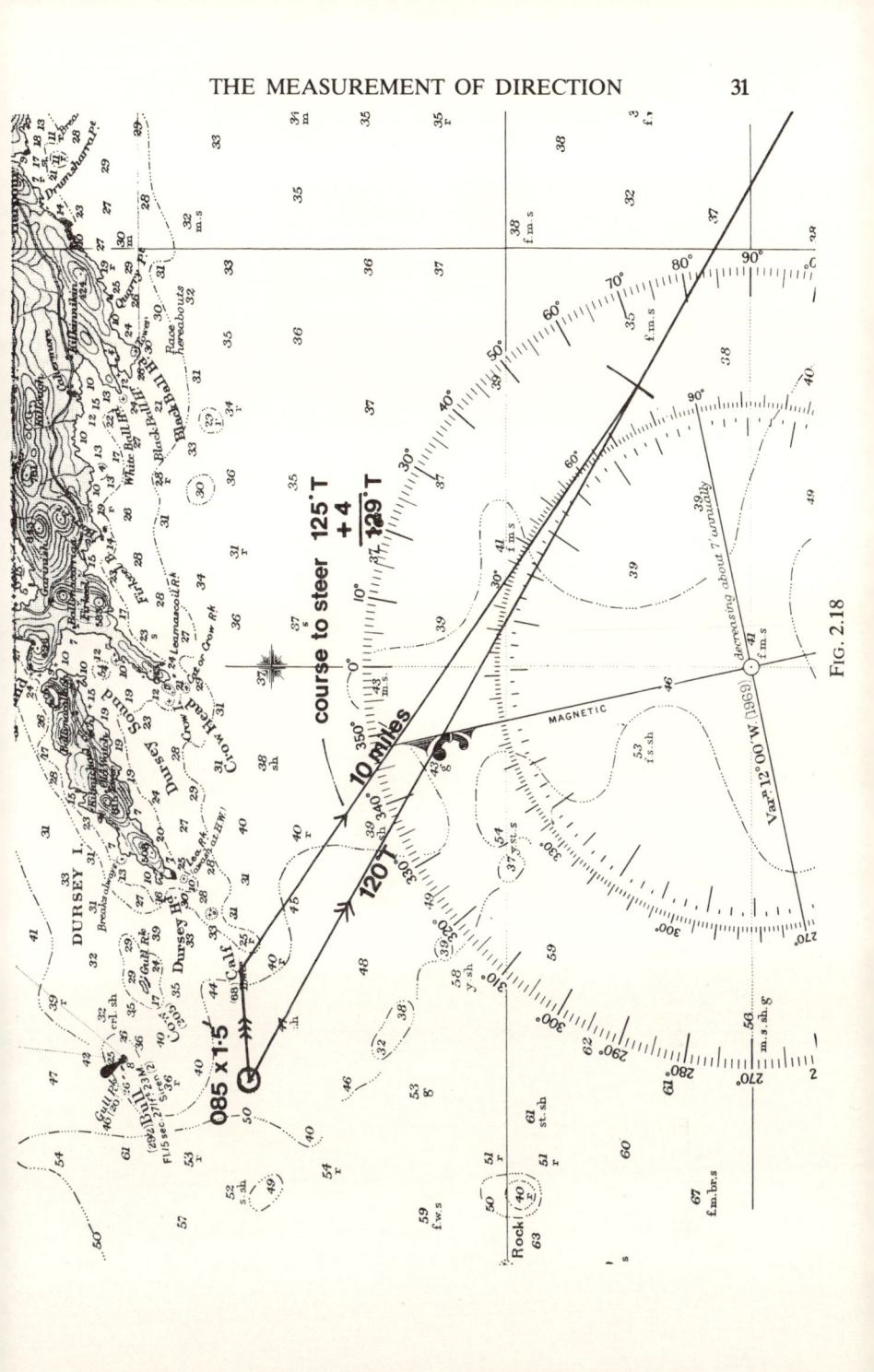

FIG. 2.18

## 32 PRACTICAL NAVIGATION FOR SECOND MATES

3. Find the set and rate of the current if a point of land is observed to bear 025° T. by 6 miles and 45 minutes later is observed to bear 300° T. by 12 miles, if in the interval the vessel was steering 095° T. at 20 knots.

4. Find the true course to steer to make good a course of 350° T. on a vessel of speed 15 knots, steaming through a current setting 005° T. at 2 knots, if a westerly wind is causing a leeway of 5°.

5. Find the course and speed made good if a vessel steaming 176° at 17 knots through a current setting 020° T. at 3 knots, is experiencing a leeway of 5° due to an easterly wind.

## CHAPTER 3

## POSITION LINES

A position line is a line on the earth's surface which represents the locus of an observer who moves such that some item of observed information remains of constant value. Position lines generally used may take the form of part of a great circle, small circle, or hyperbola, depending upon the nature of the information observed, but in general, that part of the position line which lies near to the ship's D.R. position may be considered as, and represented on a chart as a straight line without any considerable error. If such a line is drawn on a chart, the observer may be assumed to be on that line. To fix the observer's position two, non coincident position lines must be drawn to intersect, the point of intersection defining the observer's position.

**Position lines obtained from bearings**

The most commonly used method of obtaining a position line at sea is to observe the bearing of a known and charted position. The most commonly used fix is that produced from two such position lines. This is not inherently the most accurate method of fixing but bearing information is in general the easiest to obtain. This may be done visually with a compass or by observing the direction from which radio waves radiated from a shore beacon, reach the observer by means of an aerial which has directional properties. In both cases the bearing measured will be the direction of the great circle which passes through the observer and the observed position, at the observer. Both a visual line of sight and a radio wave, follow a great circle path along the earth's surface.

In the case of the visual bearing this measured bearing, corrected for any compass error, may be laid off on a mercator chart from the charted position of the point observed, as a straight line, and this line can be assumed to represent the position line. The error incurred in doing this is negligible over the short distances over which visual bearings are taken.

In the case of a bearing observed by radio direction finder, errors may be considerable if this is done due to the longer path length of the radio wave, and to the longer distances on the chart over which the bearing is laid off. To maintain errors within tolerable limits radio D/F bearings are corrected for the half convergency, which may be considered to be the difference between the great circle

34    PRACTICAL NAVIGATION FOR SECOND MATES

bearing of a point, and the mercator or rhumb line bearing which is produced on the chart by laying off a straight line. This is illustrated in figure 3.1

The half convergency may be found by the approximate formula;

$$\text{half convergency} = \frac{\text{d'long} \times \text{sine mean latitude}}{2}$$

Alternatively it may be obtained from a half convergency table provided in Burton's or Nories' nautical tables.

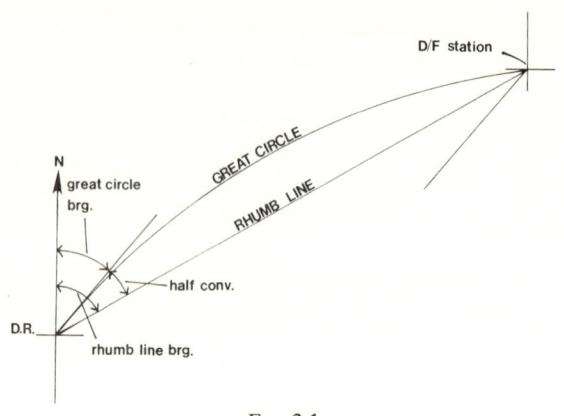

Fig. 3.1

### Direction of the half convergency correction

The appearance of a great circle on a mercator chart is a curve which is concave towards the equator. The mercator bearing always lies on the equatorial side of the great circle bearing therefore, and the correction to the great circle bearing is always applied towards the equator. This is illustrated in figure 3.2 which shows the four possible cases of;

        a. westerly bearing in north latitude,
        b. easterly bearing in north latitude,
        c. westerly bearing in south latitude,
        d. easterly bearing in south latitude.

### Example

From a ship in D.R. position 44° 10′ S. 144° 50′ E., a D/F station bore 055°. If the position of the D/F station is 42° 53′ S. 147° 14′ E., find the rhumb line bearing.

## POSITION LINES

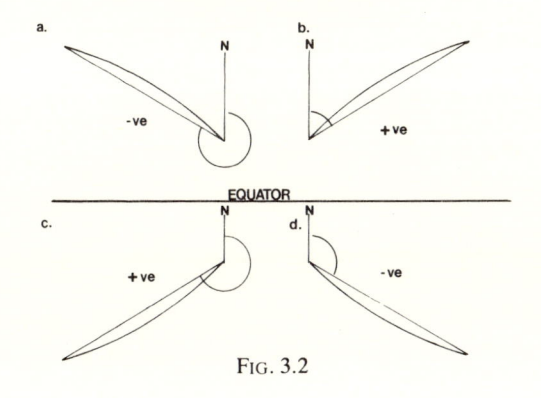

Fig. 3.2

D.R. position   44° 10' S. 144° 50' E.   Mean lat. = 43° 31' S.
D/F station    42° 53' S. 147° 14' E.

d'lat.   = 1° 17' N.   2° 24' E. = d'long.
          = 77'        = 144'

$$\text{half convergency} = \frac{144 \times \sine 43° 31'}{2}$$

$$= 49\cdot 6' = \sfrac{3}{4}° \text{ approx.}$$

great circle bearing    055°
½ conv.                  ¾°

rhumb line bearing      054¼°

Fig. 3.3

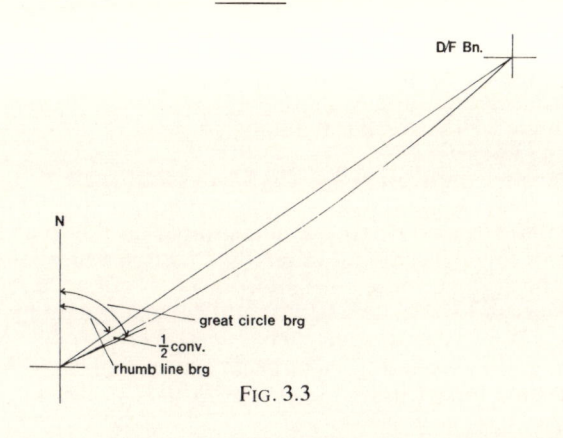

## The position circle

If the information observed is that of distance from a known and charted object or point of land then the position line will take the form of a small circle centred upon the position of the object observed and of radius the distance measured. Such information may be readily obtained by radar observation. Position lines obtained by this method are inherently more accurate than those obtained from bearing information. This is due to the limitations of the instruments used for taking bearings, be they visual or by radar, and also to the fact that whereas the effect of an error in bearing, measured in terms of a distance error, increases as the position line diverges from the position observed, an error in range is constant, and does not increase with distance from the object observed.

## The vertical sextant angle

Distance information may be obtained by observing the angle subtended by the top of a vertical structure or land formation of known height, and the sea foreshore at its base. The marine sextant is generally used for such an observation. The vertical angle subtended varies as the distance of the observer from the object, which can be readily found.

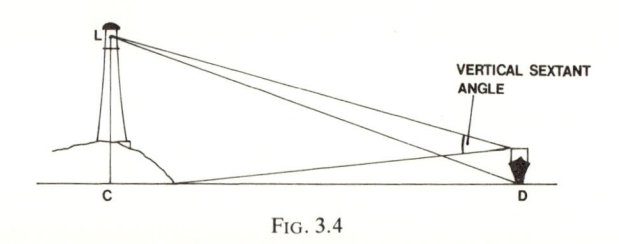

FIG. 3.4

The vertical sextant angled observed is illustrated in figure 3.4. This is considered to be equal to angle LCD without undue error. The triangle LCD, which is right angled at C can then be solved to find the distance CD.

By plane trigonometry CD=CL Tan vertical angle.

Note that the distance must be measured on the chart from the position of the highest point observed, that is in figure 3.4 from the position of the lighthouse, and not from the foreshore. Solutions for the distance off, tabulated against vertical angle and the height of the object observed, are given in nautical tables.

Heights of lights and topographical features given on Admiralty charts are expressed above Mean High Water Springs. For accurate

## POSITION LINES

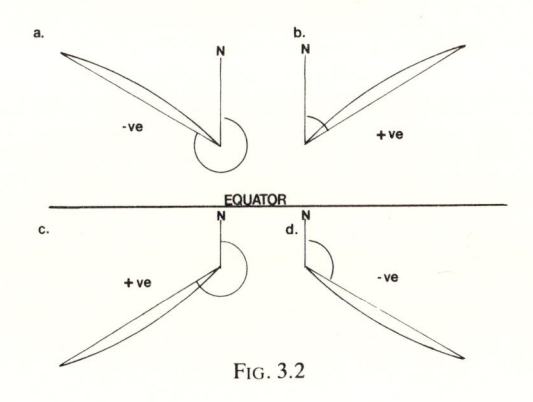

Fig. 3.2

D.R. position  44° 10′ S. 144° 50′ E.   Mean lat. = 43° 31′ S.
D/F station    42° 53′ S. 147° 14′ E.

d'lat.   =  1° 17′ N.   2° 24′ E. = d'long.
            = 77′        = 144′

$$\text{half convergency} = \frac{144 \times \sin 43° 31'}{2}$$

$= 49 \cdot 6' = \tfrac{3}{4}°$ approx.

great circle bearing    055°
½ conv.                 ¾°

rhumb line bearing     054¼°

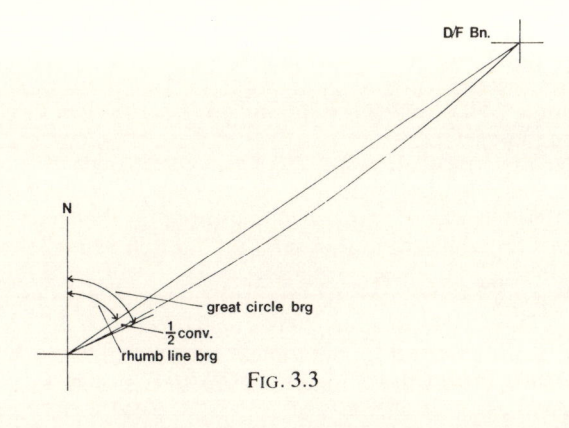

Fig. 3.3

## The position circle

If the information observed is that of distance from a known and charted object or point of land then the position line will take the form of a small circle centred upon the position of the object observed and of radius the distance measured. Such information may be readily obtained by radar observation. Position lines obtained by this method are inherently more accurate than those obtained from bearing information. This is due to the limitations of the instruments used for taking bearings, be they visual or by radar, and also to the fact that whereas the effect of an error in bearing, measured in terms of a distance error, increases as the position line diverges from the position observed, an error in range is constant, and does not increase with distance from the object observed.

## The vertical sextant angle

Distance information may be obtained by observing the angle subtended by the top of a vertical structure or land formation of known height, and the sea foreshore at its base. The marine sextant is generally used for such an observation. The vertical angle subtended varies as the distance of the observer from the object, which can be readily found.

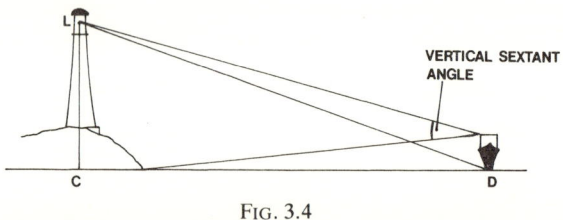

Fig. 3.4

The vertical sextant angled observed is illustrated in figure 3.4. This is considered to be equal to angle LCD without undue error. The triangle LCD, which is right angled at C can then be solved to find the distance CD.

By plane trigonometry CD = CL Tan vertical angle.

Note that the distance must be measured on the chart from the position of the highest point observed, that is in figure 3.4 from the position of the lighthouse, and not from the foreshore. Solutions for the distance off, tabulated against vertical angle and the height of the object observed, are given in nautical tables.

Heights of lights and topographical features given on Admiralty charts are expressed above Mean High Water Springs. For accurate distances by this method these heights should be adjusted for the height of the tide as shown in Chapter 15, before entering vertical angle tables. In practice the uncorrected heights are used as the unknown error will always put the vessel closer to the position observed than its true position. This will in most cases fix the vessel closer to the danger and leave the navigator with a margin of safety. This will not be the case however if the danger lies on the side of the vessel away from the position observed.

The accuracy of this method depends upon the base of the object at sea level being visible. At distances greater than that of the sea horizon for the observer's height of eye, the base will not be visible and the angle measured will be that subtended by the top of the object and the sea horizon. If this is the case vertical sextant angle tables will not be valid.

## Horizontal sextant angles

A position circle is obtained if the angular distance between two known and charted objects is measured by an observer. This information may be obtained with a marine sextant but is also readily obtained to acceptable accuracy by taking the difference between the compass bearings of the two objects. The advantage of using compass bearings in this way rather than to produce a fix by cross bearings, is that any unknown compass error will have no effect on the position line obtained. A fix may be produced by two horizontal angles obtained from compass bearings of three objects, independent of compass error, and the true bearings obtained from the chart after the fix is determined will, on comparison with the compass bearings observed, give the error of the compass.

The angle subtended by the chord of a circle at the circumference of the circle is the same at all points on the circumference. A circle may therefore be drawn a chord of which is formed by the straight

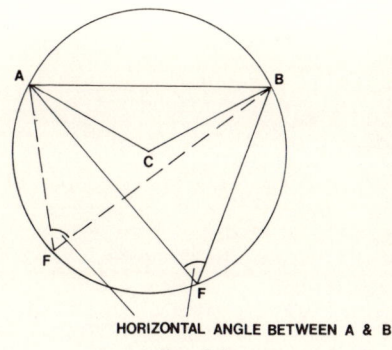

**HORIZONTAL ANGLE BETWEEN A & B**
Fig. 3.5

line between the two points whose horizontal angle is known. The observer will lie somewhere on that circle.

In figure 3.5 angle ACB = 2 × angle AFB
$$= 2 \times \text{measured horizontal angle.}$$
(the angle on a chord at the centre of a circle is twice that at the circumference.)

Also in triangle ACB, angle $\angle$CAB = angle $\angle$CBA

$$= \frac{180° - \angle ACB}{2}$$

$$= \frac{180° - (2 \times \angle AFB)}{2}$$

$$= 90° - \angle AFB$$

Thus to find the angles $\angle$CBA and $\angle$CAB, the measured horizontal angle is subtracted from 90°. (If the measured angle is less than 90°.)

*Procedure* (refer to figure 3.5). To be followed if the measured angle is less than 90°.

1. Join the charted positions of the two points between which the horizontal angle has been obtained, with a straight line. (AB)
2. Construct the angles $\angle$CBA and $\angle$CAB at the positions observed on the side of the line on which the observer lies. The values of $\angle$ACB and $\angle$CAB have been shown to be 90°— horizontal angle observed. The point of intersection of the two lines AC and BC so formed will be the centre of the required circle.
3. Draw a circle centred on C to pass through the two positions between which the horizontal angle is known, A and B. This circle is the position circle.
4. Repeat for a second horizontal angle to produce an intersection of two position circles.
5. If required, measure the true bearings of the points observed, from the chart and compare with the compass bearings to find the compass error.

**Example**

The following compass bearings were observed. Find the ship's position and also the error of the compass.
    Great Skellig lighthouse    304° C.
    Bolus Head    029° C.
    Great Hog Island (Scarriff)    074° C.    Highest point (829) observed.

# POSITION LINES

Refer to figure 3.6 and to the procedure outlined above for the construction.

Ship's position = 51° 41·8′ N. 10° 24·8′ W.

|  | Great Skellig Lt. | Bolus Hd. | Scarriff |
|---|---|---|---|
| True bearing | 310° T. | 035° T. | 080° T. |
| Compass bearing | 304° C. | 029° C. | 074° C. |
| Compass error | 6° E. | 6° E. | 6° E. |

**Note**

By checking the compass error with the three bearings a check is provided that the construction has been done correctly.

**Horizontal angle measured greater than 90°**

In this case the observer lies on the circumference of the circle on the opposite side of the chord of the circle to the circles' centre.

In figure 3.7

angle $\angle ACB = 360° - \angle AFB$ (the horizontal angle)

and angle $\angle CBA$ = angle $\angle CAB = \dfrac{180° - (360° - \angle AFB)}{2}$

$= 90° - 180° + \angle AFB$
$= \angle AFB - 90°$

Thus the angles $\angle CBA$ and $\angle CAB$ are found by subtracting 90° from the horizontal angle measured.

*Procedure*

This is the same as for the case of the horizontal angle less than 90° except that the angles $\angle CBA$ and $\angle CAB$ are laid off on the straight line joining the two positions, on the opposite side of the line to the observer's position. See example in figure 3.8.

**Example**

The following compass bearings were observed. Draw a position line by horizontal angles.

| Galley Head. Lt. | 050° C. |
|---|---|
| Castle Haven Lt. | 295° C. |
| Horizontal angle | 115° |
| Construction angles | 115° − 90° = 25° |

See figure 3.8 for construction.

40    PRACTICAL NAVIGATION FOR SECOND MATES

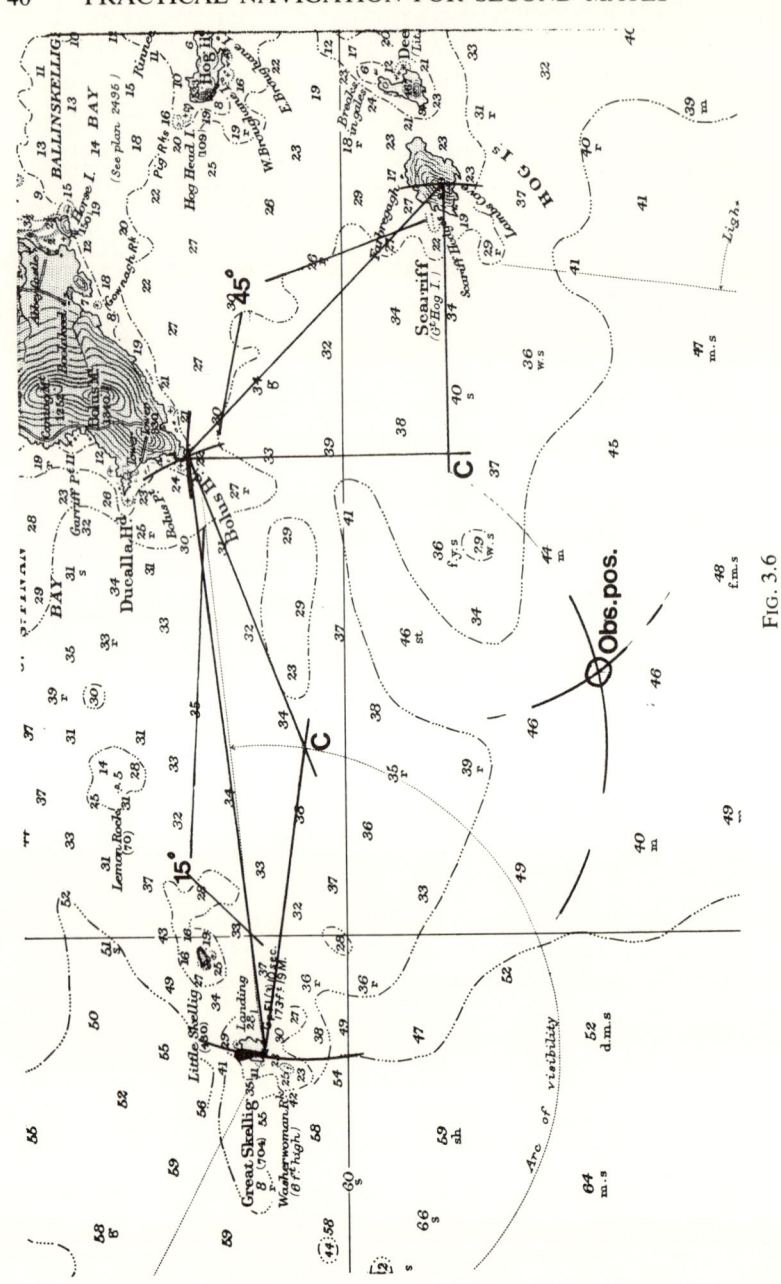

FIG. 3.6

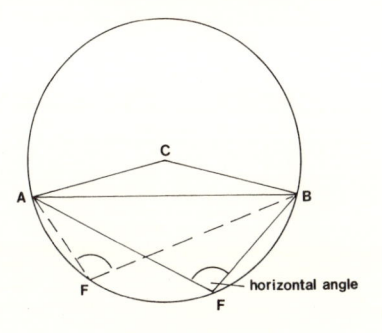

Fig. 3.7

## Notes

If the observed horizontal angle is 90°, then the centre of the circle lies on the straight line which joins the two positions between which the angle has been measured. The chord of the circle is in fact a diameter of the circle.

A poor angle of cut will result if the two position circles almost coincide. If all the observed positions and the vessel's position also, lie on the circumference of the same circle, then the two position circles constructed will coincide and a fix cannot be obtained. When choosing features to observe for horizontal angles the vessel's D.R. position should be compared with the positions to be observed to check for this condition.

## The transit bearing

One of the most useful and easily obtained position lines is that from observation of two known and charted objects which lie on the same bearing from the observer. Such objects are said to be in transit. A straight line drawn on the chart through the two positions will represent the charted position line. Observation of a transit does not require a compass but if the compass bearing is noted as the two objects come into line, a compass error may be readily obtained by comparison with the true bearing of the transit taken from the chart. If a compass bearing of some other prominent object is observed at the same time as the transit, then the accurate compass error obtained from the transit may be used to correct this compass bearing. An accurate fix may be obtained in this way very quickly, and with practice this method of fixing can be used to good effect in confined navigational waters where plenty of coastal features and beacons are available.

42     PRACTICAL NAVIGATION FOR SECOND MATES

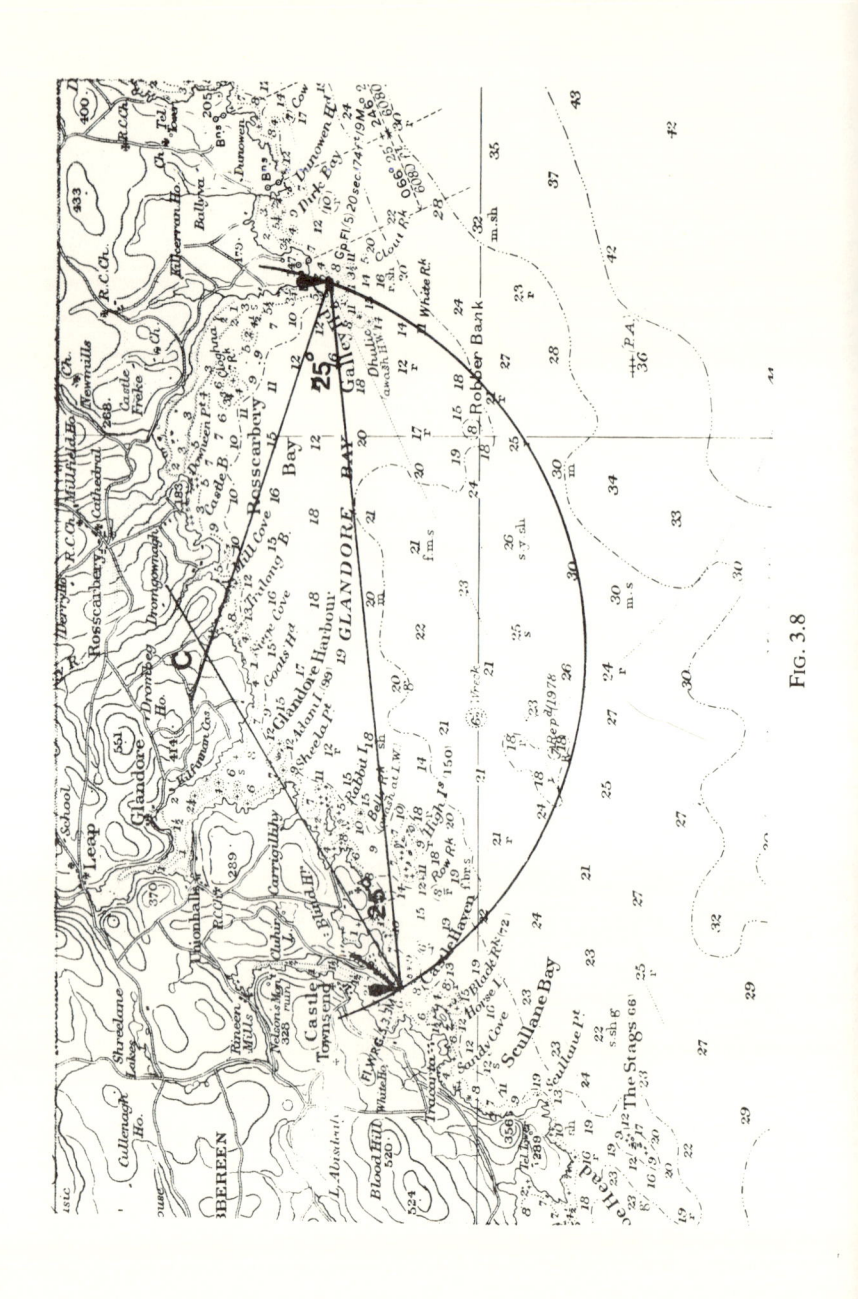

Fig. 3.8

# POSITION LINES

## Example
From a vessel entering Bantry Bay, Sheep Head Lighthouse was observed to be in transit with Three Castle Head bearing 168° by compass. At the same time Black Ball Head was observed to bear 264½° by compass. Find the ship's position.

*Procedure* (refer to figure 3.9)
1. Draw a straight line through Sheep Head Lighthouse and Three Castle Head, and produce it into Bantry Bay. Measure the true direction of this bearing from the chart compass rose.
2. Compare the true bearing of the transit with the compass bearing to obtain the compass error.
3. Using the compass error obtained in (2) correct the compass bearing of Black Ball Head to true.
4. Lay off the true bearing of Black Ball Head to cross the transit bearing at the ship's position.

## Note
A compass error obtained at the time of observation by this means or by any other means should always be used in preference to one taken from a deviation card and compass rose, or to one taken at some earlier time.

## The use of the transit bearing for leading marks
The transit bearing is used to mark the direction of approach through a navigable channel. Beacons erected to provide a transit bearing are called leading marks or leading lights. A vessel navigating in the channel maintains the leading marks in line to keep herself in the channel and on the correct approach.

Very often this technique may be used by the navigator in close waters even though leading marks are not provided, by selecting natural prominent features or charted buildings. The technique is also very useful when keeping a check on anchor bearings. At anchor there are invariably enough natural topographical features to choose transits, which tell at a glance whether the vessel is dragging or not. The swing of the vessel around her anchor should not be mistaken for a dragging.

## Position circle by rising and dipping distance
When making a landfall at night a first position may very often be obtained by observing a light which just appears above the horizon. In clear weather the loom of a light may very often be seen long before the actual light itself is seen. If a bearing is then taken when the actual light just appears or rises, then a position may be obtained by crossing this bearing with a position circle obtained

44 PRACTICAL NAVIGATION FOR SECOND MATES

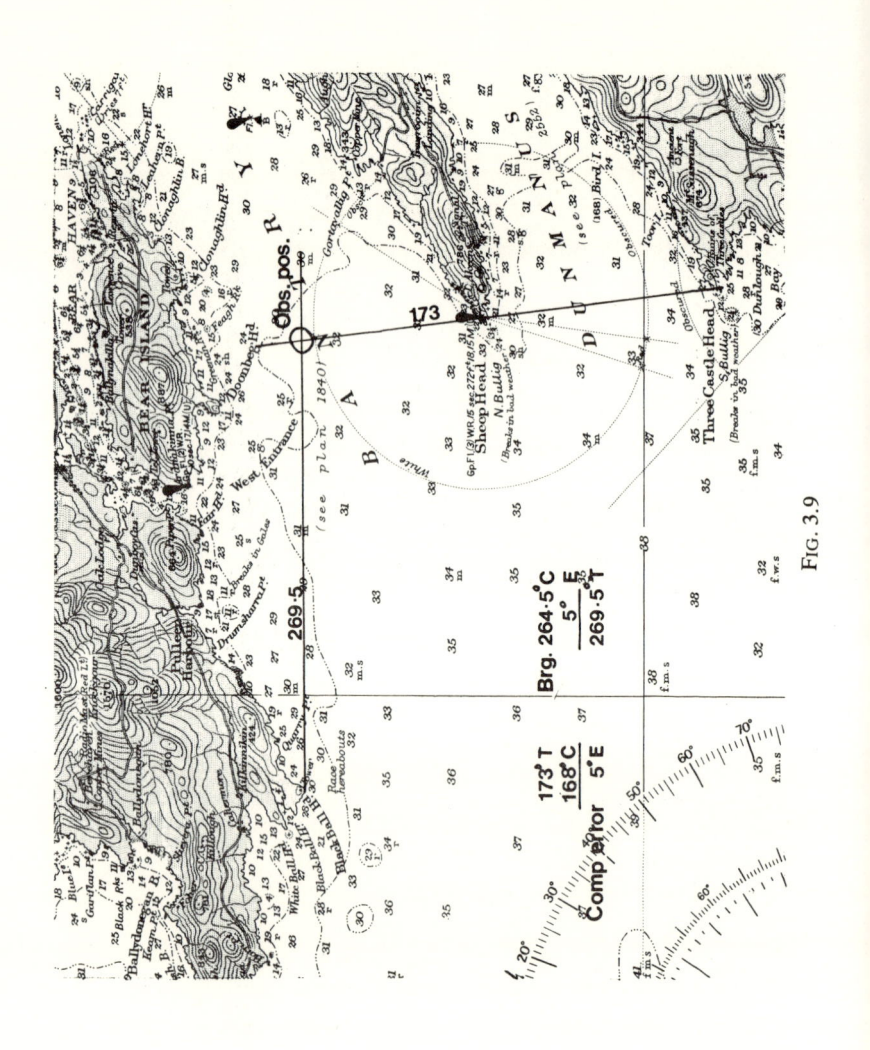

Fig. 3.9

# POSITION LINES

from a distance off. The height of the light must be obtained from the chart. The accuracy of such a position is acceptable as a first landfall position, but the distance may be approximate for a number of reasons.

The distance of the sea horizon is given by the formula $2 \cdot 08 \sqrt{h}$, where h is the height of the observer's eye in metres. The formula gives the distance in nautical miles. The constant $2 \cdot 08$ includes allowance for an estimated refraction for normal atmospheric conditions. Abnormal refraction however may cause inaccuracy.

Figure 3.10 shows that the distance at which a light will first rise above the horizon will be given by:

$2 \cdot 08 \sqrt{h} + 2 \cdot 08 \sqrt{H}$   where h = height of eye in metres.
$H$ = height of light above sea level in metres.

FIG. 3.10

The solution of this formula may be obtained from nautical tables, from the table giving the distance of the sea horizon against height of eye. The distance is taken out in two parts, one for the observer's height of eye and one for the height of the light.

It should be remembered that the heights of lights on charts are given above mean high water springs, and unless this height is adjusted for the height of tide, the distance obtained will be usually too small.

## Ranges of lights

*Nominal range*—This is the range given against the light on Admiralty charts, and also in Admiralty Lists of Lights. It is the visible range based upon its intensity, which is measured in candelas, and upon a meteorological visibility of 10 nautical miles.

*Geographical range*—This is the maximum range at which it is possible to see the light as dictated by the curvature of the earth. This will depend upon the observer's height of eye and upon the height of the light above sea level. A formula for the geographical range was given when discussing the rising and dipping of lights.

The actual range at which the light may be seen may be more or less than the nominal range depending upon the prevailing atmospheric conditions and the meteorological visibility. The range at which the light may be seen under any particular meteorological visibility is called the luminous range. This may be obtained from a luminous range diagram which is given in the introduction to each volume of the Admiralty List of Lights. This is reproduced in figure 3.11.

The nominal range obtained from the chart is entered at the top margin of the diagram. Going vertically down from this point until the cross curve which is labelled with the estimated meteorological visibility is reached, and then across to the left hand border scale gives the expected luminous range.

**Example**

A light of nominal range 25 miles is estimated meteorolgical visibility of 20 miles would be seen at a luminous range of 42 miles.

The light will only be seen at the luminous range if the observer has sufficient height of eye. It will be seen at the luminous range if the geographical range is greater than the luminous range. If the luminous range is greater than the geographical range then the light will be seen rising or dipping at the geographical range.

**Note**

The luminous range obtained from the diagram in this way is only approximate and may vary with different atmospheric conditions and conditions of background lighting from shore etc. The luminous range should only be used as a guide to when a light may be expected to be seen, and not to establish a position circle when it is seen. This may only be done when the light is seen rising or dipping.

**Danger angles**

The vertical sextant angle or the horizontal sextant angle may be used as a danger angle to enable the navigator to maintain a required distance off a navigational hazard.

The vertical sextant angle of an elevated point increases as the observer's distance from it decreases. If the distance inside which the observer does not wish to go is used to enter the distance by vertical sextant angle tables, then the vertical sextant angle corresponding to that distance may be extracted. Monitoring of the vertical sextant angle of the elevated point as the vessel passes, to ensure that it does not attain a greater value than that extracted for the minimum distance, will ensure that the vessel does not go closer than desired to the danger.

This technique may be used to advantage when rounding or passing a point which is suitable for observation of vertical sextant angle, at a small distance off. The time required to fix the ship by

## POSITION LINES

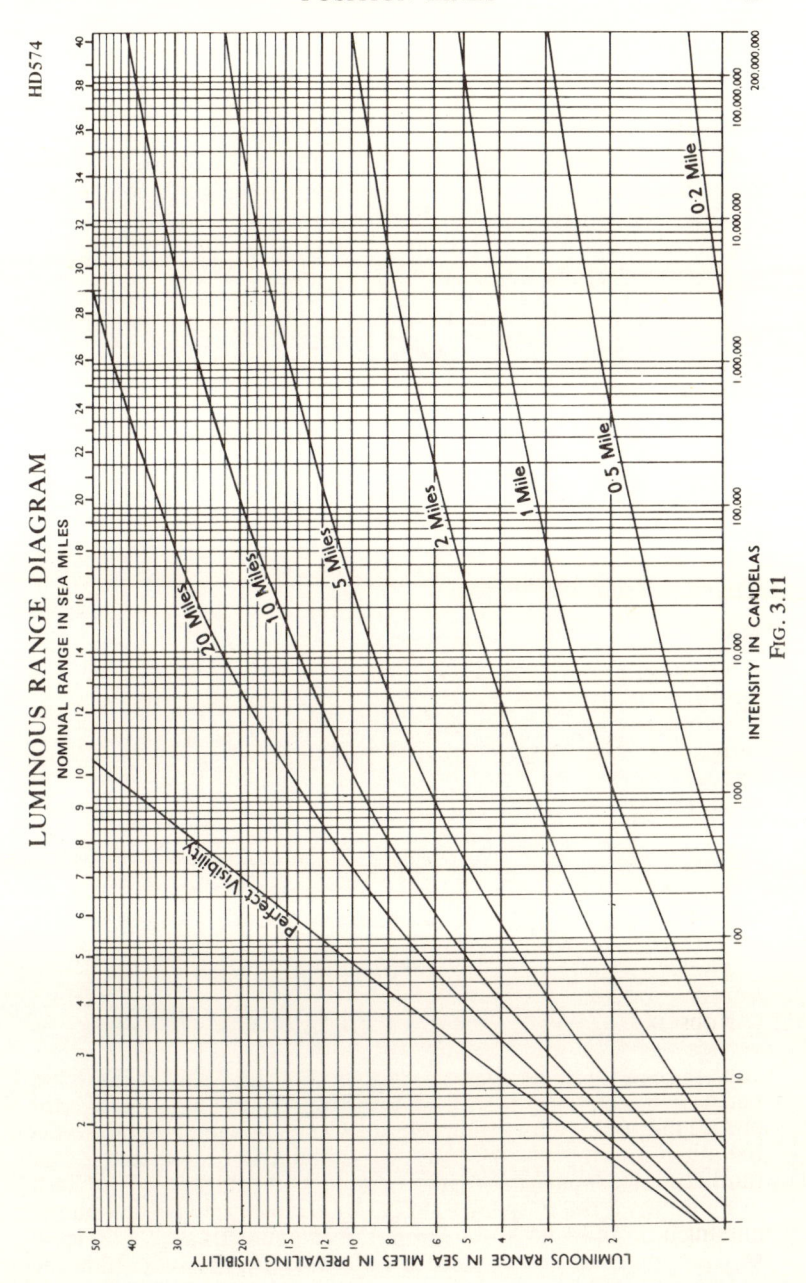

LUMINOUS RANGE DIAGRAM

Fig. 3.11

other methods to monitor the distance off may take too long to give adequate warning when navigating close to.

## Example

It is required to pass the point of land shown in figure 3.12, at a distance not less than 5 cables. A lighthouse of height 35 metres lies 4 cables inland from the outlying danger. Find the danger angle to set on a sextant to observe the vertical angle subtended by the light.

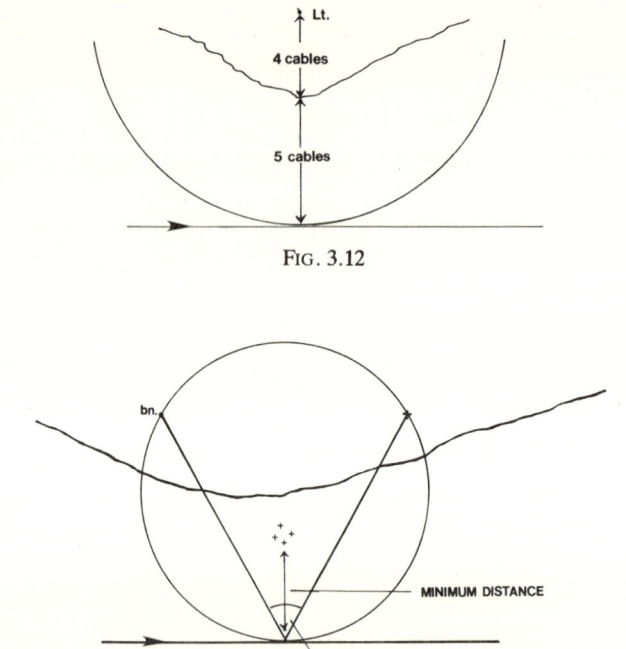

Fig. 3.12

Fig. 3.13

A horizontal angle may be used in the same way if there are beacons or charted buildings available which are suitable for horizontal angle observation. Figure 3.13 shows an outlying danger with two suitable marks, one on either side of the danger. A minimum distance may be chosen, and a point marked on the chart offshore from the outlying danger by this distance. The horizontal angle can then be measured from the chart by drawing lines from this point to the two objects to be observed, and measuring the angle between them. Monitoring of this angle as the vessel passes the danger will ensure that the vessel does not go inside the circle shown in figure 3.13.

**Hyperbolic position lines**

A hyperbolic position line is obtained from measurement of the difference in the distances from the observer to two fixed points. Invariably this informatoin is obtained by means of radio navigational aids, and the fixed points are the positions of the radio transmitters. The operator may make the necessary measurements with the receiving equipment without being aware of the underlying principles, and furthermore the position lines must be overprinted on navigational charts, as the hyperbolic shape is not easily drawn by the navigator. A series of charts is produced for each hyperbolic navigational system which are overprinted with hyperbolae which are representative of the position line at suitable intervals, between which the navigator must interpolate to find his particular position line. The charted hyperbolae are labelled with values in the same units which are displayed by the receiving equipment. Navigational aids which give hyperbolic position lines include the Decca Navigator, Loran, Omega, and Consol.

In order to qualify for a Department of Trade Class V, Class IV or Class III certificate the candidate must hold an Electronic Navigational Aids certificate which covers the use of all radio navigational aids. A description of the principles and operation of these aids is therefore not appropriate here. Candidates will however be expected to be able to plot position lines from information derived from these navigationl aids, in particular the Decca Navigator, which is the only one of the aids mentioned which give accuracy consistent with all coastal navigation requirements.

## CHAPTER 4

## THE SAILINGS

In this chapter the problems of calculating the course and distance between two positions on the earth's surface are considered.

### Parallel sailing

To be used when finding the distance to steam between two positions which are in the same latitude.

The distance measured along a parallel of latitude between any two given meridians decreases as the meridians converge towards the poles, being maximum at the equator. This distance, i.e. the distance measured along a parallel of latitude between two given meridians, is called the departure, and is expressed in nautical miles. There is therefore a relationship between departure, difference of longitude, and latitude.

The exact relationship can be seen as follows:

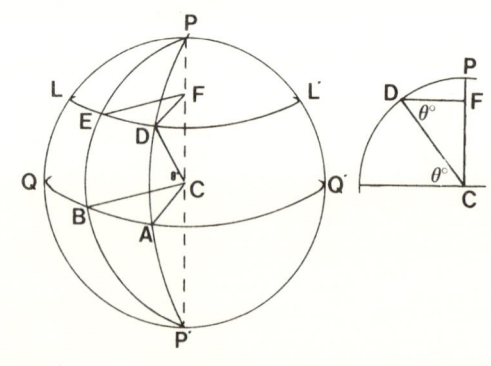

Fig. 4.1

In the figure let the circle represent the earth, C its centre, QQ′ the equator, LL′ a parallel of latitude $\theta°$, PP′ the earth's axis and F the centre of the small circle LL′.

D and E are two positions on the parallel LL′ with PAP′ and PBP′ the meridians through these two places. CA, CB and CD are radii of the earth.

# THE SAILINGS

By circular measure, the length of an arc, which subtends any given angle at its centre, is proportional to its radius.

Thus $\dfrac{DE}{DF} = \dfrac{BA}{AC}$ where DE is the departure and BA is the d. long.

Therefore $\dfrac{DE}{BA} = \dfrac{DF}{AC}$

and as DC and AC are both radii of the earth:

$$\dfrac{DE}{BA} = \dfrac{DF}{DC}$$

Thus $\dfrac{DE}{BA} = \text{Cosine } \angle FDC$

Thus $\dfrac{\text{Departure}}{\text{D. long.}} = \text{Cosine Latitude.}$

The finding of distance between any two positions on the same parallel is merely the application of this formula.

## Example 1

Find the distance to steam between the two positions:

A 51° 20′ N.   48° 30′ W.
and B 51° 20′ N.   38° 10′ W.

d. long.   = 10° 20′ E. = 620′
and Departure = d. long. × cos. latitude

Departure = 620 × cos. 51° 20′
= 387·4
Distance = 387·4 miles

| | Number | Log |
|---|---|---|
| | 620 | 2·79239 |
| | cos. 51° 20′ | $\bar{1}$·79573 |
| | | 2.58812 |

## Example 2

In what latitude will a d. long. of 3° 40′ correspond to a departure of 120 nautical miles?

sec. lat. = $\dfrac{\text{d. long. in mins.}}{\text{dep. in M.}}$

= $\dfrac{220}{120}$

= 1·8333

Latitude = 56° 56½′ N. or S.

Fig. 4.2

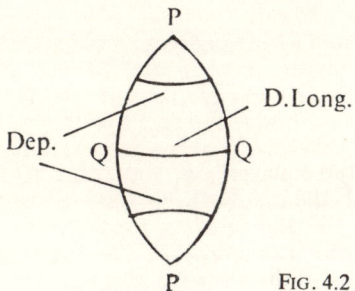

## Example 3

A vessel steams 090° T. from long. 35° 25′ W. to long. 28° 53′ W. How far did she steam if the latitude was 41° 20·5′ N.?

Initial long. = 35° 25′ W.
Final long. = 28° 53′ W

d. long. =  6° 32′ E.
       = 392′ E.

dep. in M.

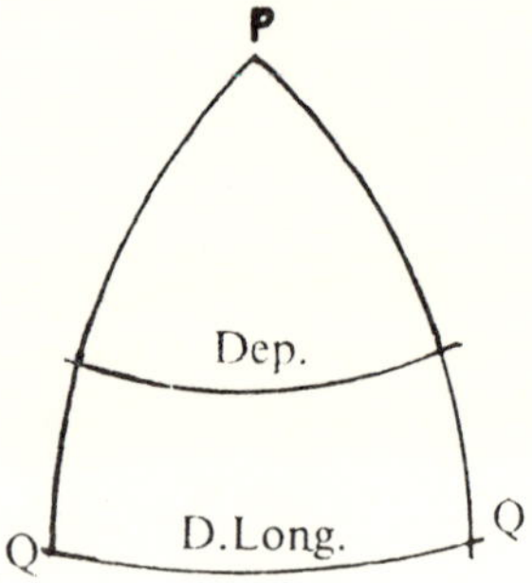

= d. long. in mins. × cos. lat.
= 392 × cos. 41° 20·5′
= 294·3

| Number | Log |
| --- | --- |
| 392 | 2·59329 |
| cos. 41° 20·5′ | 9·87552 |
| | 2·46881 |

Dist. steamed = 294·3 M.

FIG. 4.3

## Example 4

A vessel steams from a position in latitude 60°, in a direction 000° T. for a distance of 90 miles. She then steams 90 miles 090° T., 90 miles 180° T. and 90 miles 270° T. How far is she from her initial position?

**Note**

The distance steamed in a northerly direction gives a d. lat. of 90′ N. or 1° 30′ N. She will arrive therefore in latitude 61° 30′ N. at the end of the first leg. The same d. lat. is made good on the southerly leg. She will therefore arrive back in the same latitude of 60° N. A distance of 90 miles in the higher latitude will, however, give a larger d. long. than 90 miles in the lower latitude, and she will not reach her initial longitude when sailing on the westerly leg. Her distance from her initial position will be the difference in the departures for the two latitudes corresponding to the d. long. made good when sailing east.

# THE SAILINGS

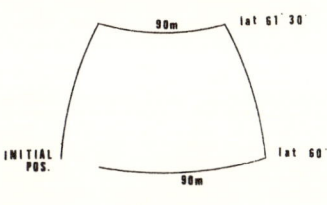

FIG. 4.4

Thus in latitude 61° 30′
 dep.   = d. long. cos. lat.
 90     = d. long. × cos. 61° 30′
 d. long. = 90 × sec. 61° 30′
         = 188·62

Thus in latitude 60°
 dep.   = 188·62 × cos. 60°
        = 94·31

Thus distance from initial position = 94·31 − 90
                                    =  4·31 miles

| Number | Log |
|---|---|
| 90 | 1·95424 |
| sec. 61° 30′ | 0·32134 |
|  | 2·27558 |
| cos. 60° | 1̄·69897 |
|  | 1·97455 |

## Example 5

Two vessels 45 nautical miles apart on the parallel of 40° 30′ N. steam 180° T., at equal speeds, until the distance between them is 55 nautical miles. How far did each steam?

d. long. in mins.
 = dep. in M. × sec. lat.
 = 45 × sec. 40° 30′

As both vessels steam 180° T., their d. long. is the same on both parallels.

sec. lat. = $\dfrac{\text{d. long. in mins.}}{\text{dep. in M.}}$

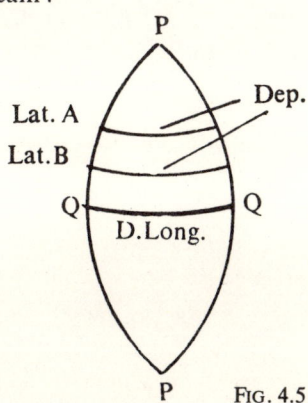

= $\dfrac{45 \times \sec. 40° 30′}{55}$

= $\dfrac{9 \times \sec. 40° 30′}{11}$

New Lat. = 21° 39·6′ N.

FIG. 4.5

## Example 6

At what rate in knots is a place in latitude 50° 56′ N. being carried around by the earth's rotation?

In 24 hours any place is carried round through 360°. This can be thought of as the d. long. Thus in one hour the d. long is 15°.

Thus distance in miles moved in one hour = departure

and dep. = 15 × 60 × cos. 50° 56′
        = 900 × cos. 50° 56′
        = 567·2
Thus speed = 567·2 knots

| Number | Log |
|---|---|
| 900 | 2·95424 |
| cos. 50° 56′ | 1·79950 |
|  | 2·75374 |

## EXERCISE 4A

1. In what latitude will a departure of 300 nautical miles correspond to a d. long. of 6° 40′?

2. On a certain parallel the distance between two meridians is 250 M., while the d. long. between the meridians is 12° 30′. What is the latitude?

3. In latitude 50° 10′ N. the departure between two meridians is 360 nautical miles. What is the d. long.?

4. A vessel steams on a course of 090° T. from $P$ in lat. 23° 30′ N., long. 59° 10′ E. to $A$ in lat. 23° 30′ N., long. 65° 30′ E. How far did she steam?

5. From lat. $X°$ N. a vessel steams 000° T. 50 M., and then 090° T. 100 M. If the difference of longitude is 185′, find lat. $X$.

6. From lat. 44° 15′ N., long. 10° 20′ W. a vessel steamed 270° T. for 550 nautical miles, and then 180° T. for 753 nautical miles. Find her final position.

7. On a certain parallel, the distance between two meridians is 150 nautical miles. On the Equator, the distance between the same two meridians is 235 nautical miles. What is the latitude of the parallel?

8. The distance between two meridians in lat. 48° 12′ N. is 250 M. What is the angle at the pole?

9. A vessel steams 470 nautical miles along the parallel of $X°$ N. from long. 15° 35′ W. to the meridian of 27° 20′ W. What is the latitude of $X$?

10. From lat. 39° 00′ N., 33° 10′ W. a vessel steamed 270° T. at 10 knots for 3 days 8 hours. In what D.R. position did she arrive?

## EXERCISE 4B

1. The distance between two meridians is 427 nautical miles in lat. 50° 20′ N. What is the angle at the pole?

2. Two ships on the parallel of 17° S. are 55 nautical miles apart. What would be their distance apart if they were on the parallel of 40° N.?

## THE SAILINGS

3. Two ports, $A$ and $B$ are in the Northern Hemisphere. On the parallel of $A$, the distance between their meridians is 250 M., on the parallel of $B$ it is 350 M., and on the Equator it is 400 M. What are the latitudes of the ports?

4. At what rate does an observer in lat. 50° 20′ rotate? (Answer to be in knots.)

5. A vessel in latitude 48° 30′ N. steams 270° T. at 10 knots for 24 hours. By how much is the longitude changed?

6. In lat. 50° 20′ N. a vessel steams from long. 15° 46′ W. to long. 31° 18′ W. What distance was made good?

7. A ship steams 090° T. for 200 nautical miles in lat. 49° 10′ N. By how much will her clocks have to be advanced?

8. The distance between two meridians in the Northern Hemisphere is 240 M. On the Equator it is 400 M., and in the Southern Hemisphere it is 360 M. What is the d. lat. between the two parallels?

9. In what latitude is the departure in nautical miles five-sevenths the d. long. in minutes?

10. In lat. 48° 30′ N. a vessel is in long. 34° 30′ W.; at noon L.A.T. the course is set 270° T., and the following day at noon L.A.T. she is in long. 40° 30′ W. What was the vessel's average speed?

11. Two vessels 200 nautical miles apart on the same parallel steam 180° T. to the parallel of 20° N., where their d. long. is found to be 5° 10′. How far did each steam?

12. A vessel leaves lat. 52° 21′ N., long. 30° 20′ W., and by steering 270° T. at 10 knots for 24 hours, arrives in lat. 52° 21′ N., long. 36° 00′ W. Find the set and drift.

**Plane sailing** (Mean Lat Sailing)

To be used to find the course and distance between two positions which are not in the same latitude, and when the distance is small.

Given the latitude and longitude of the two positions we can obtain the d. lat. and the d. long. between these positions.

The factors d. lat. and distance are measured in the same units and can be graphically represented as the two adjacent sides of a right-angled triangle. The angle between them can be made to represent the course, thus:

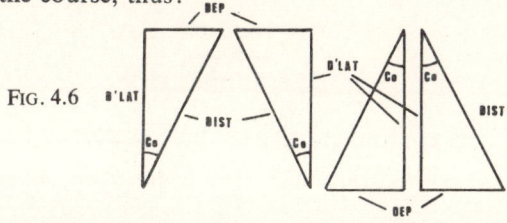

Fig. 4.6

The appropriate one of the above triangles to be used will be decided by the direction of the d. lat. and the d. long., and hence the quadrant in which the course lies.

To solve this triangle for course and distance, we need to know two other arguments of the triangle. We know d. lat., but we also need to know the length of the third side.

The length of the third side can be thought of as the departure between the two positions, and there will be one value of length which will give, when used to solve the triangle, the correct values of course and distance. We can calculate a value for departure by the parallel sailing formula:

$$\text{dep.} = \text{d. long.} \times \text{cosine latitude.}$$

But which latitude do we use in this formula. There is no readily apparent choice as the two positions are in different latitudes. The correct latitude to use would be that latitude which will give the required correct value of departure, but as yet we have no way of knowing this latitude, and as an approximation to it we use the numerical mean latitude between the two positions. (Hence the name Mean Lat. Sailing.)

**Note**

The inaccuracy due to the use of the mean latitude means that this method is only suitable for problems in which the d. lat. and therefore the distance is fairly small.

*Procedure*

1. From the two positions given calculate the d. lat. and the d. long., and also the value of mean latitude.

2. Using the mean latitude in the parallel sailing formula, find the departure.

3. Solve the plane sailing triangle, using departure and d. lat. to find course and distance, thus:

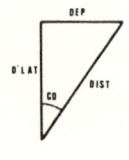

FIG. 4.7

From the triangle, $\dfrac{\text{dep.}}{\text{d. lat.}} = \tan.$ course

and distance $=$ d. lat. $\times$ sec. course

## THE SAILINGS

### Example 1

Find the course and distance between the following positions.

$A$    37° 01′ N.      9° 00′ W.
$B$    36° 11′ N.      6° 02′ W.
──────      ──────
d. lat.    50′ S.    d. long. 2  58′ E.
                              =178′

½ d. lat.    36° 11′ N.
              25′
            ──────
mean lat. 36° 36′ N.

dep. = d. long. × cos. lat.
     = 178 × cos. 36° 36′
     = 142·9

| Number | Log |
|---|---|
| 178 | 2·25042 |
| cos. 36° 36′ | 1·90462 |
| | 2.15504 |

$\dfrac{\text{dep.}}{\text{d. lat.}}$ = tan co.

tan co. = $\dfrac{142·9}{50}$
       = 70° 43′

| Number | Log |
|---|---|
| 142·9 | 2·15504 |
| 50 | 1·69897 |
| | 0·45607 |

dist. = d. lat. × sec. co.
      = 50 × sec. 70° 43′
      = 151·4

| Number | Log |
|---|---|
| sec. 70° 43′ | 0·48114 |
| 50 | 1·69897 |
| | 2·18011 |

Answer: course = S. 70¾ E., distance = 151·4 miles.

### Example 2

The course and distance from $A$ to $B$ is 055° T. 720 nautical miles. Find the d. lat. and departure made good.

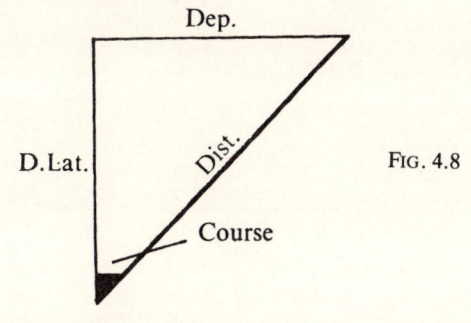

Fig. 4.8

d. lat. = dist. × cos. (course)
= 720 × cos. 55°
= 412·96
= 6° 53′ N.

| Number | Log |
|---|---|
| 720 | 2·85733 |
| cos. 55° | 9·75859 |
| | 2·61592 |

dep. = dist. × sin. (course)
= 720 × sin. 55°
= 589·8 M.

| Number | Log |
|---|---|
| 720 | 2·85733 |
| sin. 55° | 9·91337 |
| | 2·77070 |

D. lat. = 6° 53′ N., dep. = 589·8 nautical miles

**Example 3**

From lat. 50° 28′ N., a vessel steamed 156° T. 1550 nautical miles. Find the latitude in which she arrived.

d. lat. = dist. × cos. (course)
= 1550 × cos. 24°
= 1416′
= 23° 36′ S.

| Number | Log |
|---|---|
| 1550 | 3·19033 |
| cos. 24° | 9·96073 |
| | 3·15106 |

Initial lat. = 50° 28·0′ N.
d. lat. = 23° 36·0′ S.

Final lat. = 26° 52·0′ N.

**Example 4**

A vessel steers 327° T. and makes a departure of 396·7 nautical miles. How far did she steam?

Dist. = dep. × cosec. (course)
= 396·7 × cosec. 33°
= 728·4 M.
Dist. steamed = 728·4 nautical miles

| Number | Log |
|---|---|
| 396·7 | 2·59846 |
| cosec. 33° | 10·26389 |
| | 2.86235 |

THE SAILINGS 59

## EXERCISE 4C

1. Find the course and distance between the following positions;
   A  35° 12′ N.    178° 12′ W.
   B  37° 06′ N.    177° 00′ E.

2. A vessel leaves position 45° 12′ N. 161° 12′ W. and steams 213° T. for 406 miles. Find the position arrived at.

3. Find the course and distance between the following positions.
   P  5° 21′ N.    168° 17′ E.
   Q  16° 38′ S.   153° 48′ W.

4. From position 40° 30′ S. 175° 45′ E. a vessel steams 050° T. for 506 miles. Find the arrival position.

5. Find the course and distance between the following positions.
   X  7° 45′ N.    80° 30′ W.
   Y  41° 00′ S.   178° 15′ E.

**The middle latitude**

The plane sailing gives inaccurate results due to the uncertainty in the value of the departure used to solve the plane sailing triangle. The inaccuracy is acceptable over short distances of up to a few hundred miles.

The correct value of departure to use in the plane sailing triangle is that value which will give the correct value for the course between the two positions being considered. As an approximation the departure used was obtained from the parallel sailing formula, using the numerical mean latitude, thus;

departure = d'long × cosine mean latitude

It can be shown that the correct value of departure is obtained if the latitude used in this formula is the middle latitude given by the formula;

$$\text{sec. mid. lat.} = \frac{1}{\text{d'lat.}} \int_{l^s}^{l^n} \sec l \cdot dl$$

where $l^n$ and $l^s$ are the latitudes of the positions concerned.

In practice this middle latitude may be found by applying a correction to the numerical mean latitude, the correction being obtained from nautical tables. Thus the correct departure is given by;

departure = d'long × cosine middle latitude.

If this departure is used to solve the plane sailing triangle then more accurate values of course and distance are obtained. These methods are not often used however because there is an alternative method of finding course and distance which involves less calculation than plane sailing or middle latitude sailing, but gives the same accurate results as middle latitude sailing. This alternative method is called Mercator Sailing.

## Mercator sailing

To be used when finding course and distance between two positions which are in different latitudes. It is accurate for large d. lats. and distances and is in practice employed in preference to the alternative methods as it involves less calculation.

If we draw a right-angled triangle on a mercator chart, such that the hypotenuse represents the rhumb line distance between the two positions on the chart, and one side represents the meridian through one of the positions, then the third side will lie along the parallel of latitude through the other position. The angle between the meridian and the hypotenuse will represent the course. The longitude scale on a mercator chart is a constant scale, so if we express the two sides opposite and adjacent the course in units of this scale, then we can find the course by:

$$\frac{\text{opp.}}{\text{adj.}} = \tan. \text{ co.}$$

The side opposite the course, i.e. the side lying along the parallel of latitude, will be the d. long.

To express the adjacent side, i.e. the side along the meridian, a value called the meridional parts for the latitude is tabulated in nautical tables.

Meridional parts for any latitude is the length along a meridian, on a mercator chart, measured in units of the longitude scale, between the Equator and the parallel of latitude in question.

If we extract the meridional parts for each of the latitudes concerned and take the difference between them, then this 'difference of meridional parts' (d.m.p.) will be the length of the side of the triangle which lies along a meridian, adjacent to the course angle, and measured in units of the longitude scale.

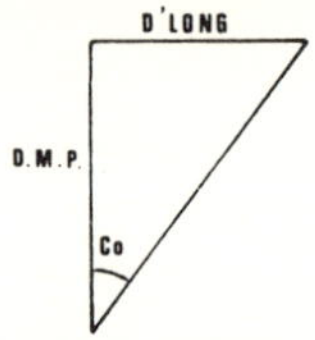

FIG. 4.9

In the triangle $\dfrac{\text{d. long.}}{\text{d.m.p.}} = \tan. \text{ course}$

# THE SAILINGS

We have thus found the course without using the factor departure and have avoided the inaccuracy which was encountered in plane sailing.

We can now revert to the plane sailing triangle with a knowledge of course and solve for distance by

$$\text{dist.} = \text{d. lat.} \times \text{sec. course.}$$

## Procedure

1. Write down the latitude and longitude of the positions, and against each latitude the meridional parts for that latitude from the nautical tables. Calculate d. lat., d. long., and the d.m.p.

**Note**

The rule for finding d.m.p. is the same as that for finding d. lat., i.e. same name take the difference and different name take the sum.

2. Calculate the course by $\dfrac{\text{d. long.}}{\text{d.m.p.}} = \tan.\ \text{course}.$

3. Calculate the distance by dist. = d. lat. × sec. course.

## Example 1

By Mercator Sailing find the true course and distance from $A$, lat. 49° 10′ N., long. 12° 30′ W., to $B$, lat. 25° 15′ N., long. 26° 50′ W.

| | | | | |
|---|---|---|---|---|
| $A$, lat. = 49° 10′ N. | M.P. | = 3379·6 | long. | = 12° 30′ W. |
| $B$, lat. = 25° 15′ N. | M.P. | = 1556·6 | long. | = 26° 50′ W. |
| d. lat. = 23° 55′ S. | D.M.P. | = 1823·0 | d. long. | = 14° 20′ W. |
| = 1435′ S. | | | | = 860′ W. |

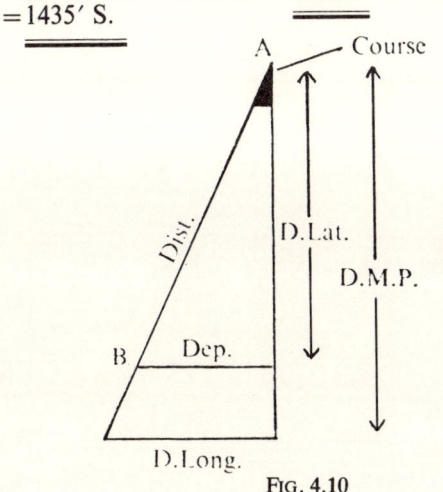

Fig. 4.10

62    PRACTICAL NAVIGATION FOR SECOND MATES

$$\tan.\text{ (course)} = \frac{\text{d. long.}}{\text{D.M.P.}}$$
$$= \frac{860}{1823}$$
Course $= \text{S. } 25° \ 15\cdot3' \text{ W.}$

| Number | Log |
|---|---|
| 860 | 2·93450 |
| 1823 | 3·26079 |
|  | 9·67371 |

**distance** = **d. lat.** × **sec. (course)**
$= 1435 \times \sec. 25° \ 15\cdot3'$
$= 1586\cdot7$ M.

| Number | Log |
|---|---|
| 1435 | 3·15685 |
| sec. 25° 15·3′ | 10·04363 |
|  | 3·20048 |

Course = 205° 15·3′ T., Dist. = 1586·7 M.

## Example 2

A vessel steams 040° T. for 2300 miles from position 39° 37′ S. 47° 28′ W. Find the arrival position.

d. lat. = dist. × cos. course (from the plane sailing triangle)
$= 2300 \times \cos. 40°$
$= 1761\cdot9$
$= 29° \ 21\cdot9'$

| Number | Log |
|---|---|
| 2300 | 3·36173 |
| cos. 40° | 1̄·88425 |
|  | 3·24598 |

| | | | |
|---|---|---|---|
| initial lat. | 39° 37′ S. | m.p. | 2577·82 |
| d. lat. | 29° 21·9′ N. | | |
| arrival lat. | 10° 15·1′ S. | m.p. | 614·25 |

d.m.p. 1963·57

d. long. = d.m.p. × tan. course (from mercator sailing triangle)
$= 1963\cdot57 \times \tan. 40°$

$= 1647\cdot6$
$= 27° \ 27\cdot6'$ E.

| Number | Log |
|---|---|
| 1963·57 | 3·29305 |
| tan. 40° | 1̄·92381 |
|  | 3·21686 |

initial long.  47° 28·0′ W.
d. long.       27° 27·6′ E.
───────────────
final long.    20° 00·4′ W.

final position 10° 15·1′ S.   20° 00·4′ W.

## EXERCISE 4D

1. Find the D.M.P. between the following pairs of latitudes:
(a) $\begin{cases} 40° \ 00′ \ N. \\ 50° \ 00′ \ N. \end{cases}$ (b) $\begin{cases} 20° \ 10′ \ N. \\ 10° \ 35′ \ S. \end{cases}$ (c) $\begin{cases} 53° \ 15′ \ S. \\ 24° \ 47′ \ S. \end{cases}$ (d) $\begin{cases} 22° \ 18′ \ S. \\ 39° \ 53′ \ N. \end{cases}$

2. Find the true course and distance from lat. 20° 14′ N., long. 22° 17′ W., to lat. 11° 35′ S., long 41° 05′ W.

3. Calculate by mercator sailing method the true course and distance from A, lat. 40° 10′ N., long. 09° 45′ W., to B, lat. 10° 15′ N., long. 18° 11′ W.

4. By using mercator sailing calculate the true course and distance from P, lat. 41° 13′ N., long. 173° 50′ W., to Q, lat. 07° 50′ S., long. 79° 55′ W.

5. A vessel steams 210° T. 750 nautical miles from 29° 30′ N., 162° 20′ E. In what position did she arrive?

6. From lat. 10° 12′ S., long. 35° 05′ W., a vessel steers 017° T. and arrives in long. 28° 29′ W. What was the distance steamed and the latitude reached?

7. A vessel steams 225° T. 800 M., and then 135° T. 800 M. from lat. 10° 00′ S., long. 00° 00′. In what position did she arrive?

8. A vessel steams 065° T. 1850 M. from lat. 20° 12′ N., long. 178° 40′ E. Find the latitude and longitude of the position in which she arrives.

9. Calculate the true course and distance from 05° 20′ N., 79° 05′ E., to 24° 20′ S., 112° 03′ E.

10. Calculate the true course and distance from 37° 03′ N., 13° 20′ E., to 31° 20′ N., 29° 55′ E.

## EXERCISE 4E

**The following problems are typical of those encountered in Class V Practical papers**

1. From the following information find the D.R. position by mercator sailing.
   initial position 50° 33′ N.  7° 25′ W.
   course 237° T.
   distance steamed 1008 miles.

2. Find by mercator sailing the true course and distance from 48° 11′ S. 169° 50′ E. to 23° 36′ S. 161° 42′ W.

3. Find the course and distance to steam by plane sailing from a position off Ushant (48° 20′ N. 5° 12′ W.) to a position off San Sebastian (42° 30′ N. 2° 00′ W.).

4. Find by plane sailing the course and distance from a position off Ushant (48° 20′ N. 5° 12′ W.), to a position off Cork (51° 44′ N. 8° 10′ W.).

5. Find by plane sailing the D.R. position if a vessel steams from a position off Esbjerg (55° 28′ N. 7° 50′ E.), on a course of 248° T. for 95 miles.

6. Find by plane sailing the D.R. position if a vessel steams 355° T. from a position off Cape Villano (43° 10′ N. 9° 30′ W.), for 18 hours 36 minutes at 9 knots.

7. Find the course and distance by mercator sailing between the following positions.

    a 52° 35′ N. 2° 38′ E.
    b 59° 15′ N. 4° 30′ E.

8. A vessel leaves a position 43° 50′ N. 9° 00′ W. and steams 328° T. for 440 miles. Find by mercator sailing the D.R. position at the end of the run.

9. Initial position 60° 40′ N. 0° 30′ W. Course 160° T. Distance steamed by log 150 miles. Find by mercator sailing the D.R. position at the end of the run.

10. A vessel steams a course of 090° T. for 145 miles from an initial position 57° 50′ N. 3° 30′ W. Find the D.R. position at the end of the run.

## CHAPTER 5

## THE TRAVERSE TABLE AND THE TRANSFERRED POSITION LINE

The traverse tables are tabulated solutions of plane right angled triangles. A table is provided for each value of the acute angles from 1° to 89° at 1° intervals, each table giving values of the three sides for a hypotenuse value from 1 unit to 600 units. By interpolation and extrapolation any right angled triangle may be solved with the traverse tables.

Traverse tables in nautical tables are specifically designed to solve the formulae associated with the parallel sailing and plane sailing problems, and columns are headed accordingly.

### Description of tables

There is one table for each whole number of degrees of the acute angles in the right angled triangle from 1° to 45°. To avoid unnecessary repetition, values of angles between 45° and 89° are listed at the foot of the table which is given for the angles complement. Separate column headings are given at the foot of each column to be used when the angle required is listed at the bottom of the page. Three columns are given with each table these being headed, hypotenuse, adjacent, and opposite. The length of the adjacent and opposite sides of the triangle are given for each value of the hypotenuse between 1 and 600. To facilitate the solution of the parallel sailing formula to solve the right angled triangle shown in figure 5.1 which corresponds to the parallel sailing formula;

$$\frac{\text{departure}}{\text{d'long}} = \text{cosine latitude.}$$

Alternative coulmn headings are given for the hypotenuse and the adjacent columns. The hypotenuse column is also headed d'long, and the adjacent column is also headed departure. In this case the table degree headings will represent degrees of latitude.

To facilitate the solution of the plane sailing problem the alternative headings distance, d'lat. and departure are given to the hypotenuse, adjacent and opposite columns respectively. In this case the table degree headings will represent the course angle in the plane sailing triangle.

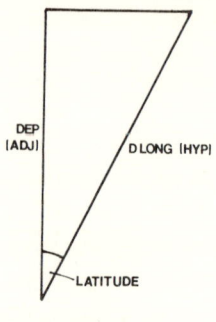

FIG. 5.1

## Solution of the parallel sailing formula

$$\frac{\text{departure}}{\text{d'long.}} = \text{cosine latitude.}$$

*Procedure*

1. Locate the table which is headed with the whole number of degrees of the latitude given.
2. In the column headed d'long. (hypotenuse column), locate the value of the d'long. given.
3. Read off the value of departure against the required value of d'long. from the column headed departure (adjacent column).
4. If the latitude given is not a whole number of degrees, repeat for the next highest value of whole number of latitude and interpolate between the two results according to the number of minutes in the latitude.

### Example 1. Find the departure for a d'long. of 138′ in latitude 38°.

1. Enter table 38°.
2. Go down the d'long. column to locate 138.
3. Against 138 extract a departure of 108·7′.

### Example 2. Find the departure in latitude 65° 40′ for a d'long. of 39·4′.

1. Enter the table headed 65° (at foot of the page).
2. Locate a d'long. of 39·4′ in the column headed d'long.
3. Against 39·4, interpolating between 39 and 40, extract a departure of 16·6. (Interpolation may be facilitated by mentally shifting the decimal place and locating a d'long. of 394. This gives a departure of 166. The decimal place can now be replaced to give departure 16·6).

# THE TRAVERSE TABLE

4. Repeat for a latitude of 66°. This gives a departure of 16·0.

5. Interpolate between 16·6 and 16·0 for a latitude of 65° 40' answer must lie two thirds of the way from 16·6 towards 16·0. The required departure is therefore 16·2'.

## To solve the plane sailing triangle

The headings distance, d'lat. and departure are used, and the table degree headings are used as the course angle. The course should be expressed in quadrantal notation. In practice the problem is usually required to be solved with the course and distance known, in order to find a D.R. position. In this case the d'lat. and departure are easily extracted against the distance steamed. The tables are a little more difficult to use if the d'lat. and departure are known and it is required to find the course and distance. In practice this problem is usually done by calculation for accuracy, but it is possible to get a quick solution by the traverse table by finding the table in which the values of d'lat. and departure appear against each other. The distance can then be taken against these values and the course from the table heading. If interpolation is required, this may take some practice. (See example 3.)

## Example 1

**Given** course 148° T., distance 520 miles. Find the d. lat. and dep.

Course 148° becomes S. 32° E. in quadrantal notation.

## Steps

1. Find the page headed 32°.

2. Move down the page in the dist. column to 520.

3. Take out the d. lat. and dep. from the appropriate columns.

*Answer.* Course S. 32° E. and dist. 520 M., d. lat.=441' S., dep.=275·6 M. E.

The course being in the S. E. quadrant indicates that the d. lat. is named S. and the departure is named E.

## Example 2

**Given** course S. 62° W., dist. 47·4 M., find the d. lat. and dep.

## Steps

1. Note that the angle is greater than 45° and will therefore be at the bottom of the page.

2. The dist. column is the same whether we are dealing with the top or bottom of the page, but the columns headed d. lat. and dep. are reversed, since we are concerned with complementary angles.

3. Turn to the page where the angle is 62°.

4. Shift the decimal point on the distance given, and look up 474 in the dist. column. This makes the task easier.

5. The d. lat. is 222·5 and the dep. is 418·5. Having multiplied the distance by 10, it will be necessary to divide these by 10 to arrive at the correct relationships for a distance of 47·4 miles.

*Answer.* Course S. 62° W. and distance 47·4 M. give d. lat. 22·25′ S. and dep. 41·85 M. W.

**Example 3**
**Given** d. lat. = 339·6′ N., dep. = 295·2 M. W., to find the course and distance.

**Steps**

1. Note that the d. lat. being greater than the dep. the angle will be less than 45°, and will therefore be found at the top of the page. Also, the values are near one another, so that the angle is approaching 45°.

2. Open the table at about 35°, and look down the d. lat. and dep. columns. The given values are found to be widely separated, so turn over a few pages, to 39°, and again look up the values. Here they are much closer, so continue to turn over the pages until they are found as near together as possible—this will be on the page headed 41°.

*Answer.* With d. lat. 339·6′ N. and dep. 295·2 M., W., course = N. 41° W. Dist = 450 M.

The values may not always be found so easily as in the examples shown. It may be necessary to (1) interpolate or (2) use aliquot parts. Interpolation for the factors dist., d. lat. and dep. can be quite accurate, since we are dealing with similar triangles; but for angles, the interpolation, though not exact, is within practical limits.

**To change d. long. into departure and vice versa**

**Example**

Find the departure corresponding to a d. long. of 58·5′ in latitude 50° 24′ N.

Under angle 50°, look up 585 in the dist. column, and this gives 376·0 in the d. lat. column.

Similarly, angle 51° and dist. 585 give 368·2 in the d. lat. column.

The dep. corresponding to the d. long. of 58·5 will therefore lie between 37·6 and 36·82. The interpolation is carried out thus, and, with practice it can be done mentally.

# THE TRAVERSE TABLE

for angle 50° and dist. 585, d. lat.=376·0
for angle 51° and dist. 585, d. lat.=368·2

$$\text{diff. for } 1° = 7·8$$
$$\text{multiplied by } 0·4 \quad 0·4$$

$$\text{diff. for } 0·4° \quad 3·12$$

∴ angle 50·4° and dist. 585 give d. lat. 376·0−3·12=372·9.

*Answer*. In lat. 50° 24′ N., d. long. 58·5′, dep.=37·29 M.

## To solve the plane sailing problem

## Example
A vessel steering 240° T. at 15 knots leaves a position 30° N. 179° 15′ W. Find the position of the vessel after 24 hours.

Course 240°=S. 60° W., distance=24 × 15=360 m.

*Procedure*

1. Turn up the page in the traverse table headed 60°.

2. Using the column names at the foot of the columns, move up the distance column to 360.

3. Extract the d. lat. and the departure from the appropriate columns named so (d. lat.=180·0′ dep.=311·8).

4. Apply the d. lat. to the initial latitude and calculate the mean lat.

5. Enter the page headed with the mean lat., and using the headings d. long. and dep. go down the dep. column to 311·8 and extract the d. long.

6. Apply d. long. to the initial longitude.

|  |  |
|---|---|
| lat. left | 30° 00·0′ N. |
| d. lat. | 3° 00·0′ S. |
| arr. lat. | 27° 00·0′ N. |
| ½ d. lat. | 1° 30·0′ |
| mean lat. | 28° 30·0′ N. |

**Note**
Direction of d. lat. and d. long taken from the name of the course

```
position left     30° 00·0′ N.    179° 15·0′ W.
d. lat.            3° 00·0′ S.      5° 54·7′ W.
                  ─────────────   ─────────────
arrival position  27° 00·0′ N.    174° 46·3′ E.
```

The solution of the mid lat. problem is exactly the same except that the correction to mean lat. is applied before taking out d. long.

### Example

Find by use of traverse table the course and distance from A lat. 46° 30′ N., long. 15° 45′ W. to B lat. 43° 50′ N., long. 25° 28′ W.

```
A lat. 46° 30′ N.      long.  15° 45′ W.      A lat. 46° 30′ N.
B lat. 43° 50′ N.      long.  25° 28′ W.      B lat. 43° 50′ N.
                                              ─────────────
d. lat.   160′ S.      d. long. 583′ W.       2)90° 20′ N.
                                              ─────────────
                                              mean lat. 45° 10′ N.
```

M. lat. 45° 00′, d. long. 583′ gives dep. 412·2
M. lat. 46° 00′, d. long. 583′ gives dep. 405·0
                                        ─────
                                   diff. 7·2

∴ For M. lat. 45° 10′, d. long. 583′ dep.=411·0.
From traverse table, with d. lat. 160′ S., dep. 411′ W.
(By inspection) co. S. 68¾° W. dist. 441 miles.

### Note

If the mid lat. had been used the distance would have been 442 miles.

If set and drift is required, this will be found by calculating the course and distance between the position by dead reckoning and the position by observation. The method is, therefore, the same as shown in the example.

### Note

If the solution of any triangle is required where the length of one of the sides is greater than the range of lengths given in the tables, then a solution can be found by dividing each known side by some convenient factor, usually 2. Then the length of any side found must be multiplied by the same factor.

# THE TRAVERSE TABLE

## EXERCISE 5A

**Traverse table**

1. True co.=N. 25° E.  dist. =238 M.   Find the d. lat. and the dep.
2. True co.=S. 10° E.  dist. =333 M.   Find the d. lat. and the dep.
3. True co.=N. 40°W.  dist. =505 M.   Find the d. lat. and the dep.
4. True co.=S. 70°W.  dist. =214 M.   Find the d. lat. and the dep.
5. True co.=306°      dist. =176 M.   Find the d. lat. and the dep.
6. True co.=065°      dep. =173·3 M.  Find the d. lat. and the dist.
7. True co.=148°      d. lat.=386·7′  Find the dep. and the dist.
8. Dist. =436 M.      dep. =262·4 M.  Find the course and the d. lat.
9. d. lat. =447·6′ N. dep. =198·3 M.E. Find the course and the dist.
10. d. lat. =351·1′ S. dep. =229·3 M.W. Find the course and the dist.
11. d. lat. = 44·6′ N. dep. = 14·5 M.E. Find the course and the dist.
12. d. lat. =312·3′ S. dep. =231·1 M.W. Find the course and the dist.
13. d. lat. =308·5′ N. dep. =367·7 M.W. Find the course and the dist.
14. d. lat. =855·0′ S. dep. =380·8 M.E. Find the course and the dist.
15. True co.=036°     dep. =723·0 M.  Find the dist. and the d. lat.

## EXERCISE 5B

**To change dep. into d. long. by inspection**
**Find the d. long, given**

1. dep. =354·8 M.    lat. =50° 00′ N.
2. dep. =261·8 M.    lat. =35° 00′ N.
3. dep. =246·0 M.    lat. =42° 30′ N.
4. dep. =197·0 M.    lat. =38° 12′ N.
5. dep. =348·4 M.    lat. =27° 00′ N.
6. dep. =361·2 M.    lat. =75° 00′ N.
7. dep. =294·6 M.    lat. =52° 00′ N.
8. dep. =326·9 M.    lat. =36° 30′ N.
9. dep. =444·4 M.    lat. =19° 15′ N.
10. dep. =258·7 M.   lat. =50° 45′ N.

## EXERCISE 5C

**To change d. long. into dep. by inspection**
**Find the dep., given**

1. d. long. =260·4′    lat. =40° 00′
2. d. long. =351·3′    lat. =48° 15′

72     PRACTICAL NAVIGATION FOR SECOND MATES

    3. d. long. = 58·1′        lat. = 56° 00′
    4. d. long. = 37·6′        lat. = 25° 00′
    5. d. long. = 667·0′       lat. = 47° 30′
    6. d. long. = 44·4′       lat. = 35° 15′
    7. d. long. = 518·5′       lat. = 36° 30′
    8. d. long. = 114·8′       lat. = 58° 30′
    9. d. long. = 534·7′       lat. = 67° 30′
   10. d. long. = 329·4′       lat. = 17° 30′

## EXERCISE 5D

**To find the course and distance**

**By inspection of the traverse table, find the course and distance**

| | | From | | | | To | |
|---|---|---|---|---|---|---|---|
| 1. | A | lat. | 50° 40′ N. | B. | lat. | 40° 50′ N. |
| | | long. | 40° 50′ W. | | long. | 50° 40′ W. |
| 2. | P | lat. | 35° 10′ N. | Q | lat. | 37° 50′ N. |
| | | long. | 27° 18′ W. | | long. | 31° 08′ W. |
| 3. | D | lat. | 25° 15′ S. | E | lat. | 22° 47′ S. |
| | | long. | 156° 44′ E. | | long. | 159° 53′ E. |
| 4. | S | lat. | 37° 53′ N. | T | lat. | 38° 10′ N. |
| | | long. | 177° 50′ W. | | long. | 177° 50′ E. |
| 5. | L | lat. | 10° 10′ N. | M | lat. | 09° 00′ N. |
| | | long. | 34° 40′ W. | | long. | 29° 10′ W. |

6. Find the set and drift, given
D.R. pos. lat. 50° 13′ N., long. 15° 15′ W. Pos. by obsn. lat. 50° 28′ N., long. 14° 44′ W.

7. Given initial position, lat. 40° 40′ N., long. 4° 04′ W.; course 214° T., dist. 100 M., find the D.R. position.

8. Find the true course and distance from 47° 06′ N., 39° 10′ W., to 48° 53·5′ N., 27° 04′ W.

9. Find the true course and distance from lat. 22° 33′ S., long. 96° 48′ E., to lat. 19° 43′ S., long. 92° 46′ E.

10. Find by inspection of the traverse table the course and distance from 18° 35·7′ N., 39° 53′ E. to 22° 45·5′ N., 37° 15·5′ E.

### Running up a D.R.

The traverse tables are used to find the D.R. position when more than one course and distance has been steamed since the last observed position. This problem is very quickly solved if the

# THE TRAVERSE TABLE

intermediate alter course positions are not required, by tabulating the d'lats. and departures for the individual courses and distances. These are then added (or subtracted if of opposite name), to find the total d'lat. and departure. The d'lat. is then applied to the initial latitude and the mean latitude found. The total departure is then converted to d'long. and applied to the initial longitude.

## Example 1

A vessel observes her position to be 40° 30′ N. 35° 15′ W. She then steams the following courses and distances:

$$056° \text{ T. distance } 45 \text{ miles}$$
$$020° \text{ T. distance } 20 \text{ miles}$$
$$335° \text{ T. distance } 35 \text{ miles}$$
$$300° \text{ T. distance } 50 \text{ miles}$$

Find the D.R. position.

| Course | Distance | D. lat. N. | D. lat. S. | Departure E. | Departure W. |
|---|---|---|---|---|---|
| N. 56° E. | 45 | 25·2 | | 37·3 | |
| N. 20° E. | 20 | 18·8 | | 6·8 | |
| N. 25° W. | 35 | 31·7 | | | 14·8 |
| N. 60° W. | 50 | 25·0 | | | 43·3 |
| | | 100·7 | | 44·1 | 58·1 |
| | | | | | 44·1 |

               d. lat. = 100·7 N.         dep. = 14·0 W.

initial latitude    40° 30·0′ N.
d. lat.              1° 40·7′ N.

arrival latitude   42° 10·7′ N.
mean latitude  = 41° 20·3′ N.  dep. 14·0 gives d. long. = 18·6′ W.
initial position   40° 30·0′ N.   35° 15′ W.
                      1° 40·7′ N.     18·6′ W.

arrival position   42° 10·7′ N.   35° 33·6′ W.

If during the steaming of the courses a current is estimated to be setting this can be treated as just another course with the drift as the distance, and the d. lat. and departure found summated with the other courses.

## Example 2

A vessel steamed the following courses and distances:

165° distance 50 miles
072° distance 63 miles
112° distance 84 miles
256° distance 58 miles

A current set 300° T., drift 10 miles. If the initial position was 46° 19′ N. 37° 47′ W., find the final position and the course and distance made good.

| Course | Distance | D. lat. N. | D. lat. S. | Dep. E. | Dep. W. |
|---|---|---|---|---|---|
| S. 15° E. | 50 | | 48·3 | 12·9 | |
| N. 72° E. | 63 | 19·5 | | 59·9 | |
| S. 68° E. | 84 | | 31·5 | 77·9 | |
| S. 76° W. | 58 | | 14·0 | | 56·3 |
| N. 60° W. | 10 | 5·0 | | | 8·7 |
| | | 24·5 | 93·8 | 150·7 | 65·0 |
| | | | 24·5 | 65·0 | |

Resultant d. lat. and dep. 69·3 S. 85·7 E.

| | | | |
|---|---|---|---|
| Initial latitude | 46° 19·0′ N. | | |
| D. lat. | 69·3′ S. | Mean lat. | = 45° 44·3′ N. |
| | | D. long. | = 121·7′ E. |
| Arrival latitude | 45° 09·7′ N. | | |
| Initial position | 46° 19·0′ N. | 37° 47·0′ W. | |
| | 69·3′ S. | 2° 01·7′ E. | |
| Arrival position | 45° 09·7′ N. | 35° 45·3′ W. | |
| From tables | with d. lat. 69·3′ S. dep. 85·7′ E. | | |
| | course = S. 51° E.  dist. = 110 miles | | |

Information required may vary somewhat, and each problem must be carefully considered. In some problems the set and drift of the current is asked for. To find this an observed position at the end of the traverse must be given and this should be compared with the D.R. calculated.

# THE TRAVERSE TABLE

## EXERCISE 5E

1. Find by traverse table the vessel's position at the end of the fourth course;

|                  |                     |
|------------------|---------------------|
| Initial position | 46° 45' N. 45° 30' W. |
| First course     | 202° T. by 72 miles |
| Second course    | 272° T. by 72 miles |
| Third course     | 337° T. by 36 miles |
| Fourth course    | 050° T. by 36 miles |

2. Find by traverse table the vessel's position at the end of the third course;

|                  |                     |
|------------------|---------------------|
| Initial position | 60° 30' N. 16° 45' W. |
| First course     | 213° T. by 64 miles |
| Second course    | 306° T. by 72 miles |
| Third course     | 082° T. by 80 miles |

3. Find by traverse table the ship's position at the end of the third course;

|                  |                     |
|------------------|---------------------|
| Initial position | 39° 25' N. 9° 38·5' W. |
| First course     | 262° T. by 9 miles  |
| Second course    | 169° T. by 146 miles |
| Third course     | 109° T. by 144 miles |

4. Find by traverse table the position at the end of the third course;

|                  |                     |
|------------------|---------------------|
| Initial position | 12° 12' S. 50° 58' E. |
| First course     | 296° T. by 60 miles |
| Second course    | 237° T. by 55 miles |
| Third course     | 215° T. by 101 miles |

5. A vessel observes a noon position 37° 54' N. 178° 29' E. The course is then 230° T. at 15 knots until 1800 hrs when an SOS is received from a position 37° 15' N. 179° 35' W. If speed is increased to 16 knots what is the course to be steered to the distress and what will be the ETA.

6. Find by traverse table the course and distance between the following positions.

|   |                       |
|---|-----------------------|
| A | 51° 30' N. 176° 42' W. |
| B | 50° 19' N. 179° 35' E. |

7. Find by traverse table the course and distance between the following positions.

|   |                       |
|---|-----------------------|
| A | 54°30' N. 37° 30' W.  |
| B | 52° 15' N. 42° 15' W. |

8. A vessel obtained a noon position 34° 06' S. 172° 09' E. She then steamed the following courses and distances;

First course    321°T. by 75 miles
Second course   037°T. by 52 miles
Third course    137°T. by 110 miles

A current was estimated to have set 260°T. by 20 miles in the interval. Find the EP at the end of the third course.

## TRANSFERRING THE POSITION LINE

If a position line is observed at some initial time, a position line valid for some later time may be found by moving the observed line in the direction made good by the vessel and by the amount of the distance steamed. The position line so found is referred to as a transferred position line and a fix may be produced by crossing it with another position line observed at the later time. The accuracy of the transferred position line depends upon the reliability of the course and distance used for running up.

Transferring the position line may be done by taking any point on the original position line and using it as a departure position. The course and distance may then be applied to this position (a) by laying off the course and distance on the chart from this point, or (b) by applying the course and distance by traverse table. The transferred position line is then drawn through the position obtained by running up, in the same direction as the original position line. The first method is normally used when coasting and navigating by the methods of chartwork, when the time intervals involved are small. The second method is normally used when out of sight of land and navigating by astronomical methods. The time intervals involved are usually longer, of the order of a few hours.

### The running fix

This is the name given to the fix produced by crossing an observed position line with a position line transferred or run up from an earlier observation.

### Example

At 0800 hrs Galley Head was observed to bear 040°T. At 0840 it was observed to bear 310°T. Find the position of the vessel at 0840 if the course and distance made good in the interval was estimated to be 075°T., 8 miles.

*Procedure* (refer to figure 5.2)

1. Lay off the two position lines given by the two bearings at 0800 (040°) and 0840 (310°), from the charted position of the point observed. Mark the lines with single arrows.

2. From any convenient point on the first position line, lay off the course and distance made good (075° by 8 miles).

# THE TRAVERSE TABLE

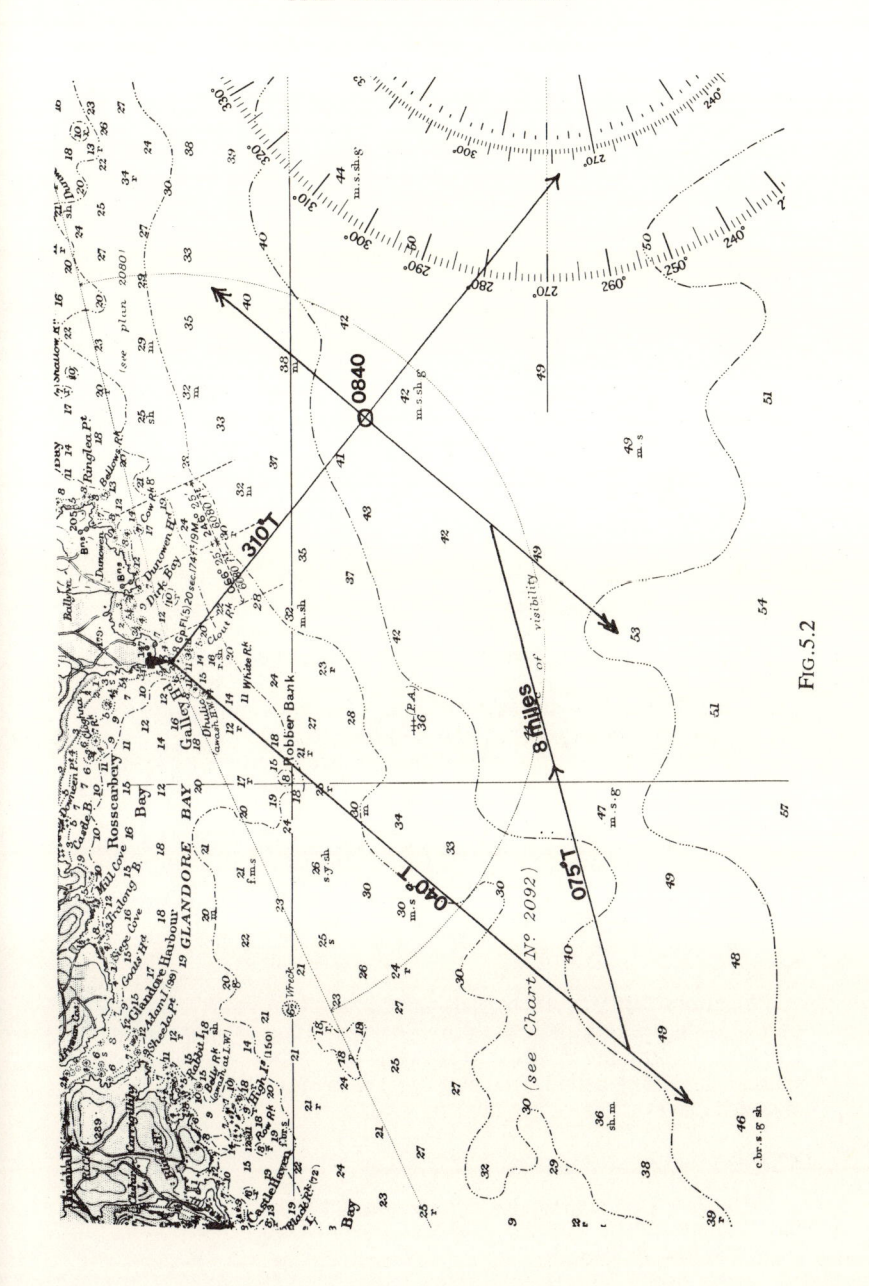

Fig. 5.2

3. Draw the transferred position line through the position reached in (2), parallel to the first position line (040), to cut the second position line (310). Mark the transferred position line with double arrows. This point of intersection gives the position of the ship at the time of the second bearing.

4. The position at the time of the first bearing may be found if required by transferring the course and distance made good, through the second position, to cut the first bearing.

### The running fix with tide

Any tide estimated to set in the duration of the running fix may be allowed for by laying off the set and drift from the end of the course and distance, before transferring the position line.

### Example

At 1300 hrs Old Head of Kinsale Lt. Hse. was observed to bear 030°T. and at 1330 hrs the same lighthouse was observed to bear 295°T. Find the position at 1330 if the vessel steered 080°T. at 16 knots in the interval and a tide was estimated to set 100°T. at 3 knots.

*Procedure* (refer to figure 5.3)

1. Lay off the two position lines given by the bearings at 1300 and 1330.
2. From any convenient position on the first bearing line lay off the course steered and the distance steamed.
3. From the end of the course and distance layed off in (2), lay off the direction of the tidal set and the amount of the drift (100°T. by 1·5′).
4. Lay off the transferred position line through the end of the tide, parallel to the first position line (030°), to cut the second position line (295°). This intersection gives the vessel's position at the time of the second observation.

### Note

The course and distance made good in the interval is given by the line joining the original departure position selected, and the end of the tide. To find the position at the time of the first observation, this course should be transferred through the position at the time of the second observation, to cut the first position line.

### The running fix with leeway

If the vessel is making leeway during the interval of the running fix, this should be applied to the course steered before laying off from the first bearing.

# THE TRAVERSE TABLE 79

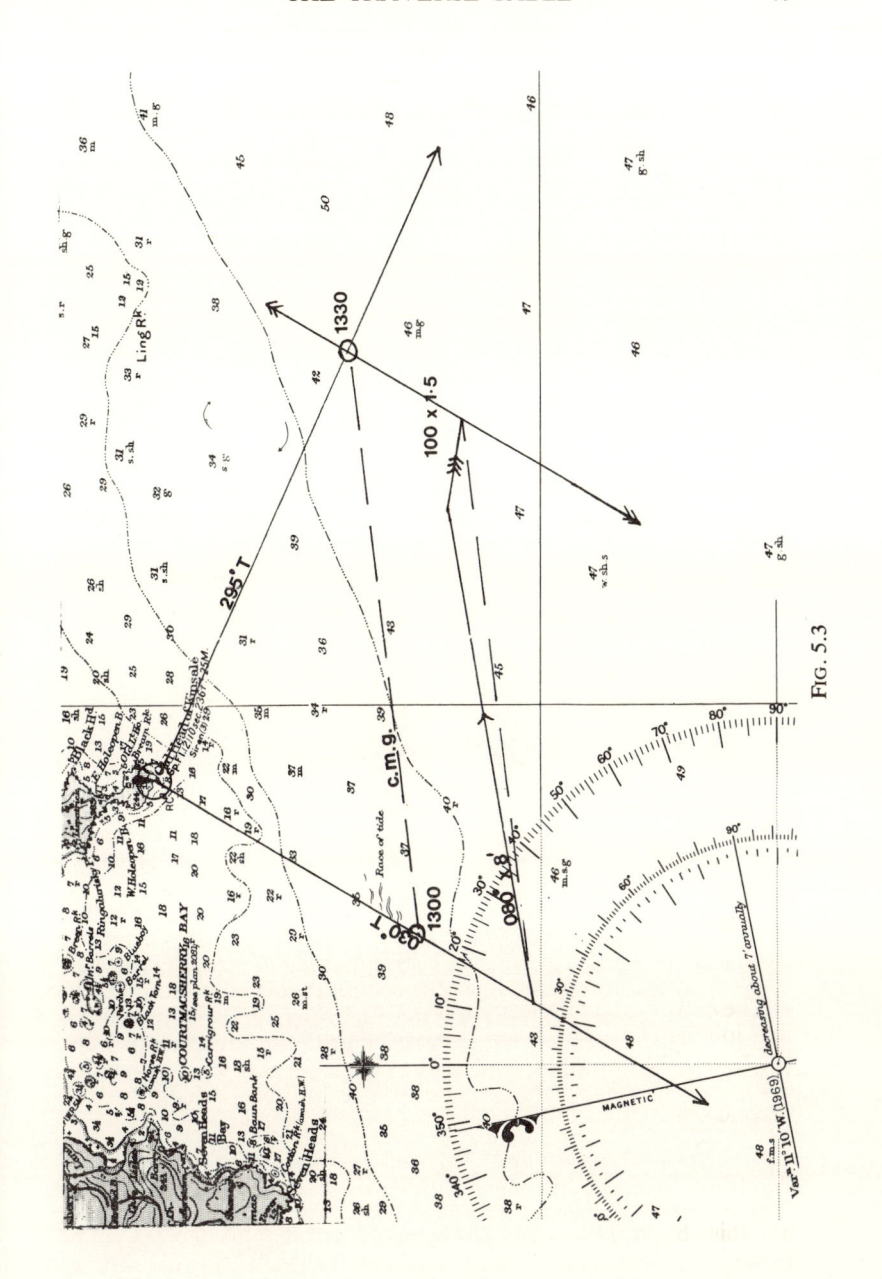

FIG. 5.3

## Transferring a position circle

The general principles of the running fix apply, irrespective of the form that the position lines take. To transfer a position circle however it is easiest to transfer the centre of the circle, that is the position whose distance has been observed. The transferred position circle is then drawn centred upon this transferred position obtained.

## Example

The distance from Wolf Rock Lt. Hse., bearing approximately north west, was observed by vertical sextant angle to be 3·0 miles. Forty-five minutes later the distance by vertical sextant angle of the same lighthouse was observed to be 3·4 miles. Find the position at the time of the second observation if the vessel was steering 030°T. in the interval and made good 4 miles by log.

*Procedure* (refer to figure 5.4)

1. Lay off the course and distance made good in the interval (030 by 4 miles), from the position of the point observed (Wolf Rock).
2. Draw the transferred position circle, radius 3·0 miles, centred upon this transferred position.
3. Draw the second position circle, of radius 3·4 miles, centred upon the position of Wolf Rock itself, to cut the transferred position circle. The intersection will give the position at the time of the second observation.

## Running up the position line by traverse table

When navigating out of sight of land, position lines are obtained by astronomical observation. The time intervals between observed position lines when using the running fix method may be as long as two or three hours. The small scale charts used for ocean navigation are not suitable for transferring the position line by plotting so that in practice the traverse table is used. The method is the same as that described for solving the plane sailing problem, that is applying the course and distance steamed to an initial position to find a D.R. position at the end of the run. The position line may be transferred by using any position on the line to which the course and distance is applied. The position obtained gives a position through which the transferred position line may be drawn.

## Example

At 0930 an astronomical observation gave a position line running 025°/205° passing through position 42° 30′ N. 32° 08′ W. Find a position through which to draw the transferred position line at 1200 hrs if in the interval the vessel steered 075°T. and made good 35 miles.

# THE TRAVERSE TABLE

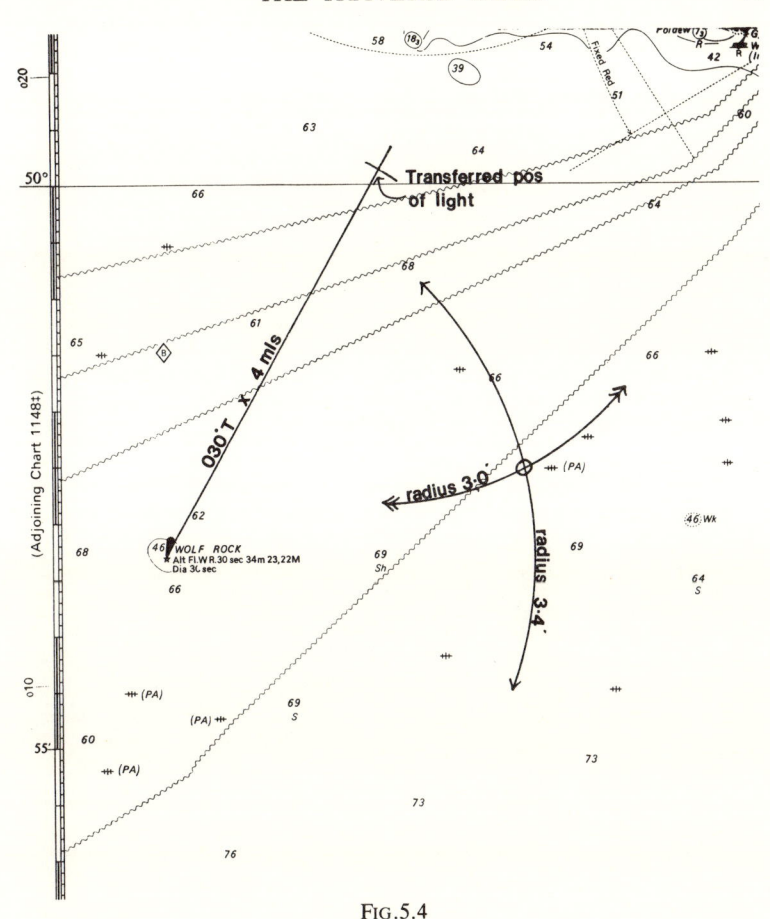

Fig. 5.4

## Procedure

1. Enter the traverse tables with the course and distance, and extract the d'lat. and departure.
2. Apply the d'lat. to the initial latitude to obtain the latitude at 1200 hrs.
3. Calculate the mean latitude and convert departure into d'long.
4. Apply the d'long. to the initial longitude to give the longitude at 1200 hrs.

Initial pos.   42° 30·0′ N.   32° 08′ W.      mean lat.=42° 34·5′ N.
d'lat.                9·1′ N.        45·9′ E. = d'long. (departure=33·8′)
hrs pos.      42° 39·1′ N.   31° 22·1′ W.

Transferred position line runs 025°/205° through 42° 39·1 N. 31° 22·1′ W.

This problem is discussed further in the chapter devoted to plotting the astronomical position line.

### Doubling the angle on the bow

This problem is a special case of the running fix method, which enables a fix to be obtained with a minimum of construction and plotting. It requires the time at which a point of land or beacon has a certain relative bearing expressed as an angle on either bow, and also the time when the same point is twice that angle on the bow, to be observed. It also requires a knowledge of the course steered and the distance run in the interval between the observations and a negligible effect from tide or wind.

In figure 5.5, $\theta$ and $2\theta$ are the relative bearings observed.

and angle BAP+angle APB=$2\theta$ (external angle of a triangle is equal to the two internal and-opposite angles).

thus angle APB=$\theta$ and the triangle is isosceles.
thus AB=PB

We can conclude therefore that the distance of the vessel from the point of land observed at the second observation, is equal to the distance run in the interval between the two observations.

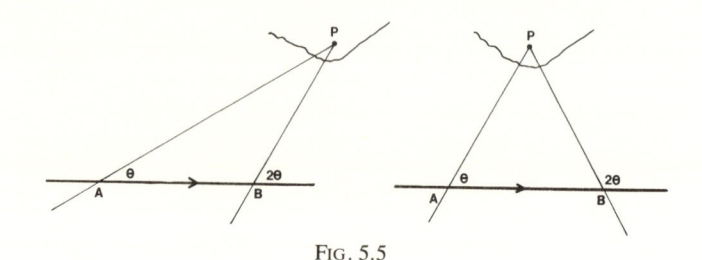

FIG. 5.5

*Procedure*

1. Note the time when a point of land or beacon is at any convenient angle on the bow.

## THE TRAVERSE TABLE

2. Note the time when the same point of land or beacon is at twice the angle on the bow.

3. Calculate the distance run between the observations.

4. Lay off the bearing of the second observation after converting to a true bearing by application of the ship's head. Mark off the observed position at a distance from the point observed equal to the distance run calculated in (3). This gives the vessel's position at the me of the second observation.

### EXERCISE 5F

1. A point of land is observed to bear 205°T. from a vessel steering 248°T. at 15 knots. Forty-eight minutes later the same point of land was observed to bear 147°T. Find the distance off the point at the time of the second observation.

2. A lighthouse was observed bearing 050°T. from a vessel steering 100°T. After running for 7 miles by log the lighthouse was observed to bear 000°T. Find the distance off the lighthouse at the time of the second observation.

3. A vertical sextant angle observation of a lighthouse gave a distance off of 6·8 miles. After steaming 174°T. for 40 minutes at 11 knots the vertical sextant angle was observed to be the same as at the first observation, while the lighthouse 054° by compass. Find the true bearing and distance from the lighthouse at the time of the second observation.

The following questions are set on the Admiralty Instructional chart No. 5051 (Lands End to Falmouth).

4. At 0800 Wolf Rock Light was observed to bear 048°T. from a vessel steering 085°T. at 16 knots. Twelve minutes later Wolf Rock was observed to bear 337°T. If a tide was estimated to set 155°T. at 3·5 knots in the interval, find the latitude and longitude of the vessel at the time of the second observation.

5. At 2000 hrs. Tater-Du light was observed to bear 338°T. from a vessel steering 250°T. at 18 knots. Thirty minutes later Wolf Rock was observed to bear 260°T. If a tide was estimated to set 145°T. at 2·5 knots in the interval, find the ship's position at 2030.

6. Lizard Pt. Light was observed to bear 015°T. from a vessel steering 270°T. at 16 knots and making 5° leeway due to a northerly wind. 1h 18m later Wolf Rock was observed to bear 335°T. If a tide was estimated to set 110°T. at 1·0 knot in the interval, find the ship's position at the time of the observation of Wolf Rock.

7. At 1200 hrs. a vessel observes Bishop Rock Lt. Hse. (position 49° 52·2' N. 6° 26·5' W.), to bear 035°T. The vessel then steams 278° for 3 hours at 15 knots. Find by traverse table a position through which to draw the transferred position line at 1500 hrs., and its direction, in order to cross it with an observation of the sun.

## CHAPTER 6

## TIDES

Tides and tidal streams are the result of gravitational attractions of astronomical bodies, mainly the sun and the moon. The tide raising forces of these bodies causes a horizontal movement of water such that tidal waves are produced directly underneath the tide raising body, and also on the opposite side of the earth to the body. Variation in the height of water at any place on the earth will occur as the earth rotates with respect to these tidal waves, producing two high waters in each rotation. The highest high waters will occur when the sun and the moon are in line with the earth, that is at new moon and at full moon. The solar tide then reinforces the lunar tide. Such tides are called spring tides occurring approximately once in two weeks. At first and third quarters the solar tide decreases the height of the lunar tide. Such tides are called neap tides.

The magnitude of tidal effects are relatively small unless they are increased by resonance in ocean basins or by the modifying effects of land and sea bed formations. This occurs to a marked extent in the North Atlantic which responds to semi diurnal components of the tide raising forces, and in which large tides are produced by the funneling effect of coastline shapes.

### Tidal streams

Tidal streams are the horizontal movements of water due to the tide raising forces. In European waters they are of a semi diurnal nature directly related to the vertical tidal variations. Their directions and rates can therefore be predicted with reference to times of high water at chosen locations. These predictions are made available to the navigator by:

    a. Tidal information on Admiralty charts.
    b. Tidal stream atlases.

### Currents

These are horizontal movements of water caused by meteorological conditions, or by flow of water from river estuaries. They are not periodic as are the tidal streams and those currents which are due to local meteorologial conditions are not included in tidal predictions. Consistent strong winds may therefore modify the streams predicted to a marked extent. The largely permanent effect of the flow of water from rivers is included in tidal stream predictions.

## Tidal information on Admiralty charts

Selected positions on Admiralty charts are chosen for which to give tidal stream information. These positions are marked by a magenta diamond with an identifying letter inside. At some convenient place on the chart a table is given for each tidal diamond, each table being headed by its appropriate identifying letter. The tables give the direction, and the spring and neap rates for each hour of the tidal cycle. The hours are referred to the time of high water at some standard port which may or may not appear on the chart. Many charts are referred to high water Dover. Information is given from 6 hours before H.W. to 6 hours after H.W. at hourly intervals. In order to relate the information to the ship's zone time the zone time of high water at the chosen standard port must be obtained from Admiralty tide tables.

To find the direction and rate at points between the tidal diamonds some interpolation between the tables is necessary together with some personal judgement as to the likely effect of the coastline shape on the direction of the stream. In this respect it should be remembered that tidal streams tend to flow parallel to coastlines and into and out of estuaries, although this may not be the case especially near the turn of the tide. To facilitate this tidal arrows are shown on charts showing the approximate mean direction of the flood (an arrow with feathers), and the ebb (an arrow without feathers), or a current (a wavy arrow). (See chart booklet 5011 for abbreviations and symbols on Admiralty charts.

## Tidal stream atlases

These are published by the Hydrographer to the Navy in a series of 11 booklets to cover the coastal waters of the British Isles. Each booklet contains chartlets of the covered area for hourly intervals from 6 hours before H.W. Dover to 6 hours after H.W. Dover. The times of high water Dover may be obtained from Admiralty Tide Tables Vol. I (N.P.200). On each chartlet the direction of the tidal stream for that hour is shown by arrows, the length and the boldness of the arrows indicating approximately the strength of the stream. Figures are given against some arrows which show the mean neap and spring rates at that place. These are shown thus:

11,24

meaning that the mean neap rate is 1·1 knot and the mean spring rate is 2·4 knots. Interpolation or extrapolation between these figures can be done by taking the range at Dover for that day and comparing it with the neap and spring ranges. An interpolation diagram is included with full instructions to facilitate this.

## Tides

The term tide refers to the variation in the level of the water

## Tidal Levels referred to Datum of Soundings

| Place | Lat N | Long W | MHWS | MHWN | MLWN | MLWS | Datum and remarks |
|---|---|---|---|---|---|---|---|
| Sennen Cove | 50° 04' | 5° 42' | 4·8 | 4·4 | — | 0·8 | 3·05 m below Ordnance Datum (Newlyn) |
| Penzance | 50 06 | 5 33 | 5·6 | 4·4 | 2·0 | 0·8 | 2·99 m below Ordnance Datum (Newlyn) |
| Porthleven | 50 05 | 5 19 | 5·5 | 4·3 | 2·0 | 0·6 | 2·90 m below Ordnance Datum (Newlyn) |
| Lizard Point | 49 57 | 5 12 | 5·3 | 4·2 | 1·9 | 0·6 | 2·90 m below Ordnance Datum (Newlyn) |
| Coverack | 50 01 | 5 05 | 5·3 | 4·2 | 1·9 | 0·6 | 2·90 m below Ordnance Datum (Newlyn) |
| Helford R Entrance | 50 05 | 5 05 | 5·3 | 4·2 | 1·9 | 0·6 | 2·90 m below Ordnance Datum (Newlyn) |
| Falmouth | 50 09 | 5 03 | 5·3 | 4·2 | 1·9 | 0·6 | 2·91 m below Ordnance Datum (Newlyn) |

## Tidal Streams referred to HW at DEVONPORT

| | ◇A 50°07·2N 5 49·5W | | | ◇B 49°58·5N 5 48·5W | | | ◇C 50°01·5N 5 27·6W | | | ◇D 49°52·2N 5 10·9W | | | ◇E 50°02·4N 5 02·3W | | | ◇F 50°02·5N 4 58·7W | | | ◇G 50°02·7N 4 54·8W | | | ◇H 50°08·0N 4 52·3W | | | ◇J 50°08·5N 5 01·5W | | | |
|---|---|---|---|---|---|---|---|---|---|---|---|---|---|---|---|---|---|---|---|---|---|---|---|---|---|---|---|---|
| Hours | Dir | Sp | Np | Dir | Sp | Np | Dir | Sp | Np | Dir | Sp | Np | Dir | Sp | Np | Dir | Sp | Np | Dir | Sp | Np | Dir | Sp | Np | Dir | Sp | Np | Hours |
| 6 | 332 | 0·4 | 0·2 | 311 | 1·7 | 0·8 | 280 | 0·9 | 0·5 | 256 | 1·8 | 0·9 | 201 | 1·0 | 0·5 | 215 | 1·0 | 0·5 | 217 | 0·7 | 0·4 | 222 | 0·4 | 0·2 | 339 | 0·2 | 0·1 | 6 |
| 5 | 002 | 1·5 | 0·7 | 327 | 2·2 | 1·1 | 290 | 0·5 | 0·3 | 254 | 1·2 | 0·6 | 309 | 0·1 | 0·0 | 220 | 0·5 | 0·2 | 232 | 0·5 | 0·2 | 249 | 0·2 | 0·1 | 005 | 0·6 | 0·3 | 5 |
| 4 | 009 | 2·4 | 1·2 | 336 | 2·0 | 1·0 | 329 | 0·3 | 0·2 | 234 | 0·4 | 0·2 | 006 | 0·1 | 0·5 | 293 | 0·1 | 0·1 | 277 | 0·3 | 0·1 | Slack | | | 022 | 0·9 | 0·4 | 4 |
| 3 | 010 | 2·5 | 1·2 | 352 | 1·6 | 0·8 | 049 | 0·4 | 0·3 | 045 | 0·4 | 0·2 | 011 | 1·4 | 0·7 | 017 | 0·5 | 0·2 | 349 | 0·4 | 0·2 | 077 | 0·2 | 0·1 | 023 | 0·6 | 0·3 | 3 |
| 2 | 009 | 1·6 | 0·8 | 023 | 1·1 | 0·5 | 065 | 0·6 | 0·3 | 054 | 1·0 | 0·5 | 015 | 1·5 | 0·8 | 029 | 0·9 | 0·5 | 014 | 0·7 | 0·3 | 037 | 0·4 | 0·2 | 022 | 0·4 | 0·2 | 2 |
| 1 | 031 | 1·1 | 0·5 | 088 | 1·0 | 0·5 | 082 | 0·8 | 0·4 | 059 | 1·8 | 0·9 | 022 | 1·5 | 0·7 | 043 | 1·2 | 0·6 | 051 | 0·9 | 0·4 | 042 | 0·5 | 0·3 | 036 | 0·2 | 0·1 | 1 |
| HW | 123 | 0·6 | 0·3 | 113 | 1·4 | 0·7 | 097 | 0·8 | 0·4 | 067 | 2·3 | 1·1 | 028 | 1·2 | 0·6 | 043 | 1·2 | 0·6 | 060 | 1·1 | 0·5 | 040 | 0·7 | 0·3 | Slack | | | HW |
| 1 | 188 | 1·7 | 0·8 | 124 | 1·8 | 0·9 | 107 | 0·7 | 0·3 | 075 | 1·8 | 0·9 | 030 | 0·5 | 0·2 | 040 | 0·7 | 0·4 | 069 | 0·8 | 0·4 | 036 | 0·5 | 0·2 | 217 | 0·3 | 0·1 | 1 |
| 2 | 181 | 2·5 | 1·2 | 137 | 1·9 | 0·9 | 124 | 0·4 | 0·2 | 082 | 0·8 | 0·4 | 202 | 0·4 | 0·2 | Slack | | | 107 | 0·3 | 0·1 | | | | 213 | 0·5 | 0·2 | 2 |
| 3 | 194 | 2·5 | 1·3 | 166 | 1·8 | 0·9 | 201 | 0·3 | 0·2 | 203 | 0·4 | 0·2 | 196 | 1·2 | 0·6 | 214 | 0·5 | 0·3 | 212 | 0·4 | 0·2 | 210 | 0·2 | 0·1 | 207 | 0·7 | 0·3 | 3 |
| 4 | 210 | 2·1 | 1·0 | 195 | 2·0 | 1·0 | 234 | 0·6 | 0·3 | 233 | 1·4 | 0·7 | 195 | 1·7 | 0·9 | 210 | 0·9 | 0·5 | 224 | 0·7 | 0·4 | 219 | 0·6 | 0·3 | 190 | 0·7 | 0·4 | 4 |
| 5 | 223 | 1·1 | 0·6 | 232 | 1·6 | 0·8 | 263 | 0·8 | 0·4 | 247 | 2·3 | 1·1 | 197 | 1·6 | 0·8 | 213 | 1·3 | 0·6 | 220 | 0·8 | 0·4 | 211 | 0·8 | 0·4 | 180 | 0·5 | 0·2 | 5 |
| 6 | 295 | 0·4 | 0·2 | 286 | 1·4 | 0·7 | 278 | 0·9 | 0·5 | 257 | 1·9 | 0·9 | 202 | 1·2 | 0·6 | 216 | 1·2 | 0·6 | 217 | 0·8 | 0·4 | 216 | 0·5 | 0·3 | 276 | 0·1 | 0·0 | 6 |

FIG. 6.1

Reproduced from British Admiralty Tide Tables with the sanction of the Controller, H.M. Stationery Office and of the Hydrographer of the Navy.

# TIDES

surface due to the tide raising forces. The following terms will be used with reference to tidal prediction.

*Chart datum*

This is an arbitrary level below which charted soundings are expressed.

*Height of tide*

This is the height of the water surface at any instant above the level of chart datum. Thus the actual depth of water is given by the sum of the charted sounding and the height of tide above chart datum. Note that it is possible to have a negative height of tide although in general chart datums are chosen such that they rarely occur.

*Mean high water springs (M.H.W.S.)*

This is the height above chart datum, which is an average of the heights of all the two successive high waters at spring tides, throughout the year. This will vary from year to year as the maximum declination of the moon varies over an 18·6 year cycle. The value of M.H.W.S. is therefore averaged over the 18·6 year cycle. The average maximum declination of the moon over this 18·6 year cycle is 23½°.

*Mean low water springs* (M.L.W.S.)

This is the height, which is an average of the two successive low waters at spring tides throughout a year when the average declination of the moon is 23½°.

*Mean high water neaps* (M.H.W.N.)

This is the height above chart datum which is an average of the two successive low waters at neap tides, throughout a year when the average declination of the moon is 23½°.

*Mean low water neaps* (M.L.W.N.)

This is the height above chart datum which is an average of the two successive low waters at neap tides, throughout a year when the average declination of the moon is 23½°.

*Height of tide above low water*

This is the height of the water surface at any instant, above the level of the nearest low water. This height can be found from the tide tables. The height of the low water is added to the height above low water to give the height of tide above chart datum.

*Drying height*

This is the height of a point on the sea bed which lies above the level of chart datum. Such a point will dry out when the height of tide above chart datum on a falling tide is equal to its drying height.

*Highest astronomical tide. Lowest astronomical tide.* (H.A.T. L.A.T.)

These are the highest and the lowest levels which can be predicted to occur under any combination of astronomical conditions, under normal meterological conditions.

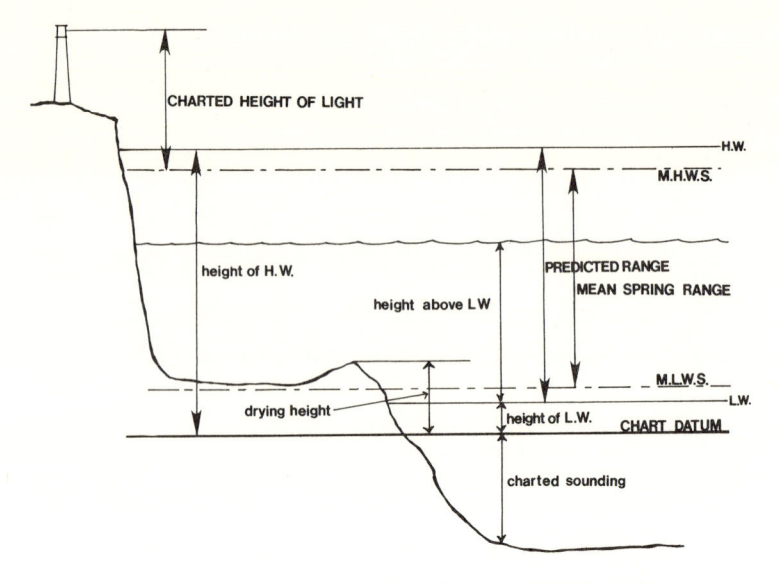

Fig. 6.2 shows the relationship between the terms defined above.

## Prediction of tidal times and heights

Soundings on Admiralty charts are expressed below chart datum. This is an arbitrary reference level chosen such that there will rarely be less water than is indicated on the chart. The level of chart datum may differ between charts, but are at present being standardised to approximate to lowest astronomical tide, which is the lowest level which can be predicted to occur under any combination of astronomical conditions under normal meteorological conditions. The relationship between chart datums at various places and the L.A.T. are shown in Table V in the front of the Admiralty Tide Tables Vol. I. For comparison of chart datums between charts of different areas Table III in the tide tables is consulted. This gives the height of chart datums at various places relative to the ordnance datum (Newlyn) which is the datum for the land levelling system of England, Scotland and Wales. For chart datums at places outside these countries the reference is the datum used in the respective countries. Because of differences in chart datums there may be differences in soundings on different charts of the same area. The level of the chart datum is shown on Admiralty charts in the titles.

In all cases however tidal predictions for ports are referred to the chart datum established at that port and which is used on the largest scale chart of it. The total depth of water at any point is therefore the sum of the sounding shown on the chart, and the tidal height above chart datum predicted from the tide tables.

**Tidal calculations**

Candidates for Class V and Class IV certificates are required to be able to use the Admiralty tide tables Volume I (European Waters), in order to predict times and heights of high and low waters, and to predict the height of tide at times between high and low water.

The tides of European waters are of a semi diurnal nature, that is there are two high and two low waters each lunar day. Part I of ATT Vol. I gives the predictions of the times and heights above chart datum of these high and low waters for a number of selected ports which are called Standard Ports. For each standard port there is also a tidal curve plotting the tidal height between high and low water, against the interval of time from the nearest high water. Part II of the tide tables gives tidal predictions for a large number of ports which lie between the chosen standard ports. These are called secondary ports as their tidal information is given in the form of time and height differences between the secondary port and one of the standard ports.

**To find times and heights of high and low water at a standard port**

These may be extracted directly from Part I of ATT Vol. I for the required standard port, and for the required date (see extracts from ATT Vol. I). Note that the times given are in the zone time for that area in which the port lies. The difference between the zone time used and G.M.T. is given at the top of each page. The sign attached to this time difference is appropriate to correct the tabulated zone times to G.M.T. Thus if the time zone is given as −0100, then the times tabulated are 1 hour ahead of G.M.T. The time zone used for the British Isles is G.M.T., but care must be taken when British Summer Time is being kept. Similarly care must be taken that the time kept in any other country is in fact the time zone used in the tables.

**Example** (refer to extract from ATT Vol. I)

Find the times of high and low water at Avonmouth on the morning of 29th January 1980, and the depth of water at these times at a place off Avonmouth where the charted sounding is 4·2 metres.

From ATT   H.W. 0505   Ht. 11·4m
               L.W. 1140   Ht.  2·0m

Depth of water = charted sounding + height of tide
depth at H.W. = 11·4 + 4·2 = 15·6m
depth at L.W. =  2·0 + 4·2 =  6·2m

**To find the height of tide at times between high and low water** (Standard Port)

This is done with the aid of the tidal curves given with each standard port. There is one curve for neap tides and one for springs. For times between springs and neaps interpolation between the curves must be done (see examples).

*Procedure*

1. Extract from ATT Vol. I Part 1 the times and heights of the high and low waters that 'straddle' the time for which the prediction is required. This time should be expressed in the same zone time as the tidal predictions for the standard port.
2. Subtract the height of low water from the height of high water to obtain the predicted range.
3. Take the difference between the time required for prediction and the time of high water. This is the interval from high water. Note whether the interval is positive (falling tide) or negative (rising tide).
4. Compare the predicted range with the mean spring and neap ranges given on the tidal curve. This will determine whether the spring curve or the neap curve should be used or whether interpolation between the two is necessary.
5. Enter tidal curve or curves with the interval from high water along the horizontal axis and go vertically to meet the tidal curve. From this point go across to obtain the factor.
6. Multiply the factor by the predicted range found in (2). If the predicted range is between the spring and neap ranges the factor is found by interpolating between the spring and neap factors. If the predicted range is above the spring range then the spring factor should be used. If the predicted range is below the neap range then the neap factor should be used. The factor multiplied by predicted range gives the height above low water.
7. Add on the height of the low water to obtain the height of tide above chart datum.

**Example**

Find the height of tide at Avonmouth at 1530 G.M.T. on 9th April 1980, and hence the depth of water at a place where the charted sounding is 2·0 metres.

## TIDES

```
From ATT H.W. 1245   Ht. 10·0m
         L.W. 1901   Ht.  3·5m    ..................................... 1
                          6·5m    = predicted range .......... 2

    time H.W. 1245
 required time 1530
interval from H.W. 0245 + (falling tide) ............................... 3
From tidal curve      Spring range = 12·3m
                       Neap range =  6·5m
                    Predicted range =  6·5m (use neap curve) ....... 4
From neap curve factor = 0·59 (see tidal curve for Avonmouth) .. 5
Height above L.W. = 0·59 × 6·5    = 3·8 ................................ 6
            L.W.                  = 3·5
Height above chart datum          = 7·3 metres ....................... 7
Charted sounding                  = 2·0
Depth of water                    = 9·3 metres
```

**To find the time at which there will be a given depth of water between high and low waters at a standard port, on a given tide**

With this problem it is first necessary to determine the height of tide which corresponds to the given depth of water. This will require consideration of the charted sounding. The problem is often given in the form of a required clearance under a vessels keel. In this case the ship's draft must be given. The draft and the clearance will be the required depth of water.

*Procedure*

1. Extract the times of high and low waters for the tide specified, and the heights.
2. Subtract the height of low water from the height of high water to obtain the predicted range.
3. Ascertain the height of tide above chart datum required to produce the given depth of water (subtract the charted sounding).
4. Subtract the height of L.W. to give the height above L.W.
5. Find the factor from:

$$\text{factor} = \frac{\text{height above L.W.}}{\text{predicted range}}$$

6. Compare the predicted range with the spring and neap ranges to determine which curve to use or whether interpolation between both curves is necessary.
7. Enter appropriate curve or curves with the factor and extract the interval from H.W. by the reverse process to that described in the previous example.
8. If necessary interpolate between the intervals from H.W. from spring and neap curves, as indicated by the comparison between the predicted range and the spring and neap ranges.

9. Apply interval from H.W. to the time of high water found in (1).

**Example**

Find the time when a vessel of draft 6·5 metres will have a clearance of 1·0 metre over a shoal of charted depth 1·0 metre off Avonmouth, on the rising tide of the morning of 23rd February 1980.

From ATT 23rd Feb. H.W. 1219      Ht.11·0m  
                        L.W. 0611      Ht. 2·2m ................. 1  
                                  pr. range    8·8m ................. 2

Draft                 6·5m  
Clearance          1·0m  
Required depth    7·5m  
Sounding           1·0m  
Ht. above cht. datum 6·5m ................................. 3  
Height of L.W.      2·2m  
Height above L.W.   4·3m ................................. 4

$$\text{Factor} = \frac{\text{ht. above L.W.}}{\text{pdctd. range}}$$

$$= \frac{4\cdot 3}{8\cdot 8} = 0\cdot 49 \quad\quad\quad\quad\quad ................................. 5$$

Spring range   12·3  
                      predicted range 8·8 ................................. 6  
Neap range     6·5

From spring curve    interval from H.W.    =−2h 50m  
From neap curve      interval from H.W.    =−3h 20m ............ 7

interpolating between springs and neaps for a predicted range of 8·8m                    interval from H.W.=−3h 09m ................ 8

Time of H.W.         12h 19m  
Interval from H.W.    −03h 09m  
Time required        09h 10m ................................. 9

**Secondary ports**

Part II of ATT Vol. I gives tidal information for a large number of secondary ports. This information is given as time and height differences between the secondary port and some convenient standard port.

# TIDES

## High and low water time differences

The time differences between high or low water at the secondary port and high or low water at the standard port varies between springs and neaps. There are two values given which should be taken as the maximum and minimum differences. These are tabulated against the time of high or low water at the standard port which will depend mainly on the neap-spring cycle. If the time of high or low water at the standard port falls between the times tabulated then the time differences must be interpolated.

Figure 6.3 shows an extract from Part II of ATT Vol. I. The standard port is Milford Haven.

## Example

Find the time of high water at Ilfracombe if the time of high water at Milford Haven is 0330.

|  |  |  | 0100 and 1300 | 0700 and 1900 | 0100 and 1300 | 0700 and 1900 |  |  |  |  |
|---|---|---|---|---|---|---|---|---|---|---|
| 496 | **MILFORD HAVEN** | (see page 110) |  |  |  |  | 7·0 | 5·2 | 2·5 | 0·7 |
| 535 | Ilfracombe | 51 13   4 07 | −0030 | −0015 | −0035 | −0055 | +2·2 | +1·7 | +0·5 | 0·0 |
|  | *Rivers Taw and Torridge* |  |  |  |  |  |  |  |  |  |
| 536 | Appledore | 51 03   4 12 | −0020 | −0025 | +0015 | −0045 | +0·5 | 0·0 | −0·9 | −0·5 |
| 537 | Yelland Marsh | 51 04   4 10 | −0010 | −0015 | +0100 | −0015 | −0·4 | −0·9 | −1·7 | −1·1 |
| 538 | Fremington | 51 05   4 07 | −0010 | −0015 | +0030 | −0030 | −0·5 | −1·2 | −1·6 | +0·1 |
| 539 | Barnstaple | 51 05   4 04 | 0000 | −0015 | −0155 | −0245 | −2·9 | −3·8 | −2·2 | −0·4 |
| 540 | Bideford | 51 01   4 12 | −0020 | −0025 | 0000 | 0000 | −1·1 | −1·6 | −2·5 | −0·7 |

FIG. 6.3

Produced from British Admiralty Tide tables with the sanction of the Controller, H.M. Stationery Office and of the Hydrographer of the Navy.

Time difference for $\left. \begin{array}{l} 0100 \\ 0330 \end{array} \right\}$ 2·5 hrs.   $\left. \begin{array}{l} -0030 \\ -0015 \end{array} \right\}$ 15 mins.

Time required

Time difference for 0700

Thus for 0330 the time difference is:

$$-\left(0030 - \frac{2 \cdot 5 \times 15}{6}\right)$$
$$= -(0030 - 6)$$
$$= -0024 \text{ mins.}$$

|  |  |
|---|---|
| Time H.W. Milford Haven | 0330 |
| Time difference | −0024 |
| Time H.W. Ilfracombe | 0306 |

## Height differences

Differences in tidal height between secondary port and standard port are tabulated for M.H.W.S., M.H.W.N., M.L.W.S., and

94    PRACTICAL NAVIGATION FOR SECOND MATES

M.L.W.N. The differences between the mean spring and mean neap levels should be assumed to vary linearly and can be found by interpolation. For levels outside the mean range the height differences are found by extrapolation. The level of the required tide at the standard port should be compared with the mean spring and neap levels given for the standard port in Part II for interpolation purposes.

**Example** (refer to figure 6.3)

The height of H.W. at Milford Haven is 6·0 metres. Find the height of H.W. at Ilfracombe.

Difference for M.H.W.S.

M.H.W.S. level 7·0 ⎫ 1·0 ⎫ difference +2·2 ⎫
Required H.W. 6·0 ⎭     ⎬ 1·8  ⎬ 0·5
M.H.W.N. level 5·2       ⎭     difference +1·7 ⎭

$$\text{difference} = +(2\cdot 2 - \frac{1\cdot 0 \times 0\cdot 5}{1\cdot 8})$$
$$= +(2\cdot 2 - 0\cdot 3)$$
$$= +1\cdot 9$$

H.W. Milford Haven    6·0 metres
Difference            +1·9
H.W. Ilfracombe       7·9 metres

## Example

Find the times and height of high and low waters at Watchet (ATT 531) on 12th February 1980.

| Standard Port (Avonmouth) differences | H.W. 0242 −0037 | L.W. 0859 −0049 | H.W. 1524 −0039 | L.W. 2148 −0057 |
|---|---|---|---|---|
| Secondary Port (Watchet) | 0205 | 0810 | 1445 | 2051 |
| Standard Port heights differences | 9·8 −1·5 | 3·5 +0·1 | 10·3 −1·5 | 3·2 +0·1 |
| Watchet heights | 8·3 | 3·6 | 8·8 | 3·3 |

## Note

Care should be taken in problems like this that a high or low water at the standard port on the preceeding or the following day does not produce a high or low water at the secondary port on the day in question, after the differences are applied.

# TIDES

**To find the height of tide at a secondary port at a time between high and low water**

Unless indicated otherwise in the tide tables the tidal movements at a secondary port are similar enough to those at the standard port for the tidal curve given for the standard port to be used for the secondary port also. After finding the times and heights of the high and low waters at the secondary port the problem is similar to that for a standard port. The spring and neap ranges for the secondary port must be found to compare with the predicted range to determine which curve to use.

*Procedure*

1. Extract the times of the high and low water on either side of the required time for the standard port, and the heights.
2. Apply the time differences to obtain the times of high and low water at the secondary port. Apply the height differences, interpolating between spring and neap ranges, by comparing the heights at the standard port with the mean spring and neap heights given in Part II for the standard port, as explained previously. This gives the heights at the secondary port.
3. Subtract the height of low water at the secondary port from the height of high water to obtain the predicted range.
4. Take the difference between the required time and the time of the high water to find the interval from H.W.
5. Apply the secondary port spring and neap height differences to the heights of M.H.W.S., M.H.W.N., M.L.W.N and M.L.W.S. for the standard port to obtain these values for the secondary port.
6. Subtract the height of M.L.W.N. from that of M.H.W.N. to obtain the neap range, and subtract the height of M.L.W.S from that of M.H.W.S. to obtain the spring range. Compare these with the predicted range to determine whether to use the spring curve or the neap curve or whether interpolation is necessary between the two.
7. Enter the tidal curve given for the standard port with the interval from high water and extract the factor. Interpolate if necessary between the spring and neap factors.
8. Multiply the factor by the predicted range to obtain the height above L.W.
9. Add on the height of low water to the height above L.W. to obtain the height above chart datum.

## Example

Find the height of tide at Clevedon (ATT 528) at 1000 G.M.T. on 16th March 1980. What will be the under keel clearance of a vessel of draft 8·1 metres, when passing over a shoal of charted depth 3·4 metres?

96    PRACTICAL NAVIGATION FOR SECOND MATES

```
Standard port        H.W.    Height      L.W.   Height
(Avonmouth)          0654    13·9        1417   0·4      .....1
Differences         −0018   −0·4        −0024  −0·0      .....2
                     ─────   ─────       ─────  ─────
Clevedon             0636    13·5        1353   0·4
                             0·4
                             ─────
                             13·1 = predicted range ............3
```

High Water Clevedon 0636
Required time        1000
                     ─────
Interval from H.W. + 0324 .............................................4

|  | M.H.W.S. | M.H.W.N. | M.L.W.N. | M.L.W.S. |
|---|---|---|---|---|
| Avonmouth mean levels | 13·2 | 10·0 | 3·5 | 0·9 |
| Clevedon differences | −0·4 | −0·2 | +0·2 | 0·0 |
| Clevedon mean levels | 12·8 | 9·8 | 3·7 | 0·9 .........5 |
|  | 0·9 | 3·7 |  |  |
| Spring range = 11·9 |  | 6·1 = neap range ..............6 |

Predicted range = 13·1 use spring curve

From spring curve for Avonmouth factor = 0·46 ..............7
Height above L.W. = 0·46 × 13·1         = 6·0 ..............8
Height of L.W.                          = 0·4
                                          ─────
Height of tide above chart datum        = 6·4 ..............9
Charted sounding                        = 3·4
                                          ─────
Depth of water at 1000 G.M.T.           = 9·8
Vessel's draft                          = 8·1
                                          ─────
Underkeel clearance                     = 1·7 metres

**To find the time between high and low water at a secondary port, when there will be a given depth of water on a given tide**

*Procedure*

  1. Extract the times and heights of the high and low water of the given tide for the standard port.
  2. Apply the time and height differences to obtain the times and heights of the high and low water at the secondary port. Find the predicted range.
  3. Apply the secondary port spring and neap height differences to the heights of M.H.W.S., M.H.W.N., M.L.W.N. and M.L.W.S. for the standard port to obtain these values for the secondary port.
  4. Subtract the height of M.L.W.N from that of M.H.W.N. to

# TIDES

obtain the mean neap range, and subtract the height of M.L.W.S. from that of M.H.W.S. to obtain the mean spring range for the secondary port. Compare these with the predicted range at the secondary port to determine which tidal curve to use or whether to interpolate between the two.

5. Ascertain the required height of tide above chart datum.
6. Subtract the height of L.W. to obtain the height above L.W.
7. Find the factor from factor = $\dfrac{\text{ht. above L.W.}}{\text{predicted range}}$

8. Enter the tidal curves for the standard port and extract the interval from H.W. Interpolate if necessary between the spring interval and the neap interval.

9. Apply interval from H.W. to the time of high water at the secondary port to obtain the required time.

## Drying heights

A drying height is a 'sounding' on a chart of a point which lies above the level of chart datum. The height of the point above chart datum will give the height of tide above chart datum when the point dries on a falling tide or covers on a rising tide.

## Example

A vessel is berthed at Watchet alongside a quay with a drying height of 1·5 metres. Find the time when the vessel will take the ground on the falling P.M. tide on 28th January 1980 if the vessels draft is 3·8 metres.

|  | H.W. | Height | L.W. | Height |  |
|---|---|---|---|---|---|
| Standard port (Avonmouth) | 1638 | 11·1 | 2301 | 2·2 | ......1 |
| Differences | −0042 | −1·6 | −0108 | +0·1 |  |
| Watchet | 1556 | 9·5 | 2153 | 2·3 | ......2 |
|  |  | 2·3 |  |  |  |
|  |  | 7·2 = predicted range |  |  | ...........2 |

|  | M.H.W.S. | M.H.W.N. | M.L.W.N. | M.L.W.S. |  |
|---|---|---|---|---|---|
| Avonmouth mean levels | 13·2 | 10·0 | 3·5 | 0·9 |  |
| Watchet differences | −1·9 | −1·5 | +0·1 | +0·1 |  |
| Watchet mean levels | 11·3 | 8·5 | 3·6 | 1·0 | ......3 |
|  | 1·0 | 3·6 |  |  |  |
| Spring range = | 10·3 | 4·9 = neap range |  |  |  |
| Predicted range = | 7·2 | interpolation necessary |  |  | .................4 |

Vessel's draft            3·8 metres (depth of water when taking ground)
Drying height             1·5
Ht. above chart datum     5·3 .................................................. 5
Ht. of L.W.               2·3
Ht. of tide above L.W.    3·0 .................................................. 6

Factor = $\dfrac{3\cdot 0}{7\cdot 2}$ = 0·42 ................................................. 7

From spring curve interval from H.W. = + 3h 37m
From neap curve interval from H.W. = + 3h 30m reqd. interval
Required interval from H.W.           + 3h 33m ................... 8
Time of high water                     15h 56m ................... 9
Time when vessel will take the ground  19h 29m G.M.T.

## Charted heights

Heights of terrestrial objects such as lighthouses and topographical features are expressed above Mean High Water Springs. If these heights are required accurately above the water level, such as for observation of accurate distance by vertical sextant angle, then a correction must be applied equal to the height of M.H.W.S. above or below the water surface.

## Example

Find the correction to apply to the charted height of a lighthouse at a place where the level of M.H.W.S. is 13·2, and the height of tide above chart datum is found to be 9·5 metres.

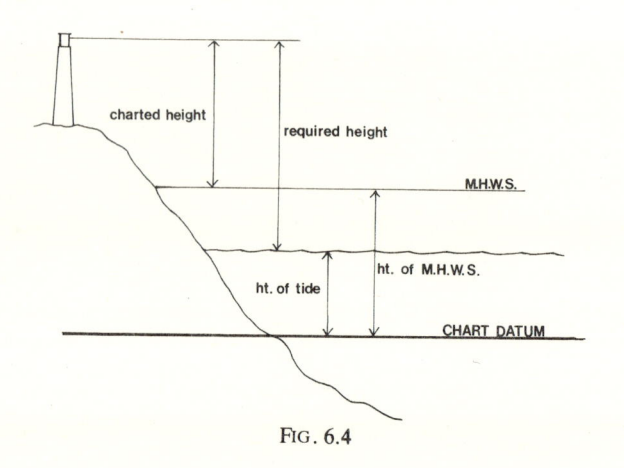

FIG. 6.4

# TIDES

From figure 6.4 the correction may be seen to be the height of M.H.W.S. minus the height of tide.

Thus:
   Actual height = Charted height + M.H.W.S. − Height of tide.

If the height of tide is greater than the height of M.H.W.S then

   Actual height = Charted height − (Height of tide − M.H.W.S.)

## EXERCISE 6A (Standard Port)

1. Find the times (G.M.T.) and heights of high and low waters at Avonmouth on 27th February 1980.

2. Find the height of tide above chart datum at a place off Avonmouth at 0924 B.S.T. on 31st March 1980.

3. Find the depth of water beneath the keel of a vessel of draft 5·8 metres when passing over a shoal of charted depth 2 metres, at 1715 B.S.T. on 29th April 1980 off Avonmouth.

4. Find the depth of water over a rock of drying height 1·5 metres at the P.M. high water on 26th January 1980. Will this rock dry during the following tide?

5. Find the earliest time (B.S.T.) that a vessel of draft 6·5 metres can pass over a shoal of charted depth 2·5 metres with a clearance of 2·0 metres on the rising tide of the morning of 20th April 1980.

6. A vessel is aground off Avonmouth with her for'd section on a sandbank of charted drying height 1·0 metre. At what time (G.M.T.) can she expect to float off on the P.M. rising tide of 9th March 1980 if the for'd draft is 8·0 metres?

7. Find the height of a lighthouse near Avonmouth, above the water surface, at 0800 G.M.T. on 11th January 1980, if the charted height of the light is 48 metres.

## EXERCISE 6B (Secondary Ports)

1. Find the times and heights of all high and low waters at Sharpness Dock on 29th January 1980.

2. Find the depth of water at a place off Sharpness Dock where the charted depth is 2·5 metres, at the high water on the afternoon of 12th March 1980, and the time G.M.T.

3. Find the clearance under the keel of a vessel at anchor off Watchet at 0830 (G.M.T.) on 27th April 1980 if the charted depth is 3·2 metres, and the vessel's draft is 5·5 metres.

4. Find the correction to charted soundings at a place off Weston-super-Mare at 1200 G.M.T. on 1st February 1980.

5. Find the time (B.S.T.) when there will be 13·0 metres of water

over a place where the charted depth is 5 metres, at Beachley on the rising tide on the morning of 19th April 1980.

6. Find the latest time that a vessel can pass over a shoal off Watchet of charted depth 1 metre on the falling tide of the evening of 1st January 1980, if the vessel's draft is 6 metres and a clearance of 0·5 metres is required.

7. Find the height of tide above chart datum at Bristol (Sea Mills) at 1500 G.M.T. on 15th January 1980. What would be the charted sounding at a place which was just drying at this time?

# CHAPTER 7

# THE CELESTIAL SPHERE AND THE NAUTICAL ALMANAC

## The celestial sphere

The concept of the celestial sphere is one in which all astronomical bodies are considered to lie on the surface of a sphere of infinite radius, which is concentric with the terrestrial sphere. This concept is acceptable to the navigator as he is concerned only with the measurement of angles subtended at the centre of the sphere, and the fact that astronomical bodies lie at different distances from the earth is of little consequence in the measurement of these angles. The motion of the closer bodies and the motion of the earth will cause these bodies to exhibit a movement on the celestial sphere relative to the bodies which can be considered to lie at infinitely great distances, that is the stars.

An astronomical body may have its position defined with reference to the celestial sphere, or with reference to the terrestrial sphere. The latter is required in order to use the body for observation to find an observer's terrestrial position. The terrestrial position of the body is defined by the latitude and longitude of the point on the earth's surface where a line joining the body to the earth's centre cuts the surface. This point is called the body's Geographical Position (G.P.). The position of a G.P. changes rapidly with the rotation of the earth so that it is convenient to define position on the celestial sphere by coordinates which change relatively slowly.

## Definition of position on the celestial sphere

### Equinoctial

This is the great circle on the celestial sphere which lies in the same plane as the earth's equator. It may sometimes be referred to as the celestial equator.

### Celestial poles

These are the points on the celestial sphere at which the earth's axis of rotation when produced, meet the celestial sphere. The north and south celestial poles will be 90° removed from all points on the equinoctial.

## Celestial meridians

These are semi great circles which terminate at the celestial poles, cutting the equinoctial in a right angle in the manner of terrestrial meridians. They are sometimes referred to as Hour Circles.

## Declination

This defines the position of a celestial body with respect to the equinoctial. It may be defined as the arc of the celestial meridian which passes through the body, contained between the equinoctial and the body. It is measured north or south of the equinoctial in the manner of the terrestrial latitude. The declination of a celestial body will be equal to the latitude of its geographical position at all times.

The declination of the sun varies between 23° 27' N. and 23° 27' S. approximately. The period of the variation is one year, the period of the earth's revolution around the sun. The variation is caused by the earth's axis of rotation being inclined to the plane of its annual orbit around the sun at an angle of 66° 33' approximately. The plane of the equator and equinoctial is therefore inclined to the plane of the earth's orbit by 23° 27'. The plane of the earth's orbit is called the ecliptic plane and the inclination of the equinoctial to the ecliptic plane is called the obliquity of the ecliptic. The intersection of the two planes, both of which are orientated in space in a constant plane (except for small long term changes), defines a constant direction in space. The point of their intersection which is occupied by the sun when going from south to north declination is called the First Point of Aries. The passage of the sun through the first point of Aries marks the beginning of the Spring season on earth.

## Sidereal hour angle

This is the arc of the equinoctial measured westwards from the first point of Aries to the meridian which passes through the body being considered.

The declination and the sidereal hour angle (S.H.A.) define a position on the celestial sphere in the same way as latitude and longitude define a position on the terrestrial sphere. (Note that although the declination is equal to the latitude of the G.P., the S.H.A. is not equal to the longitude of the G.P. This is because, due to the earth's rotation within the celestial sphere, the meridian of Greenwich does not always coincide with the meridian of Aries, whereas the plane of the equinoctial is always coincidental with the plane of the equator by definition.)

If small long term movements of the first point of Aries are neglected the S.H.A. and declination of a fixed point on the celestial sphere remain constant. Stars lie at such vast distances that they almost constitute fixed directions in space. The S.H.A. and

# THE NAUTICAL ALMANAC

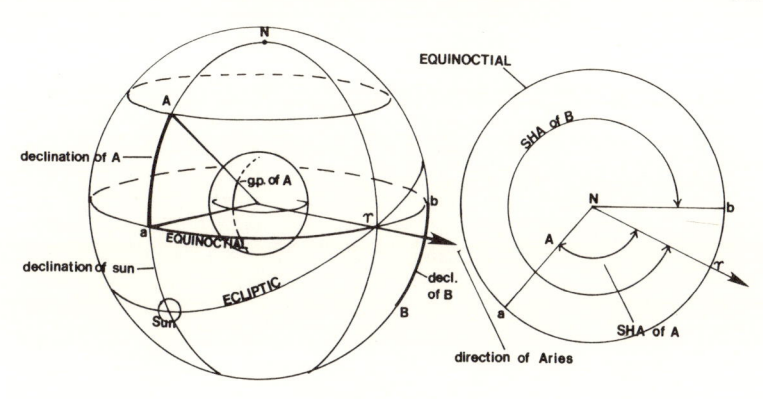

Fig. 7.1

declination of stars exhibit slow changes due to the stars own movement in space and due to the long term movements of Aries referred to previously. The changes are in the order of minutes of arc over a period of months.

The sun appears to move completely around the celestial sphere, relative to the stars and relative to the first point of Aries due to the earth's annual motion in orbit around the sun. This motion causes the S.H.A. of the sun to decrease by approximately 1° per day, that is by 360° per year. The inclination of the earth's axis causes the declination to vary from 23° 27' S. to 23° 27' N. and back again in the same period. The maximums are reached on December 22nd and June 21st respectively, and the sun's declination goes to zero on March 21st and September 23rd.

Planets exhibit movements on the celestial sphere relative to the stars due to their own orbital motion around the sun and also due to the earth's motion around the sun. Planets S.H.A. and declinations therefore have slow changes throughout the year.

The moon, because of it proximity to the earth exhibits, a relatively rapid decrease in S.H.A. This is due to the moon's orbital motion around the earth in the same direction as the earth's orbital motion around the sun. This amounts to approximately 13° per day. The inclination of the moon's orbit to the equinoctial also produces a rapid change of declination. This varies between limits which are themselves variable, but may be as much as 28¾°S. and 28¾°N., and as little as 18¼°S. to 18¼°N. The declination will change between these limits in approximately two weeks.

## Definition of a celestial body's position on the terrestrial sphere

The longitude of the geographical position of a celestial body changes by 360° each day as the earth rotates. The longitude of the

104    PRACTICAL NAVIGATION FOR SECOND MATES

G.P. is called the body's Greenwich Hour Angle (G.H.A.). This may be defined as the arc of the earth's equator measured westwards from the meridian of Greenwich to the meridian through the G.P. This is the same as the longitude except that whereas longitude is measured from 0° to 180° E. and W. of Greenwich, the G.H.A. is measured continually westwards from 0° to 360°.

The G.H.A. may be found if the S.H.A. is known and the longitude of the first point of Aries is known (G.H.A. Aries). The G.H.A. of Aries will also change by 360° each time the earth rotates within the celestial sphere, and may be found at any instant from the nautical almanac. The G.H.A. of a celestial body may therefore be found from the relationship:

G.H.A.=S.H.A.+G.H.A. Aries (see figure 7.2).
(if greater than 360° subtract 360°)

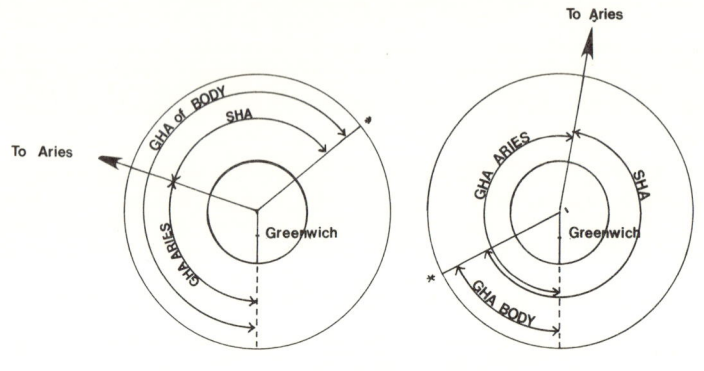

FIG. 7.2

### The Nautical Almanac

The Nautical Almanac is published by Her Majesty's Stationery Office as publication N.P. 314 and provides all astronomical data necessary for use in marine astronomical navigation.

The arrangement of the information in the almanac may be divided into sections as follows.

1. Altitude correction tables. These will be explained in full under the chapter heading of 'Correction of Altitudes'.
2. The list of selected navigational stars, giving the S.H.A. and declination of each star for each month of the year.
3. The Pole Star Tables. These will be described fully under the chapter heading 'The Pole Star Problem'.
4. The daily pages and increment tables. These are the most important pages of the almanac and allow the navigator to pick out

# THE NAUTICAL ALMANAC

the G.H.A. and the declination of the sun, moon, four navigational planets, and each of the selected stars, for any second of Greenwich Mean Time throughout the year.

## Arrangement of the daily pages

Each double page contains data for three days. This is arranged in columns headed Aries, Venus, Mars, Jupiter, Saturn, Sun and Moon. The column for each body contains the G.H.A. and declination (except for Aries) for each hour of G.M.T. for the three days contained on the double page. The increment tables allow easy interpolation between the hourly figures to obtain the values for each second of G.M.T. These are found by adding an increment to the value given for the hour of G.M.T. The increment is extracted from the increment tables by entering the table headed with the number of minutes in the G.M.T., and extracting the increment from that table against the seconds of the G.M.T. down the page. There are three tables given, one for sun and planets, one for Aries and one for the moon. These are given because of slight variations in the rate of change of the G.H.A. Care should be taken to use the appropriate column.

## Corrections to the increment. The 'v' correction

The mean rate of increase of G.H.A. of the sun is 15° per hour, and it is upon this value that the sun's increment table is based. Variations from the mean rate of change are small and are allowed for in the hourly figures given in the daily pages. The sun's increment requires no correction.

The rate of increase of G.H.A. of the planets, is with the exception of Venus always greater than 15° per hour. That of Venus may be slightly greater or slightly less. The sun's increment table is used however for the increment to planets G.H.A. The increment may be in error however if the hourly increase in G.H.A. is not exactly 15° as assumed by the increment table figures. The difference is allowed for by the 'v' correction. The difference between the assumed value of 15° per hour in the increment table and the actual hourly change as indicated by the hourly figures in the daily pages is given in the daily pages, at the foot of each of the planet's columns as 'v'. This is a mean figure for the three days. A more accurate figure may be obtained by noting the actual change in G.H.A. over the hour in which the required G.M.T. lies. The actual 'v' correction to apply is a proportion of this value depending upon the number of minutes in the G.M.T. (If the minutes of the G.M.T. given is 30, then only half of the 'v' value is applied.) The 'v' correction may be found by entering the 'v' correction table given with each of the increment tables using that one which is given with the table headed with the minutes in the G.M.T. The 'v' value from the daily pages is located and the correction is extracted against it. (Correction is in bold type.) This correction is added to the increment, unless in the case of Venus, the 'v' value from the daily pages is given a negative sign, in which case it is subtracted from the increment.

The rate of change of the moon's G.H.A. varies between 14° 19' and 14° 37' per hour. The moon's increment table is based upon the least value of 14° 19'. Any excess over this value is taken care of by a 'v' correction in the same manner as the planets. The correction is always positive for the moon. The variation in the hourly change of G.H.A. for the moon makes it necessary to give a 'v' value in the daily pages for each hour of G.M.T. This is given against each hourly value of G.H.A.

The rate of change of G.H.A. of Aries is constant of value 15° 02·46' per hour, and this is the value upon which the Aries increment table is based. There is therefore no 'v' correction required for Aries.

## Declination

The declination for each body is given for each hour of G.M.T. against the value of G.H.A. The rate of change of declination is usually small enough to allow for mental interpolation between the hourly values. However this may be done by the interpolation tables (the same table as for interpolating the 'v' value). The difference between hourly values of declination is given as value 'd' in the daily pages. 'd' is given once for the three days for planets at the foot of each planet column, and once for the three days for the sun. These are mean values and a more accurate value may be obtained by inspecting successive hourly values of the declination. In the case of the moon the change of declination over one hour is large enough and variable enough to require a 'd' value for each hour of G.M.T. and this is given against the declination value in a column headed 'd'.

The correction to apply to the hour value of declination is obtained by entering the increment interpolation tables ('v' or 'd') with the value from the daily pages and extracting the correction against it. The sign of the 'd' correction must be inferred by noting whether the declination is increasing or decreasing over the hour in which the given G.M.T. lies.

## Example

Required the G.H.A. and declination of the sun on 5th January 1980, at 18h 45m 17s G.M.T.

| | | | |
|---|---|---|---|
| G.H.A. at 18h | 88° 41·4' | dec. | 22° 38·7' S. |
| Incr. | 11° 19·3' | 'd' corr. | −0·2' |
| G.H.A. | 100° 00·7' | dec. | 22° 38·5' S. |

# THE NAUTICAL ALMANAC

**Notes**

1. The G.H.A. for 18h 00m is obtained from the daily page for the date of G.M.T.

2. The declination is taken out at the same time.

3. The increment is for 45m 17s, and is taken from the Interpolation Tables. Turn to the page headed 45m, and proceeding down the page to 17s, the increment will be found under the column headed Sun. The increment is always added.

4. The correction to the declination is found by proceeding down the column headed *v* or *d* correction to the value of *d*, as found at the foot of the column headed 'Sun' on the daily page and taking out the quantity abreast of it. Note from the values of the declination whether *d* is plus or minus.

## To find the G.H.A. and declination of the moon

The method is similar to that for the sun, except that the *v* correction must be applied to the G.H.A. Follow the steps in the example.

## Example

Find the G.H.A. and declination of the moon on 7th January 1980 at 19h 34m 36s G.M.T.

| | | | | | |
|---|---|---|---|---|---|
| G.H.A. for 19h | 222° 28·9' | dec for 19h | 6° 24·4' N. | 'v' = 15·2 |
| Incr. | 8° 15·4' | 'd' corr. | −5·3' | 'd' = 9·2 |
| 'v' corr. | 8·7 | | 6° 19·1' | |
| G.H.A. | 230° 53·0' | | | |

**Note**

The *v* value of 15·2 is extracted against the G.H.A. value for 19h and the actual correction to apply is obtained from the *v* or *d* interpolation table for 34m against 15·2. Similarly the increment to declination of −5·3 is obtained against 9·2.

(*d* correction is particularly important in the case of the moon as the declination is changing rapidly.)

## To find the G.H.A. and declination of a planet

Follow the same steps as for the moon, but the increment for minutes and seconds of time is found in the interpolation tables by using the sun table.

## Example

18th September 1980 at 21h 35m 45s G.M.T. Find the G.H.A. and declination of the planet Venus.

| | | | | | |
|---|---|---|---|---|---|
| G.H.A. for 21h | 178° 48·7′ | dec. for 21h | 16° 18·2′ N. | 'v' | −0·3 |
| Incr. | 8° 56·3′ | | −0·4 | 'd' | 0·6 |
| | 187° 45·0′ | dec. | 16° 17·8′ N. | | |
| | −0·2′ | | | | |
| G.H.A. | 187° 44·8′ | | | | |

Note that the *v* is minus.

## To find G.H.A. of a star

In order that G.H.A. information need not be given for every star a value of sidereal hour angle (S.H.A.) is given for selected stars once on each three day page. This is the angular distance of the star west of the first point of Aries, and as the direction of the first point of Aries and the direction of the stars are, over short periods of time, constant then the S.H.A. of a star is fairly constant.

The list of selected navigational stars down the right hand side of the left hand page contains the value for S.H.A. If the star is not contained in this list it will be found in the full list at the back of the almanac, where a value is given for each month.

Now, if the S.H.A. of the star is the angular distance west of the first point of Aries, if we knew the angular distance of the first point of Aries west of Greenwich, then we could add them together and obtain the angular distance of the star west of Greenwich, i.e. the G.H.A. of the star.

Thus: S.H.A.* + G.H.A.♈ = G.H.A.*

The G.H.A. of the first point of Aries is given every hour in the extreme left hand column of the almanac and an interpolation table is given to find values at times between the hours exactly as for the sun.

**Note**

If the G.H.A.* as found exceeds 360° then this amount should be subtracted.

## Example

Required the G.H.A. of the star *Canopus* 17, at 16h 50m 10s G.M.T. 7th January, 1980.

| | |
|---|---|
| G.H.A. for 16h | 346° 22·9′ |
| Incr. | 12° 34·6′ |
| G.H.A. | 358° 57·5′ |
| S.H.A.* | 264° 06·9′ |
| | 623° 04·4′ |
| | −360° |
| G.H.A.* | 263° 04·4′ |

**Note**

*Canopus* is listed on the daily pages. The declination and the S.H.A. are taken from that list.

# THE NAUTICAL ALMANAC

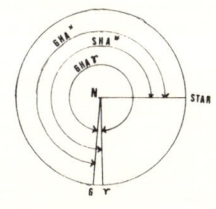

Fig. 7.3

## Exercise
### Find the G.H.A. and declination of the sun on:
1. 7th January, 1980, at 10h 50m 00s G.M.T.
2. 18th September, 1980, at 15h 40m 00s G.M.T.
3. 19th December, 1980, at 11h 58m 25s G.M.T.
4. 26th June, 1980, at 17h 58m 34s G.M.T.
5. 30th September, 1980, at 04h 15m 47s G.M.T.

*Answers*

|    | G.H.A.       | Dec.          |
|----|--------------|---------------|
| 1. | 341° 00·4′   | 22° 26·7′ S.  |
| 2. | 56° 30·0′    | 1° 38·6′ N.   |
| 3. | 000° 17·5′   | 23° 25·3′ S.  |
| 4. | 88° 55·8′    | 23° 20·1′ N.  |
| 5. | 246° 26·7′   | 2° 50·4′ S.   |

## Exercise 1
### Find the G.H.A. and declination of:
1. The moon, 8th January, 1980, at 02h 50m 20s G.M.T.
2. The moon, 19th September, 1980, at 08h 50m 40s G.M.T.
3. The moon, 19th December, 1980, at 06h 35m 42s G.M.T.
4. Venus, 18th September, 1980, at 11h 45m 10s G.M.T.
5. Mars, 18th September, 1980, at 21h 51m 20s G.M.T.
6. Jupiter, 8th January, 1980, at 20h, 31m 20s G.M.T.

*Answers*

|    | G.H.A.       | Dec.          |    | G.H.A.       | Dec.          |
|----|--------------|---------------|----|--------------|---------------|
| 1. | 336° 41·8′   | 5° 11·8′ N.   | 4. | 40° 09·3′    | 16° 23·3′ N.  |
| 2. | 202° 06·1′   | 19° 31·8′ S.  | 5. | 104° 41·3′   | 16° 30·6′ S.  |
| 3. | 134° 12·5′   | 13° 51·5′ N.  | 6. | 253° 26·7′   | 8° 54·6′ N.   |

## LOCAL HOUR ANGLE (L.H.A.)

This is the arc of the equinoctial contained between the meridian of the observer and the meridian of the celestial body. It is always measured westwards from the observer to the body and expressed in degrees and minutes of arc.

Clearly the difference between the L.H.A. and the G.H.A. of any body will be the longitude of the observer.

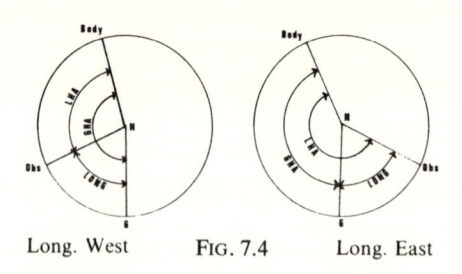

Long. West    FIG. 7.4    Long. East

Thus from the figure:
L.H.A. = G.H.A. − W'ly longitude.
L.H.A. = G.H.A. + E'ly longitude.

### To find the Local Hour Angle (L.H.A.) of a body

First find the G.H.A. of the body, and then apply the longitude of the observer.

**Note**

If the longitude is west and the G.H.A. of the body is less than the longitude then 360° should be added to the G.H.A. before subtracting the longitude.

**Example**

Find the L.H.A. of the sun at 23h 10m 48s G.M.T. on 7th January 1980, if the longitude of the observer is 153° 20' E.

| | |
|---|---|
| G.H.A. 23h | 163° 26·8' |
| Incr. | 2° 42·0' |
| G.H.A. | 166° 08·8' |
| Long. E.+ve | 153° 20·0' |
| L.H.A. | 319° 28·8' |

**Example 2**

Find the L.H.A. of the moon on 1st October 1980 at 15h 31m 29s G.M.T. to an observer in longitude 150° 42·0' W.

# THE NAUTICAL ALMANAC

|  |  |  |
|---|---|---|
| G.H.A. 15h | 130° 04·4' | 'v'=8·8 |
| Incr. | 7° 30·7' | |
| 'v' corr. | 4·6' | |
| G.H.A. | 137° 39·7' | |
|  | +360 | |
|  | 497° 39·7' | |
| Long. W.−ve | 150° 42·0' | |
| L.H.A. | 346° 57·7' | |

## Exercise
**Find the L.H.A. in each case**

| Date | G.M.T. | Body | Longitude |
|---|---|---|---|
| 1. 19th Dec. 1980 | 08h 35m 30s | Sun | 125° 10·0' E. |
| 2. 18th Sept. 1980 | 21h 58m 57s | Sun | 72° 18·3' W. |
| 3. 18th Sept. 1980 | 03h 50m 41s | Aries | 140° 10·2' W. |
| 4. 17th Dec. 1980 | 20h 10m 40s | Arcturus | 164° 16·2' E. |
| 5. 18th Dec. 1980 | 21h 10m 14s | Kochab | 38° 20·2' W. |
| 6. 19th Sept. 1980 | 18h 30m 40s | Sun | 162° 20·0' W. |
| 7. 26th June 1980 | 20h 00m 12s | Aries | 17° 33·0' W. |
| 8. 30th Sept. 1980 | 20h 31m 20s | Betelgeuse | 162° 00·0' W. |

*Answers*

1. 74° 44·6'  2. 78° 57·4'  3. 274° 43·4'  4. 339° 50·8'
5. 144° 10·9'  6. 296° 56·0'  7. 197° 35·5'  8. 67° 02·3'

## TO FIND TIMES OF MERIDIAN PASSAGE

Meridian passage of a heavenly body occurs when the body, in its movement across the heavens due to the daily rotation of the earth, crosses the observer's meridian. At this point the body bears due north or due south and the altitude reaches a maximum. Observation of the altitude at meridian passage provides a quick and easy method of finding the observer's latitude (see Chapter 11, The Meridian Altitude Problem). It is necessary to calculate the G.M.T. of the meridian passage in order to extract the declination from the almanac, and also to know approximately at what time to take the observation. The exact time is not necessary for observation as the practice is to watch the altitude with a sextant until it reaches a maximum and starts to decrease. The maximum altitude is then taken as the meridian altitude. The time is not required for the calculations. The accuracy to which the time is required in order to extract the declination will depend upon how rapidly the declination is changing. In all cases however the time to the nearest minute is adequate.

Information is given in the daily pages of the Nautical Almanac to enable the G.M.T. of meridian passage of any body to be found.

**To find the time of meridian passage of the sun**

At the bottom right hand corner of each double page in the daily pages section of the Nautical Almanac is a box headed 'SUN'. In this box is given a figure headed 'mer pass', one figure given for each of the three days on the page. This may be taken as the Local Mean Time (L.M.T.) of the meridian passage of the true sun for any observer's meridian. It must be converted into Greenwich Mean Time (G.M.T.) by application of the observer's longitude in time as follows:

G.M.T.=L.M.T.+westerly longitude in time.
or G.M.T.=L.M.T.−easterly longitude in time.

The conversion of longitude in degrees to time at 15° per hour may be easily done by using the arc to time conversion table which is given in the Nautical Almanac immediately preceding the increment tables (the first of the 'yellow pages').

**Example**

Find the time of meridian passage of the sun to an observer in longitude 75°W. on the 6th January, 1980.

```
L.M.T. mer. pass.   1206
Long. in time       0500
                    ————
G.M.T. mer. pass.   1706   6th
```

Note that longitude is west and therefore the G.M.T. is later than the L.M.T.

(The conversion of longitude in arc to longitude in time is best done by the conversion table on the page immediately preceding the increment tables in the almanac.)

**Example**

Find the time of the meridian passage of the sun to an observer in longitude 168° 50′ E. on 29th September, 1980.

```
L.M.T. mer. pass.   1150
Long. in time       1115
                    ————
G.M.T. mer. pass.   0035   29th
```

# THE NAUTICAL ALMAMAC

## To find the times of meridian passage of the moon

At the bottom of the right hand side of each double daily page there is a box headed 'MOON'. This gives figures for the times of upper and lower meridian passage of the moon, the age of the moon, and its phase. We are mainly concerned with the time headed 'mer pass'—'upper'. This is the Local Mean Time of the passage of the moon over the meridian of Greenwich. Unlike the sun we cannot take this figure as being an L.M.T. for any observer's meridian, due to the rapid motion of the moon on the celestial sphere. The L.M.T. given for Greenwich must be corrected by an amount referred to as the 'longitude correction'. This correction produces the L.M.T. for the observer's meridian. (Note it does not convert to G.M.T.)

The longitude correction is found by taking the difference between the times of meridian passage given for the day in question, and for the following meridian passage, which will usually be on the next day, if in westerly longitude, or the preceding meridian passage which will usually be on the preceding day, if in east longitude. This difference in minutes of time is then multiplied by the longitude and divided by 360°. Thus:

$$\text{longitude correction} = \frac{\text{daily difference} \times \text{longitude}}{360°}$$

This calculation is easily done by use of Table II at the back of the Nautical Almanac immediately after the increment tables (the last of the 'yellow pages'). The table is entered with the longitude and the daily difference and the correction may be extracted to the nearest minute by simple interpolation.

The longitude correction is added if in westerly longitude, and subtracted if in easterly longitude. Full instructions are given with Table II in the almanac.

## Note

There is on average 24h 50m between two successive meridian passages of the moon. Once in each month there will therefore be one day which does not have a meridian passage. In this case the time given in the Nautical Almanac is for the meridian passage of the following day given in the form 2430. This means that the meridian passage occurs at 0030 hrs. of the next day. The time given for the next day is 0030. The same meridian passsage is therefore given twice. Care must be taken in a case like this to take out the time of the following meridian passage and not the same meridian passage when finding the daily difference.

Although the times given may indicate that there is no meridian passage at Greenwich on a particular day, there may be one on that day over other meridians. If a problem involves finding the time of

114    PRACTICAL NAVIGATION FOR SECOND MATES

meridian passage on a day for which there is no passage at Greenwich, then the figure given for that day (the time of mer pass on the following day), should be used in the manner described. The longitude correction will bring the time back into the day in question.

**Example**

**To find the G.M.T. of the moon's meridian passage**

7th January, 1980. Find the G.M.T. of the moon's meridian passage to an observer in longitude 150° 10′ W.

| | |
|---|---|
| L.M.T. mer. pass. 7th long. 0° | 03h 44m |
| L.M.T. mer. pass. 8th long. 0° | 04h 26m |
| difference | 42m |
| corr. from table II = | 18m |
| L.M.T. mer. pass. 7th long. 0° | 03h 44m |
| Long. corr. | 18m |
| L.M.T. mer. pass. 7th long. 150° 10′ W. | 04h 02m |
| Long. in time | 10h 01m |
| G.M.T. | 14h 03m (7th) |

**Example**

29th September, 1980, find the G.M.T. of the moon's meridian passage to an observer in longitude 94° 37′ E.

| | |
|---|---|
| L.M.T. mer. pass. 29th long. 0° | 04h 09m |
| L.M.T. mer. pass. 28th long. 0° | 03h 13m |
| difference | 56m |
| corr. from table II = | 15m |
| L.M.T. mer. pass. 29th long. 0° | 04h 09m |
| Long. corr. | 15m |
| L.M.T. mer. pass. 29th long. 94° 37′ | 03h 54m |
| Long. in time | 06h 18m |
| G.M.T. | 21h 36m (28th) |

## THE NAUTICAL ALMANAC

### To find the time of meridian passage of a planet

At the bottom right hand corner of the left hand page in the daily pages the L.M.T. of meridian passage of the four planets is given once for the three days. The figure refers to the middle day of the page. The difference in the times for successive days may be several minutes so that to obtain the time of meridian passage for any meridian other than the Greenwich meridian, to the nearest minute, a longitude correction must be applied in the same way as described for the moon. This correction will never amount to more than a few minutes. Inspection of the declination figures for the planets will show that the changes are slow and will never amount to any significance over a few minutes. In practice therefore it is sufficient to take the L.M.T. given for the day in question, interpolating between the three days on the page, and, ignoring the longitude correction, apply the longitude in time to obtain the G.M.T.

### Example (with longitude correction)

Find the G.M.T. of meridian passage of Saturn on 7th January, 1980 to an observer in longitude 179°W.

| | |
|---|---|
| L.M.T. mer pass 8th | 04h 45m |
| L.M.T. mer pass 5th | 04h 57m |
| By inspection | |
| L.M.T. mer pass 7th | 04h 49m |
| L.M.T. mer pass 8th | 04h 45m |
| Difference | 4m |

$$\text{Longitude correction} = \frac{4 \times 179°}{360°} = 2m$$

| | |
|---|---|
| L.M.T. mer pass 7th | 04h 49m |
| Longitude correction | 2m |
| L.M.T. mer pass 7th 179°W. | 04h 47m |
| Longitude in time | 11h 56m |
| G.M.T. 7th | 16h 43m |

(correction negative—west longitude with times becoming earlier. See instructions with Table II).

### Example without longitude correction (same example is used)

| | |
|---|---|
| By inspection from almanac | |
| L.M.T. mer pass 7th long. 0° | 04h 49m |
| Longitude in time | 11h 56m |
| G.M.T. 7th | 16h 45m |

## To find the time of meridian passage of a star

The method which is recommended here for finding the time of meridian passage of a star may also be used with any of the other bodies, if preferred.

At meridian passage the body concerned is on the same meridian as the observer. The G.H.A. of the body, which it will be remembered is the longitude of the body's geographical position, must therefore be equal to the longitude of the observer. As the G.H.A. is expressed from 0° to 360° westwards from Greenwich, then if the longitude is east it may be subtracted from 360° and then said to be equal to the G.H.A. It only remains to extract from the Nautical Almanac the exact G.M.T. at which that particular G.H.A. occurs, from the daily pages which list G.H.A. against G.M.T. In dealing with stars it will be necessary to apply the S.H.A. of the star to its G.H.A. to give the G.H.A. of the first point of Aries, which is listed in the daily pages. This is given by:

$$G.H.A. \; \Upsilon = G.H.A.* - S.H.A.*$$

## Note

When using this method a problem arises as to what will be the date at Greenwich when the required meridian passage occurs on the observer's meridian, on the date given in the question. The date given always refers to the date at the vessel. The Greenwich date may be the same, or it may be one day later if the observer is in west longitude, or one day earlier if in east longitude. When extracting the G.M.T. from the almanac for the appropriate value of G.H.A. at meridian passage, it must be extracted on such a date that when the longitude in time is applied to it to get the L.M.T., then the date at the observer is the date required by the problem. In practice this will not arise as the Greenwich date will be known, but in the context of an examination question this point should be given careful consideration (see Example 3).

## Example 1

Find the G.M.T. and L.M.T. of meridian passage of the star Capella to an observer in longitude 45° 18′ W. on 2nd October, 1980.

| | | |
|---|---|---|
| G.H.A. Capella | 45° 18′ | |
| S.H.A. Capella | 281° 11·1′ | |
| G.H.A. $\Upsilon$ | 124° 06·9′ | |
| G.H.A. $\Upsilon$ 07h 2nd | 116° 09·1′ | (by inspection of almanac) |
| Increment | 7° 57·8′ | =31m 46s (from Aries increment tables) |

# THE NAUTICAL ALMANAC

G.M.T. mer pass     07h 31m 46s 2nd
Longitude            03h 01m 12s
L.M.T.                04h 30m 34s 2nd

**Note**

By inspection of the almanac the value of G.H.A. next less than the required G.H.A. should be extracted, mentally checking at this point that the application of the longitude in time will give a local date required by the question.

## Example 2

Find the time of meridian passage of the moon over the meridian of 78° 45′ E. on 18th December, 1980.

G.H.A. moon = E. Long. − 360°    281° 15′
G.H.A. moon 15h 18th          269° 01·0′
Increment                       12° 14·0′   ('v' value from daily
'v' corr.                            −6·6′    pages 7.8 correction
Increment corrected           12° 07·4′    from increment table for 50 minutes = 6·6)

From increment table for moon an increment of 12° 07·4′ corresponds to 50m 49s.

G.M.T.    15h 50m 49s   18th
Long.       5h 15m
L.M.T.     21h 05m 49s   18th

**Note**

Adequate accuracy is obtained by neglecting the 'v' correction. In this case the increment of 12° 14·0′ would have given 51m 16s. The G.M.T. would therefore have been 15h 51m 16s. In practice the error of a few seconds would be negligible. The other method described for the moon in fact only gives the times to the nearest minute.

## Example 3

Find the G.M.T. and L.M.T. of meridian passage of the star Antares to an observer in longitude 175° 30′ E. on 8th January 1980.

G.H.A. Antares     184° 30′
S.H.A. Antares     112° 57·7′
G.H.A. ♈            71° 32·3′
G.H.A. ♈ 21h 7th   61° 35·3′
Increment             9° 57·0′    = 39m 42s

```
G.M.T.   21h 39m 42s  7th Jan.
Long.    11h 42m
L.M.T.   33h 21m 42s
   =      9h 21m 42s  8th Jan.
```

**Note**

The date at Greenwich is one less than the date at ship in east longitude in this example. Had the G.H.A. for 21h on 8th been extracted, the L.M.T. would have fallen on the 9th. By extracting the G.H.A. for 21h on 7th the L.M.T. is on the required date.


### EXERCISE

Find to the nearest second the G.M.T. of meridian passage in the following examples.

1. Jupiter, 26th June, longitude 50°W.
2. Canopus, 19th September, longitude 40°E.
3. Moon, 18th December, longitude 105°W.
4. Procyon, 28th June, longitude 169° 50′ E.
5. Spica, 5th January, longitude 124° 30′ W.

*Answers*

1. 19h 30m 09s 26th June
2. 03h 50m 41s 19th Sept.
3. 04h 34m 17s 19th Dec.
4. 01h 53m 38s 28th June
5. 14h 44m 41s 5th Jan.

**To find times of lower meridian passage (Lower transit or meridian passage below the pole)**

If the body is visible above the horizon when it crosses the observer's lower meridian, latitude may be found readily by observation of altitude at this occurrence (see Chapter 11). The declination is required and therefore the G.M.T. of lower meridian passage.

**Sun and planets**

It is sufficiently accurate for practical purposes to add 12 hours to the time of upper meridian passage given in the almanac, and to proceed as described for upper meridian passage.

**Moon**

The L.M.T. of lower meridian passage is extracted from the almanac and treated in the same way as described for upper meridian passage.

# THE NAUTICAL ALMANAC

**Stars**

At lower meridian passage the G.H.A. of the body is 180° different from the observer's longitude. The G.M.T. when such G.H.A. occurs is extracted from the daily pages in the same way as for upper meridian passage.

## TO FIND TIMES OF SUNRISE, SUNSET, MOONRISE AND MOONSET

The G.M.T. of sunrise and sunset, and moonrise and moonset is required to solve the amplitude problem in which the compass error is obtained by observation of the sun or moon at rising or setting. The declination is required in the problem hence the requirement for the G.M.T. (see Chapter 8, The Amplitude Problem).

**To find times of sunrise or sunset**

The time of sunrise and sunset is given on the right hand side of the right hand page in the daily pages. It is given once for the three days on the page, the figure referring to the middle day. In moderate latitudes the times will change little over three days, so that the figure given for the page may be used without interpolation. In higher latitudes the daily change may be such that interpolation between the three days may be necessary for accuracy. No significant error will be caused in practice if this is not done. The times vary with latitude so that the argument latitude must be used to extract the time, interpolating between the latitudes tabulated. The interpolation is not linear and may be done with the aid of Table 1 on the page immediately following the increment tables. Full instructions are given with this table. The change of the times with latitude are usually small and in practice no significant error will be caused by interpolation mentally assuming linear changes.

The times contained in these columns are local mean times (L.M.T.), and may be taken to be for any meridian. The longitude in time must be applied in order to obtain G.M.T. (see Times of Meridian Passage).

**Example**

Find the G.M.T. of sunrise on 19th September to an observer in D.R. position 50° N. 165° 24′ W.

|  |  |  |
|---|---|---|
| L.M.T. sunrise 50°N. 19th | 05h 42m |  |
| Longitude in time (W) | 11h 02m |  |
| G.M.T. | 16h 44m | 19th |

## Example

Find the time of sunset on 26th June 1980 to an observer in position 55° S. 172° 30′ E.

| | | |
|---|---|---|
| L.M.T. sunset 55°N. 27th | 17h 12m | (interpolating between |
| L.M.T. sunset 55°N. 24th | 17h 17m | 54° and 56°) |
| L.M.T. sunset 55°N. 26th | 17h 14m | (interpolating between |
| Longitude in time (E.) | 11h 30m | 27th and 24th) |
| G.M.T. | 05h 44m | 26th |

## To find times of moonrise and moonset

The times of moonrise and moonset are tabulated against latitude in the same manner as those for sunrise and sunset. The times are given for each day however due to the large differences between daily figures and the variations in the daily differences. For the same reasons the times, which are local mean times, cannot be taken as L.M.T. for any meridian, but only for the meridian for which they were calculated, the Greenwich meridian. In order to find the L.M.T. for any other meridian a longitude correction must be applied as described for the time of meridian passage of the moon. Again this longitude correction is given by:

$$\text{daily difference} \times \frac{\text{longitude}}{360}$$

After correction for longitude, the longitude in time must be applied in order to give the G.M.T. Thus the procedure is:

1. Extract the L.M.T. tabulated for the date in question, interpolating for latitude using Table 1.

2. Extract the L.M.T. for the following day if in west longitude, or the preceding day if in east longitude, and thus find the daily difference.

3. Find the longitude correction from Table II or by the formula given above.

4. Apply the longitude correction to the L.M.T. extracted for the day in question as explained in (1). This is normally added if in west longitude and subtracted in east longitude. This only applies if the times are getting later each day, as is usually the case. If the times are getting earlier each day then this rule is reversed, and the correction subtracted if in west longitude and added if in east longitude. In all cases the result will lie between the times extracted in (1) and (2).

5. Apply longitude in time, +ve for west longitude and −ve for east longitude, to obtain the G.M.T.

## Example

Find the G.M.T. of moonrise on 9th January to an observer in position 21° 30′ S. 100° E.

| | | |
|---|---|---|
| L.M.T. moonrise 9th Long. 0° | 23h 37m | (interpolating for latitude between 20° and 30°) |
| L.M.T. moonrise 8th Long. 0° | 23h 01m | |
| Difference | 36m | |

Longitude correction $= 36 \times \dfrac{100}{360} = 10\text{m}$

| | |
|---|---|
| L.M.T. moonrise 9th Long. 0° | 23h 37m |
| Longitude correction | 10m |
| L.M.T. moonrise 9th Long. 100° E. | 23h 27m |
| Longitude in time (E.) | 06h 40m |
| G.M.T. | 16h 47m 9th |

## Example

Find the G.M.T. of moonset on 26th June 1980 to an observer in position 33° N. 170° W.

| | | |
|---|---|---|
| L.M.T. moonset 26th Long. 0° | 03h 24m | (using Table I for interpolating for latitude) |
| L.M.T. moonset 27th Long. 0° | 04h 08m | |
| Difference | 44m | |

Longitude correction $= 44 \times \dfrac{170}{360} = 21\text{m}$

| | |
|---|---|
| L.M.T. moonset 26th Long. 0° | 03h 24m |
| Longitude correction | 21m |
| L.M.T. moonset 26th Long. 170° W. | 03h 45m |
| Longitude in time (W.) | 11h 20m |
| G.M.T. | 15h 05m 26th |

## CHAPTER 8

## COMPASS ERROR BY ASTRONOMICAL OBSERVATION

### (The Azimuth Problem and the Amplitude Problem)

The compass error may be found by observing the compass bearing of an astronomical body and comparing it with the true bearing found for the instant of observation, by calculation. The Azimuth problem may be used with any body which is visible above the horizon, except for bodies which are close to the zenith, when the observation of bearing is inaccurate. The Amplitude problem is used when the compass bearing of a body is observed at the moment of rising or setting.

The calculation of the true bearing requires the solution of the PZX triangle for the angle Z (see Chapter 12 for an explanation of the PZX triangle). The solution may be obtained quickly and easily with the use of the ABC tables contained in nautical tables (Nories' or Burton's). These are tables which can be used for the solution of any spherical triangle, just as the traverse tables may be used for the solution of any plane right angle triangle. The ABC tables are specifically designed however for the solution of the astronomical triangle for the angle Z, and are headed accordingly. The arguments are the known values of latitude declination and hour angle (L.H.A.).

The angle Z in the PZX triangle is the angle contained between the observer's meridian and the direction of the body. This is also a definition of the bearing (see Chapter 2). The angle Z however is called the azimuth and is measured from 0° to 180° from the direction of the elevated pole (north if in north latitude and south if in south latitude), to the east or west depending upon whether the body is rising or setting. The azimuth therefore is merely the bearing of the body expressed and named according to a different set of rules.

Converting azimuth to bearing is therefore an easy matter. If the azimuth is named north then the bearing will be the same as the azimuth if named E, and 360° − azimuth if named west. If the azimuth is named south, the bearing will be 180° − azimuth if named E, or 180° + azimuth if named W.

## Examples

| North Lat. | | South Lat. | |
|---|---|---|---|
| Azimuth | Bearing | Azimuth | Bearing |
| N. 40° E. | 040° | S. 40° E. | 140° |
| N. 140° E. | 140° | S. 140° E. | 040° |
| N. 40° W. | 320° | S. 40° W. | 220° |
| N. 140° W. | 220° | S. 140° W. | 320° |

When a body is on the observer's meridian the L.H.A. is 000° and when it lies to the west of the meridian the L.H.A. is between 000° and 180°. As the body passes to the eastwards of the meridian and commences to rise the L.H.A. increases from 180° to 360°, when it is again on the observer's meridian.

Figure 8.1 shows a body with a N. declination and an observer in a higher N. latitude. As the body rises the azimuth or bearing of the body is N.E'ly and increases, bearing 090° when it crosses the Prime Vertical (WZE). It will pass to the south of the observer, bearing 180° when the L.H.A. is 360° (or 000°). Subsequently, as the body moves to the west, the bearing continues to increase, being 270° when it again crosses the Prime Vertical and finally it sets bearing N.W'ly.

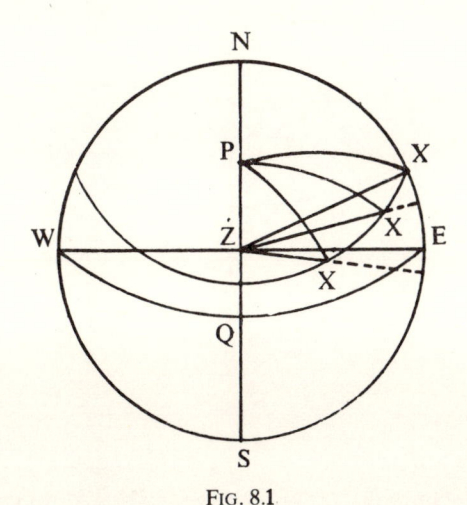

Fig. 8.1

Figure 8.2 shows an observer in N. latitude and a body with a S. declination. It will be noted that the body rises, bearing S.E'ly and sets, bearing S.W'ly, so that the range of azimuth is less than in the previous case.

Figure 8.3 shows an observer in N. latitude and a body with a

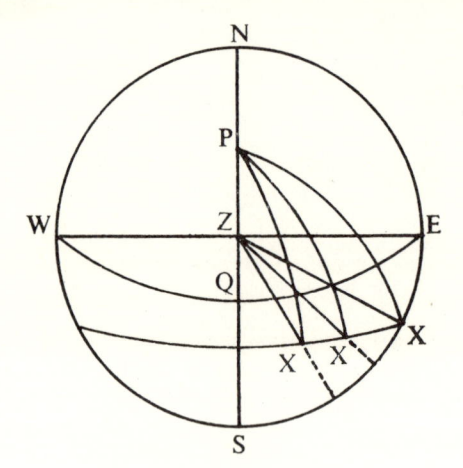

FIG. 8.2

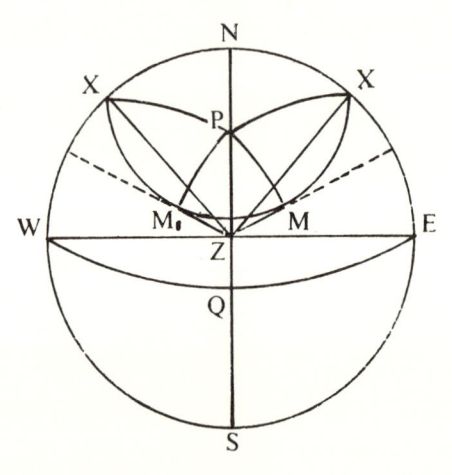

FIG. 8.3

higher N. declination. The body bears N.E'ly when rising and N.W'ly when setting, but in this case the body passes to the N. of the observer bearing 000° when the L.H.A. is 360° (or 000°). It will be noted that as the body rises it moves first to the right, reaching a maximum azimuth at $M$ after which the bearing moves to the left until it reaches a maximum westerly azimuth at $M_1$. It then moves to the right again before setting.

# COMPASS ERROR

If a body has a declination of 0°, it will rise bearing due east and set, bearing due west. The L.H.A. on rising will be 270° and on setting the L.H.A. will be 90°.

Inspection of the figures will show, therefore, that when the latitude and declination are of the same name, the L.H.A. ($E$) of the body on rising and the L.H.A. on setting will be greater than 90°.

If the latitude and declination are of opposite names the L.H.A.($E$) on rising and L.H.A. when setting will be less than 90°.

## Steps in the problem

1. Ascertain the G.M.T. and date from the time given.
2. Take out the necessary elements from the *Nautical Almanac*:
   for the sun—the declination and G.H.A.
   for a star—the declination, *S.H.A. and G.H.A. ♈
3. Using the appropriate time formula, derive the L.H.A. of the body. If the L.H.A. is greater than 180° it may be found more convenient to subtract this from 360° to obtain L.H.A.($E$).
4. Using ABC tables, with L.H.A. and latitude take out the quantity $A$, interpolating as necessary. If the L.H.A. is less than 90° name this opposite to the latitude. With L.H.A. and declination take out the quantity $B$ interpolating as necessary. Name this the same as the declination.

Add the two quantities, $A$ and $B$, if they are the same name, otherwise take the lesser from the greater to obtain the quantity $C$. Name this according to the greater.

From the $C$ table against the latitude take out the azimuth. Name this according to the quantity $C$ and the L.H.A.

An example should make this clear.

L.H.A. 48° 45′   Latitude of observer 40° 42′ N.
Declination of body 16° 20′ S., to find the True Bearing.

L.H.A. 48°, lat. 40° 42′ N., A = 0·774
L.H.A. 49°, lat. 40° 42′ N., A = 0·751
∴ L.H.A. 48° 45′, lat. 40° 42′ N.
A = 0·757 S.                                 A

| Lat. | Hour Angle | |
|---|---|---|
|  | 48° | 49° |
| 40° | 0·76 | 0·73 |
| 41° | 0·78 | 0·76 |

L.H.A. 48°, dec. 16° 20′ S., B = 0·396
L.H.A. 49°, dec. 16° 20′ S., B = 0·390
∴ L.H.A. 48° 45′, dec. 16° 20′ S., B = 0·391 S.

i.e. A  0·757 S. (named opposite to lat.)
    B  0·391 S. (named same as dec.)       B
    ─────
    C  1·148 S.

| Dec. | Hour Angle | |
|---|---|---|
|  | 48° | 49° |
| 16° | 0·39 | 0·38 |
| 17° | 0·41 | 0·41 |

C 1·14, lat. 40° 42′, Az.=49·2°
C 1·16, lat. 40° 42′, Az.=48·7°
∴C 1·148, lat. 40° 42′, Az.=49·0°
i.e. True bearing S. 49° W.
  or 229°

|   | 1·14 | 1·16 |
|---|---|---|
| C Lat. | Azimuth ||
| 40° | 48·9° | 48·4° |
| 41° | 49·3° | 48·8° |

**Notes**

1. Had the L.H.A. been greater than 90°, then A would have been named the same as the latitude. The B factor is always named the same as the declination.

2. With practice the interpolation can be done mentally.

3. When using *Burton's Tables* the method is exactly the same except that (*a*) + and − signs are used instead of N. and S., (*b*) the factors A and B are given to 3 decimal places, (*c*) the azimuth is given for every full degree. This may make interpolation a little more awkward, but this can be overcome by using the interpolation table at the end of the ABC tables, and by following the concise instructions given there.

**Example**

30th September, 1980, in D.R. position lat. 45° 22′ N., long. 125° 10′ E., where the variation was 24° E., the sun bore 229° C. at 07h 51m 06s G.M.T. Find the sun's true azimuth, and thence the deviation of the compass.

G.M.T. 30th 07h 51m 06s

From N.A.
G.H.A.   287° 30·5′
Incr.      12° 46·5′

G.H.A.   300° 17·0′
Long. E. 125° 10·0′

L.H.A.   425° 27·0′
         360°

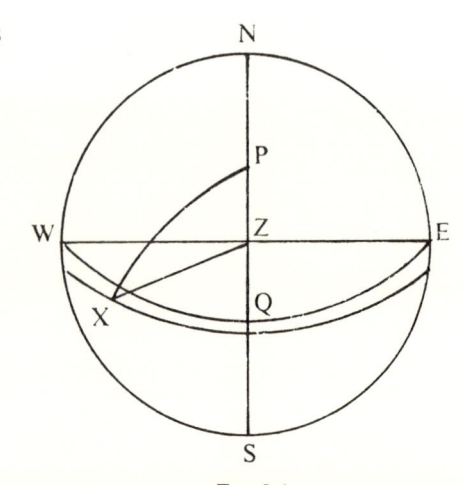

L.H.A.    65° 27·0′

Fig. 8.4

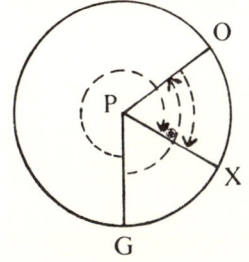

Fig. 8.5

| Dec. | 2° 53·1′ S. |
| d. corr. | + 0·9′ |
| Dec. | 2° 54·0′ S. |

A  0·464 S.
B  0·054 S.
―――――
C  0·518 S.
―――――
T. Az. S. 70·0° W.

| True bearing | 250·0° |
| Comp. bearing | 229·0° |
| Comp. error | 21·0° E. |
| Var. | 24·0° E. |
| Dev. | 3·0° W. |

## EXERCISE 8A

### SUN AZIMUTHS

1. 19th September, 1980, in D.R. position lat. 42° 50′ N., long. 46° 10′ W. at 11h 40m 19s G.M.T., the sun bore 149° C. Find the true azimuth and the deviation, the variation being 24·5° W.

2. 6th January, 1980, in E.P. 48° 20′ S., 96° 30′ W., at 20h 40m 30s G.M.T., the sun bore 286° C. Find the deviation, the variation being 23° E.

3. 19th December, 1980, in D.R. position 46° 15′ N., 168° 35′ W., the observed azimuth of the sun was 122° C. at 20h 31m 10s G.M.T. Find the sun's true azimuth and the deviation, the variation being 12° E.

4. 27th June, 1980, in D.R. position, lat. 38° 10′ S., long. 124° 10′ E., a.m. at ship, when the chronometer showed 11h 58m 10s, the observed azimuth of the sun was 057° C. Find the deviation, the variation being 2° W.

5. 19th September, 1980, at 15h 20m 00s L.M.T., the sun bore 262·5° to an observer in D.R. position lat. 19° 20′ N., long. 149° 50′ E., where the variation was 1° E. Find the deviation.

## STAR AZIMUTHS

**Example**

18th September, 1980, at 06h 14m 09s, G.M.T. in D.R. position 37° 36′ N., 47° 50′ W., the observed bearing of *Alpheratz* 1 was 289·5°C. Find the true azimuth and the deviation, the variation being 22·5° W.

G.M.T. 18th 06h 14m 09s

| From N.A. | |
|---|---|
| G.H.A. ♈ | 87° 18·6′ |
| Incr. | 3° 32·8′ |
| G.H.A. ♈ | 90° 51·4′ |
| *S.H.A. | 358° 09·0′ |
| | 449° 00·4′ |
| | 360° |
| *G.H.A. | 89° 00·4′ |
| Long. W. | 47° 50·0′ |
| L.H.A. | 41° 10·4′ |

*Dec. 28° 59·1′ N.

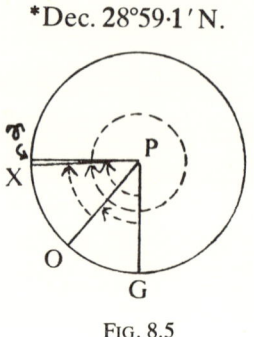

Fig. 8.5

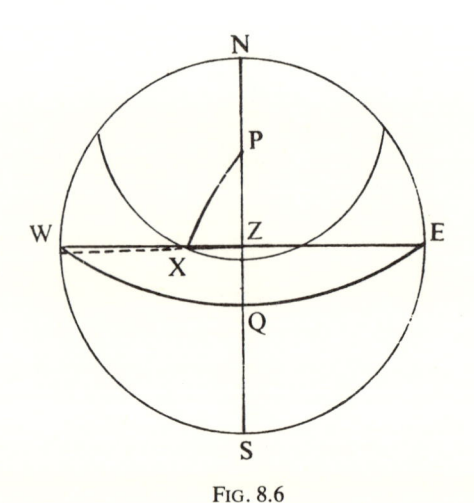

Fig. 8.6

|   |   |   |   |
|---|---|---|---|
| A | 0·88 S. | True bearing | 267·1° |
| B | 0·84 N. | Comp. bearing | 289·5° |
| C | 0·04 S. | Comp. error | 22·4° W. |
|   |   | Var. | 22·5° W. |
| True Az. S. 87·1° W. |   | Dev. | 0·1° E. |

### EXERCISE 8B

### STAR AZIMUTHS

1. On 19th December, 1980, in position lat. 46° 40′ N., long. 168° 20′ W. the observed bearing of the star *Gienah* 29 was 134° C. at 15h 15m 27s G.M.T. If the magnetic variation in the locality was 13° E., find the deviation for the ship's head.

2. On 26th June, 1980, at 02h 11m 43s G.M.T. the star *Rasalhague* 46 bore 247° C. to an observer in lat. 38° 20′ N., long. 5° 40′ E. If the variation was 6·5° E., find the deviation for the ship's head.

3. On 1st October, 1980, a.m., at ship in lat. 41° 15′ N., long. 145° 26′ E., when the chronometer, which was correct on G.M.T., showed 5h 48m 19s, the star *Procyon* 20 was observed bearing 125° C. If the variation was 7° W., find the deviation for the ship's head.

4. On 6th January, 1980, at ship in lat. 46° 20′ N., long. 47° 52′ W., the star *Schedar* 3 bore 336° C. when the chronometer which was correct on G.M.T. indicated 01h 41m 28s. If the variation was 27° W., find the deviation for the ship's head.

5. On 19th September, 1980, at about 03.30 at ship in lat. 32° 24′ S., long. 80° 15′ E. the star *Peacock* 52 was observed bearing 250° C. when the chronometer, which was correct on G.M.T., showed 10h 14m 20s. If the variation was 33° W., find the deviation for the ship's head.

### THE AMPLITUDE PROBLEM

**Definition**

The amplitude of a body is the angle between the direction of the body when rising or setting, and the direction of east or west respectively.

Thus the amplitude is merely another way of expressing the bearing of a body at the moment it rises or sets.

By observing the bearing of a body by compass (magnetic or gyro) when the centre is on the rational horizon and comparing this with the calculated true bearing, the compass error, and if a magnetic compass, the deviation for the ship's head, are very simply found.

A body is on the rational horizon at theoretical rising or setting, and the true altitude at this instant is 00° 00′. Because of refraction and dip, etc., the visible rising or setting will occur earlier and later respectively. As a general rule of thumb, theoretical sunrise or sunset can be taken to occur when the sun's lower limb is about one semidiameter above the visible horizon. In lower and medium latitudes the bearing will be changing slowly at this time and any small error in time will not make any difference to the bearing calculated. In very high latitudes, however, when the bearing is changing quickly at sunrise and sunset more care must be taken.

The true amplitude can be calculated from the formula:

$$\text{Sine amplitude} = \text{sine declination} \times \text{secant latitude}.$$

This is given in tabulated form in *Norie's*, *Burton's* and other nautical tables.

The amplitude is named east when rising, west when setting and either north or south according to the name of the declination, e.g. a body with a declination of 20° N. will rise to an observer in latitude 30° N. with an amplitude of E. 23° 16′ N. This can then be converted to the usual three figure notation of 066° 44′.

If the declination had been 20° S. then, in the same latitude, the amplitude would have been E. 23° 16′ S. and this would have given a true bearing of 113° 16′.

**Note**

This problem usually involves the use of the sun. Sometimes the moon may be used, but stars and planets are rarely visible at their rising or setting due to horizon haze. If the moon is used care must be taken to obtain an accurate G.M.T. when taking out the declination, as this may be changing rapidly.

*Procedure*

1. Obtain the L.M.T. and hence the G.M.T. of rising (or setting) from the almanac as explained in Chapter 7.

2. From the *Nautical Almanac* find the declination.

3. Obtain the true amplitude either by calculation or from tables.

4. Convert the true amplitude into a true bearing in three figure notation.

# COMPASS ERROR

5. Compare the true bearing and the compass bearing and obtain the compass error.

6. If the variation is known, find the deviation for the ship's head.

**Example**

1st October, 1980, in D.R. position, lat. 36° 10′ N., long. 28° 20′ W., at 05h 57m 22s L.M.T., the sun rose bearing 112° C. Find the true amplitude, and if the variation was 18° W., find the deviation for the direction of the ship's head.

| L.M.T. sunrise 1st | 05h 55m | Decl. 07h | 3° 16·3′ S. |
|---|---|---|---|
| Long. (W.) | 01h 53m | 'd' corr. | +0·8′ |
| G.M.T. | 07h 48m | Decl. | 3° 17·1′ S. |

Sin ampl. = sin 3° 17·1′ sec. 36° 10′
  ampl. = 4° 04·2′

| True ampl. | E. 4° 04·2′ S. |
|---|---|
| True brg. | 094° T. |
| Comp. brg. | 112° C. |
| Comp. error | 18° W. |
| Variation | 18° W. |
| Deviation | 0° |

**Example**

On 8th January, 1980, at ship in lat. 30° 45′ S., long. 166° 15′ W., the sun set bearing 230° by compass. If the variation for the place was 16° E., find the deviation for the ship's head.

| L.M.T. sunset, lat. 30° S. 8th | 19h 06m |
|---|---|
| Diff. 12m Table I corr. | + 01m |
| L.M.T. sunset, lat. 30° 45′ S. 8th | 19h 07m |
| Long. W. | 11h 05m |
| G.M.T. 9th | 06h 12m |

| Dec. 9th 06h | 22° 12·8′ S. |
|---|---|
| Corr. | − 0·1′ |
| Dec. | 22° 12·7′ S. |

```
Sin ampl.      = sin 22° 12·7' sec. 30° 45'
               = 26° 05·7'
True ampl.     = W. 26° S.
True brg.      = 244° T.
Comp. brg.     = 230° C.
Comp. error    =  14° E.
Variation      =  16° E.
Deviation          2° W.
```

## EXERCISE 8C

1. 30th September, 1980, in D.R. position lat. 20° 52' N., long. 153° 10' W., at 06h 03m 14s L.A.T., the sun rose bearing E. 11·5° N. by compass. Find the true amplitude and the deviation, the variation in the locality being 11° E.

2. 18th September, 1980, at 05h 52m 03s L.A.T., the observed amplitude of the sun to an observer in lat. 39° 53' N., long 51° 00' E., was E. 5° N. Find the true amplitude and the deviation. The variation was 5° E.

3. 27th June, 1980, at ship in D.R. positon, lat. 40° 20' S., long. 00° 00', the sun set bearing 301·5° C. Find the sun's true amplitude and the deviation, the magnetic variation being 26° W.

4. 18th December, 1980, at ship in D.R. position 37° 30' N., 32° 15' W. the sun rose bearing 138° C. Find the true amplitude and thence the deviation, the variation being 21° W.

5. 5th January, 1980, the sun set bearing 258° C. to an observer in E.P. lat 49° 10' S., long 98° 45' W., where the variation was 24° E. Find the deviation for the direction of the ship's head.

6. 26th June, 1980, the sun rose bearing 062° C. to an observer in D.R. position lat. 42° 30' N., long 142° 30' W.. Find the deviation of the compass, the variation in the locality being 20° E.

**Revision papers**

The following revision papers are similar in structure to the Department of Trade Class V (Chartwork and Practical Navigation) paper, and the Class IV (Chartwork) paper.

The chartwork papers are set on charts of British coastal waters in the hope that they will be available to navigators studying at sea. Those to whom the charts are not available may wish to obtain them from Admiralty chart agents. The charts used are published at minimal price as practice charts, which are full size charts containing the same information as navigational charts, but are printed on strong thin paper and may not be corrected to date. They must not be used for navigation but are entirely adequate for practice purposes. The following charts are used:

Lands End to Falmouth        Chart No. 5051
Falmouth to Plymouth         Chart No. 5050
Bristol Channel
(Worms Head to Watchet)      Chart No. 1179

The practical navigation papers may be done with a set of nautical tables (Nories' or Burton's), and the extracts from the Nautical Almanac.

## CLASS V CHARTWORK AND PRACTICAL NAVIGATION

### PAPER 1 (3 hours)

#### Chartwork

Chart. Lands End to Falmouth No. 5051
Use variation 8° W. throughout.

**Use deviation card given in Chapter 2**

1. At a time 5 hours after H.W. Devonport (spring tides), a vessel was in a position with Lizard Lt. Hse. bearing 000° T. distant 5 miles. Find the compass course to steer to pass 3 miles to the south of Wolf Rock Lt., making allowance for any tide you may expect. Estimate the time of arrival off Wolf Rock. Vessel's log speed is 12 knots.

2. From a vessel leaving Falmouth steering 212° C. at 8 knots, Black Head was observed in transit with Lizard Lt. Hse. bearing 232° C. After maintaining this course for 1 hour Lizard Lt. Hse. bore 309° C. If a current set 230° T. at 1·5 knots in the interval find the position of the vessel at the time of the second observation.

3. It is required to round Lizard Point maintaining a minimum distance off Men Hyr Rocks of 1·5 miles. What would be the vertical danger angle to set on a sextant to observe Lizard Pt. Lighthouse?

4. A vessel steering 125° C. at 10 knots observed Longships Lighthouse bearing 345° C. and Tater Du Lighthouse bearing 035° C. 1½ hours later Lizard Pt. Lt. Hse. was bearing 065° C. while Mullion Island was bearing 029° C. Find the set and drift of the tide in the interval.

## Practical navigation

1. Find by traverse table the vessel's position at the end of the third course.

        Initial position    49° 30' N. 8° 00' W.

        First course       261°T. distance steamed 70 miles
        Second course   210°T. distance steamed 72 miles
        Third course     166°T. distance steamed 65 miles

2. Find by plane sailing the course and distance between the following positions.

        A 50° 15' N. 5° 25° W.
        B 52° 10' N. 7° 05' W.

3. From the following information find the compass error and the deviation for the ship's head.

Date: June 26th 1980        D.R. position 50° 30' N. 6° 30' W.
Sun rose bearing 053°C.     Variation 8° W.


## CLASS V CHARTWORK AND PRACTICAL NAVIGATION

### PAPER 2 (3 hours)

### Chartwork

Chart. Falmouth to Plymouth No. 5050.
Use variation 8° W. throughout

## Use deviation card provided in Chapter 2

1. From a position where Eddystone Rock Lt. bears 360°T. distant 3 miles find the compass course to steer to a position where Dodman Point bears 307°T. distant 2·4 miles, in order to counteract a tidal stream estimated to set 133°T. at 2 knots, and allowing for a 10° leeway due to a SW'ly wind. Ship's speed by log 8 knots.

2. The following compass bearings were obtained from a vessel:

        Chapel Point          246°C.
        Black Head           009°C.
        Gribbin Hd. daymark  064°C.

Find the latitude and longitude of the vessel's position and the compass course to steer to arrive at a position where Rame Head chapel ruins is 30° on the port bow distant 3 miles.

# COMPASS ERROR

3. A vessel steering 330°C. has the buoy (Fl.R.10 sec.) in position 50° 07' N. 4° 30' W. approximately, bearing 015°C. distant 1·2 miles. After steaming for 40 minutes at 12 knots the vessel's position was fixed by three bearings:

| Udder Rock buoy | 015°C. |
| Cannis Rock buoy | 326°C. |
| Yaw Rock buoy | 273°C. |

Find the set and rate of the current, the course made good and the speed made good.

4. Find the rising and dipping distance of Eddystone Light. Ht. of eye 12·5m.

## Practical navigation

1. Find by mercator sailing the position at the end of the run.
   Initial position 55° 55' N. 7° 18' E.
   Course 257°C. Variation 8°W. Deviation 3°E. Distance run 120 miles.

2. Find by traverse table the vessel's position at the end of the third course:

| Initial position | 49° 30' N. 8° 00' W. |

| First course | 265°T. distance 132 miles |
| Second course | 347°T. distance 97 miles |
| Third course | 180°T. distance 40 miles |

3. From the following information find the compass error and the deviation for the ship's head.

Time at ship 1608 2nd October. D.R. 43° 30' N. 9° 40' W. Bearing of sun by compass 262°. Chronometer 05h 08m 02s. Chronometer error 2m 18s slow on G.M.T. Variation 12°W.

## CLASS IV CHARTWORK

### PAPER 1

Chart. Bristol Channel Chart 1179.
Use variation 9°W. throughout

**Use specimen deviation card Chapter 2**

1. From a vessel the following simultaneous bearings were taken:

| Hartland Point Lt. Hse. | 204°C. |
| Lundy Is. South Lt. | 287°C. |
| Bull Point Lt. Hse. | 067°C. |

Find the ship's position, the compass error and the deviation for the ship's head.

136     PRACTICAL NAVIGATION FOR SECOND MATES

2. At 1135 from a vessel steaming at 12 knots, a navigator observed St. Gowan Lt. V/L in transit with Warren Church spire, bearing 009°C. At the same time Caldy Is. Lt. Hse. bore 059°C. Find the latitude and longitude of this position and the course to steer by compass to a position with Worms Head bearing 000°T. distant 9·9 miles in order to counteract a tide setting 115°T. at 3 knots, and allowing for 6° leeway due to a northerly wind. Estimate the distance off Helwick Lt. V/L when it is abeam and the time at this position.

3. At 0830 from a vessel steering 259°C. at 12 knots Breaksea Lt. V/L was observed bearing 352°C. distant 1 mile. If 1 hour later Foreland Point Lt. Hse. and Selworthy Beacon bore 192°C. and 121°C. respectively find the set and rate of the tide. Assuming that the same tidal conditions continue find also the course to steer by compass and the speed required to reach a position with Bull Point Lt. Hse. bearing 172°T. distant 5·4 miles in 50 minutes.

4. From a vessel in D.R. position 51°20′ N. 4°58′ W. and steering 113°C. at 7 knots in reduced visibility, the relative bearing of Lundy Island radio beacon was 050°. Two hours later the relative bearing of Breaksea Lt. V/L radio beacon was 357°. If during the interval the vessel was making 8° leeway due to a southerly wind, and the tidal stream set as indicated at position E. 3h before high water Swansea at spring rate, find the latitude and longitude of the vessel at the time of the second bearing.

5. On 25th February 1980 at 1200 G.M.T. a vessel off Watchet passed over a shoal of charted depth 4 metres. If the vessel's draft was 10·5 metres what was her clearance underkeel?

### CLASS IV CHARTWORK

#### PAPER 2

Chart. Lands End to Falmouth No. 5051.
Use variation 9°W. throughout.

**Use deviation card provided in Chapter 2**

1. At 0800 hours in poor visibility, from a vessel steering 181°G. at 5 knots, the Longships Lt. Hse. bore 148°G. Gyro error 1° High. The vessel continued on this course and at 0915 Wolf Rock Lt. was observed to bear 238°G. If a tide set 127°T. at 2 knots throughout, find the vessel's position at 0915.

2. From a position with Wolf Rock Lt. bearing 350°T. distant 2·4 miles, find the true course and distance to a position latitude 49° 46′ N. longitude 5° 26·6′ W. Find also the compass course to steer to counteract a tide setting 084°T. at 3 knots and 6° leeway due to a SW'ly wind, if the vessel's speed by log is 6 knots. What will be the E.T.A. at the position given?

3. On a vessel at anchor off Falmouth the following compass bearings were taken;

| | |
|---|---|
| Hill (102), (50° 02′ N. 5° 05′ W.) | 226°C. |
| Mawnan House (Conspic) (50° 06′ N. 5° 05′ W.) | 280°C. |
| St. Anthony Hd. Lt. | 010°C. |

Find the vessel's position and the compass error.

4. At 1000 hrs. the Runnel Stone buoy was observed in transit with Longship Lt. Hse. bearing 318°C. At the same time Wolf Rock bore 251°C. Find the vessel's position. If course is now set 095°C., and 20 minutes later Tater Du Lt. bore 321°C. and St. Michael's Mount bore 028°C., estimate the set and drift of the tide if the distance run by log in the interval was 5 miles.

5. Find the times and heights of high and low waters at Sharpness Dock on 14th January 1980.

# SECTION 2

The following section contains work required for Class IV Practical Navigation paper in addition to the work contained in Section 1.

## CHAPTER 9

## POSITION CIRCLES AND POSITION LINES

If a distance off a charted point of land is obtained, either by radar or by vertical sextant angle, then a circle can be drawn centred on the point of land and with radius the distance off, and this circle will represent a line, any point on which the ship might be from the observed information. It is in fact a position circle. The intersection of two such circles will give an observed position, as will the intersection of two position lines.

We employ the same procedure when navigating with astronomical bodies. The charted point of land is replaced in this case with the geographical position of the heavenly body, i.e. the G.H.A. and the declination.

When we observe the altitude of a body, we correct it and subtract it from 90° and obtain the zenith distance. Now this zenith distance is the angular distance of the observer's zenith from the position of the body on the celestial sphere, and as the centres of the celestial sphere and the earth are coincident, this will be the same as the angular distance on the earth, of the observer from the geographical position of the body. Furthermore, when we measure an angular distance on the surface of the earth and express it in minutes of arc, it becomes, by definition of the nautical mile, a distance measured in those units.

Thus the zenith distance becomes the radius of our position circle which is centred on the geographical position of the body.

In this manner it would be possible to navigate by plotting the geographical positions of two or more bodies on a chart and drawing circles of radius the bodies' zenith distances and obtain a fix at the intersection of the circles. However, to do this we would need a chart of a very large area, and because of the large radii the plot would be on a very small scale. Far too small for the required accuracy. Moreover the circles would not appear on the mercator chart as true circles and would thus be difficult to plot.

To get around this problem we only draw that part of the position circle that passes near to the D.R. position, and because of the large radius of the circle such a small part of it can be taken as a straight line without material error. Thus our position circle now becomes a position line, which strictly speaking is a tangent to the position circle.

The direction in which the line runs near to the D.R. position is found by calculating the true bearing of the body. The line representing this bearing will be a radius of the position circle and therefore the position line, being a tangent to the circle is at right angles to this bearing.

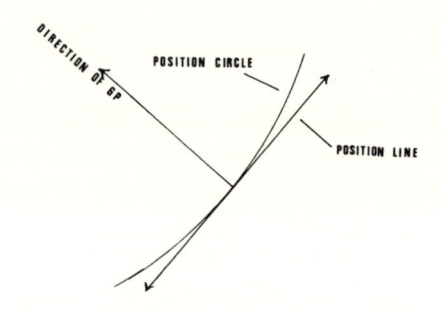

The position line is marked at each end with an arrow

FIG. 9.1

Thus we can draw this position line on the chart without plotting the geographical position as long as we know some point through which to draw it. All methods of sight reduction are means of finding the direction of such a position line and a position through which it passes. The calculations involved in finding this information are dealt with in the following chapters, but first we will see how we obtain the fix once we have calculated the necessary position line information from two or more observations.

## THE MARCQ ST. HILAIRE METHOD (INTERCEPT METHOD)

This is a universal method. Any sight can be worked this way.
We assume a D.R. position and calculate a zenith distance using this position. The bearing is also calculated. Thus we could draw a position line through the D.R. position at right angles to the bearing. We could call this the calculated position line. We now take the observed altitude and find the TRUE zenith distance from it. We compare this true zenith distance with the calculated zenith distance and the difference between the two in minutes will be the distance in nautical miles between the calculated and the true position lines. This distance we call the INTERCEPT. It should be named 'towards' or 'away' depending upon whether the true position line is nearer the geographical position of the body than the D.R. position or farther away from it.

# PLOTTING POSITION LINES

**Note**

True Z.D. less than Calc. Z.D.—towards.
True Z.D. greater than Calc. Z.D.—away.

The true position should be close enough to the D.R. position to assume that the true bearing is the same as that which was calculated. The calculated and the true position lines are therefore parallel.

In practice we need only draw the true position line. This can be done by measuring the intercept from the D.R. position either towards or away from the direction of the bearing, as the intercept is named and drawing the position line at right angles to this direction through the intercept terminal point (I.T.P.). Thus the plot of one position line looks like this.

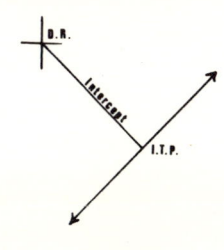

FIG. 9.2

If this is plotted to scale for two or more observations we can take the observed position as being the intersection of the position lines and can measure the d. lat. and the departure between the D.R. and the observed position and thus find the latitude and longitude of the observed position.

## To find the position by plotting the position lines from two simultaneous sights

*Procedure*

1. Plot a convenient point to represent the D.R. position.

2. From this point draw the directions of the intercepts either towards the direction of the bearing of the body, or away from it depending on how the intercept is named.

3. Mark off to a scale of nautical miles the lengths of the intercepts from the D.R. position.

4. Through the intercept terminal points draw the position lines at right angles to the intercepts. Where the two position lines cross is the observed position.

5. Measure the d. lat. and the departure between the D.R. position and the observed position, and apply to the D.R. position after having converted departure into d. long.

**Example 1**

Using D.R. position, lat. 47° 56′ N., long. 27° 50′ W., simultaneous observations of two stars gave:
(1) bearing 148° T., intercept 5′ away.
(2) bearing 065° T., intercept 4′ towards.
Find the ship's position.

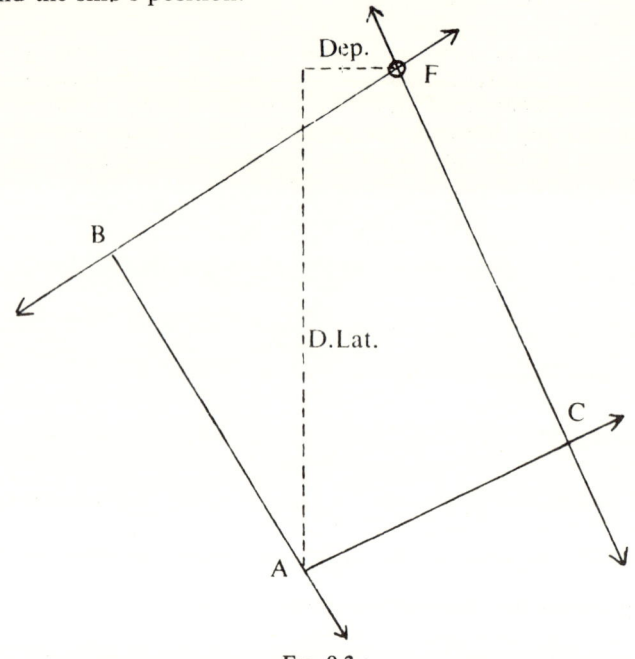

FIG. 9.3

**By measurement from *A* to *F***

d. lat. =   6·7′ N.   dep. =      1·3′ E.
*A* lat. 47° 56·0′ N.   long. 27° 50·0′ W.
d. lat.       6·7′ N.   d. long.   2·0′ E.
──────────   ─────────────
*F* lat. 48° 02·7′ N.   long. 27° 48·0′ W.

**Description of plot**

*A*—the D.R. position
*B*—the intercept terminal point for P.L. 058° T. −238° T.
*C*—the I.T.P. for P.L. 155° T.−335° T.
*F*—Position by observation.

**To find the position from two position lines by Marcq St. Hilaire, with a run in between them**

A running fix using celestial position lines gives an observed position at the time of the second sight, by crossing the second

# PLOTTING POSITION LINES

sight with an earlier observation, which position line is transferred up to the time of the second sight by applying the course and distance run.

*Procedure*

1. Take the D.R. position at the time of the first sight and apply the course and distance run between the sights by the traverse table. This gives a D.R. position at the time of the second sight, which is used to calculate the second intercept and position line, and from which the two intercepts are plotted.

2. Plot the transferred position line and the second position line from this D.R. in the same manner as in the previous section.

**Example 2**

In D.R. position 23° 40′ S. 98° 50′ E. an observation of a star gave an intercept of 10′ towards with a bearing of 117° T. The ship then ran 254° for 27 miles, when a second observation gave an intercept 3·8′ away with a bearing of 226° T. Find the position at the time of the second sight.

| D.R. at first sight | 23° 40·0′ S. | 98° 50·0′ E. |
|---|---|---|
| Run 254°, 27m. | 7·4′ S. | 28·3′ W. |
| D.R. at second sight | 23° 47·4′ S. | 98° 21·7′ E. |

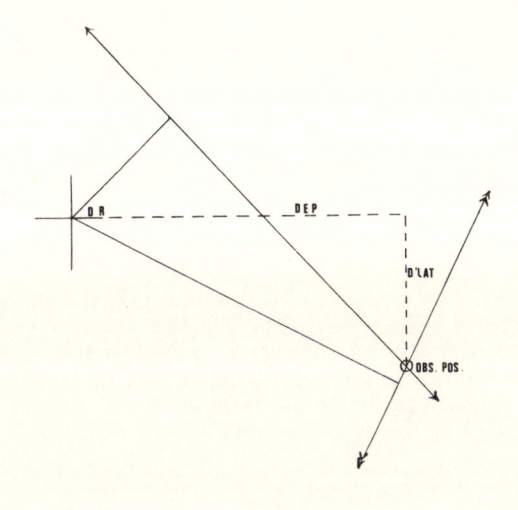

Fig. 9.4

Note that any transferred position line is marked with double arrows

By measurement:

d. lat. = 4·2′ S.   departure = 9·2′ E.
D.R. at second sight   23° 47·4′ S.   98° 21·7 E.
                         4·2′ S.         10·1′ E.
                       ─────────────────────────
Observed position      23° 51·6′ S.   98° 31·8′ E.

**Note**

A D.R. for the second sight may be given that is different from the D.R. obtained from running up the first D.R. In this case each sight must be plotted from its own D.R. Thus the transferred position line is plotted from the first D.R. after the run is applied, and the second position line is plotted from the D.R. given, which will be the D.R. which has been used to calculate the intercept and bearing. (See example 4.)

**The longitude by chronometer method**

In this type of problem we assume only a D.R. latitude, and then calculate the longitude at which the position line crosses this latitude. It should be realised that if any one observation is worked by two different methods of sight reduction then exactly the same position line should result. Only the position that we calculate, through which to draw the position line differs.

In the longitude by chronometer method we calculate the true bearing of the body to give us the direction of the position line and then draw the line through the position given by the D.R. latitude and the longitude found by calculation. Thus there is no intercept involved.

**To find position by simultaneous position lines by longitude method**

*Procedure*

1. Plot the TWO positions through which the position lines pass. (There will be one position to be plotted for each observation, as, although probably the same D.R. latitude will have been used to work the sights, a different longitude will result from each. Note that the distance between the positions in an east–west direction on the plot should be the departure and not the d. long.)

2. Draw each position line through its respective plotted position. The position where they cross is the observed position.

3. Measure the d. lat. and the departure between the observed position and one of the known positions, and apply to this position after having converted departure into d. long.

# PLOTTING POSITION LINES

## Example 3

By using D.R. latitude 25° 20′ N., simultaneous observations of two stars gave:
(1) Longitude 36° 05′ W., bearing 060° T.
(2) Longitude 35° 57′ W., bearing 300° T.
Find the ship's position.

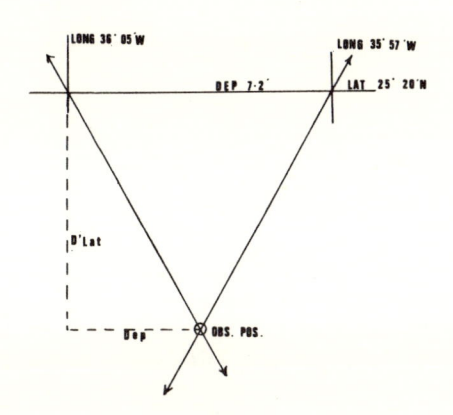

FIG. 9.5

### By measurement from *A* to observed position

d. lat.=6·2′ S., dep.=3·6′ E.

| | | | |
|---|---|---|---|
| Pos. *A* lat. | 25° 20·0′ N. | long. | 36° 05′ W. |
| d. lat. | 6·2′ S. | d. long. | 4·0′ E. |
| Obs. pos. lat. | 25° 13·8′ N. | long. | 36° 01′ W. |

## Example 4

At 1100 hours ship's time an observation of the sun by longitude by chronometer, using a D.R. latitude of 30° 14′ N. gave a longitude of 36° 18′ W., with the bearing of the sun 150° T. At 1300 hours ship's time, the sun was bearing 215° T., and a second observation gave a longitude of 36° 55′ W., using a D.R. latitude of 30° 00′ N. If the ship's course and speed between the observations was 250° T., 16 knots, find the position at the time of the second observation.

| | | |
|---|---|---|
| D.R. at first sight | 30° 14·0′ N. | 36° 18·0′ W. |
| Run 250°, 32m. | 10·9′ S. | 34·8′ W. |
| D.R. to plot transferred P.L. | 30° 03·1′ N. | 36° 52·8′ W. |

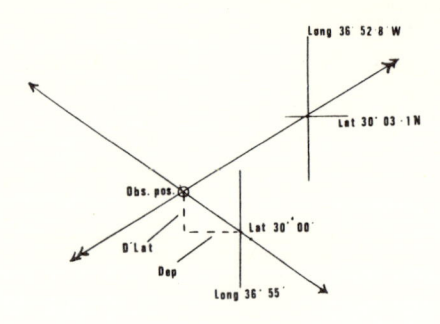

FIG. 96

## By measurement

d. lat. between observed position and D.R. of second sight =
1·1′ N.
departure = 1·6′ W.

| D.R. second sight | 30° 00·0′ N. | 36° 55·0′ W. |
|---|---|---|
|  | 1·1′ N. | 1·9′ W. |
| Observed position | 30° 01·1′ N. | 36° 56·9′ W. |

## Note

The east–west distance on your plot between the two D.R. positions should be the departure and not the d. long.

## Latitude by meridian altitude method

The meridian altitude is the altitude when the body is on the same meridian as the observer. Under these circumstances the bearing of the body is either north or south, and therefore the position line will always run in an east–west direction. It therefore coincides with the parallel of latitude upon which the observer lies.

The standard method of finding the noon position at sea is to obtain a position line during the forenoon and transfer it up to the time of noon. The transferred position line can then be crossed with the noon latitude.

## Example 5

In D.R. position 30° 15·0′ N. 26° 40·0′ W., an observation of the sun gave a bearing of 110° T. intercept 6·5′ towards. The ship then steamed 245° T. 20 miles, when the latitude by meridian altitude of the sun was 30° 00·0′ N. Find the ship's position at noon.

A complete picture of the problem would look as follows. In fact only the part inside the dotted lines need be plotted.

# PLOTTING POSITION LINES

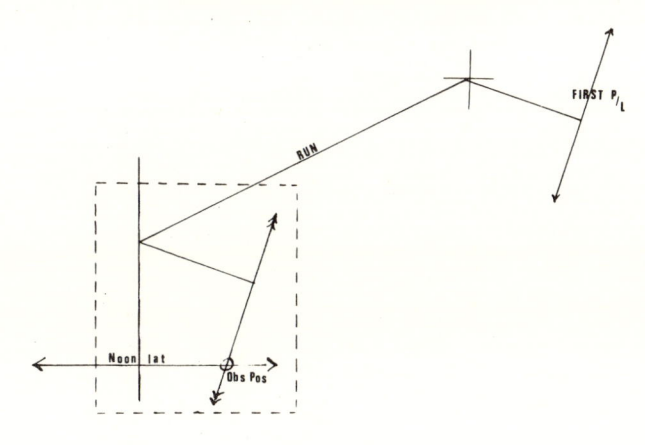

Fig. 9.7

The accuracy of the position obtained depends largely upon the length of the run up to noon. Hence this should be kept as short as possible consistent with a good angle of cut between the forenoon position line and the noon latitude.

The course and distance should be applied to the morning D.R. by use of the traverse table to find the position through which to draw the transferred position line.

*Procedure*

1. Apply the course and distance to the morning D.R. to give a noon D.R.

2. Plot this position and draw in the morning intercept and the transferred position line.

3. Take the difference between the noon D.R. and the observed latitude and hence plot the observed latitude, drawing in the position line running east–west.

4. The point where the transferred position line cuts the noon latitude is the position of the observer at noon.

5. Measure the departure between the observed position and the D.R. longitude, and convert it to d. long.

6. Apply this d. long. to the D.R. long. to get the observed longitude.

| | | |
|---|---|---|
| D.R. at morning sight | 30° 15·0′ N. | 26° 40·0′ W. |
| Run 254° T. 20m. | 8·5′ S. | 20·9′ W. |
| D.R. at noon | 30° 06·5′ S. | 27° 00·9′ W. |

150    PRACTICAL NAVIGATION FOR SECOND MATES

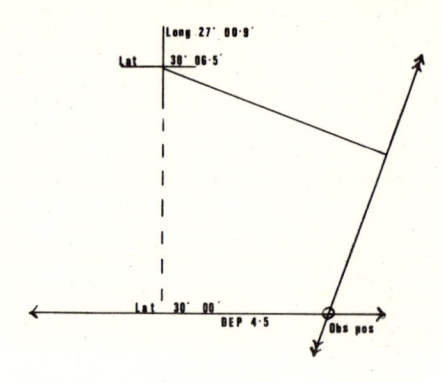

FIG. 9.8

**By measurement departure between observed longitude and D.R. long. = 4·5′ E.**

    Longitude of D.R. noon    27° 00·9′ W.
    D. long.                       5·2′ E.
                              ─────────────
    Observed longitude        26° 55·7′ W.
    Observed position 30° 00·0′ N. 26° 55·7′ W.

**Example 6**

An observation in D.R. latitude 42° 30′ N. gave a longitude of 32° 08′ W. bearing of the observed body 050° T. The ship then ran 075° T. for 35 miles when a meridian altitude gave a latitude of 42° 42′ N. Find the ship's position at the time of the meridian altitude.

    D.R. at first sight          42° 30·0′ N.   32° 08·0′ W.
    Run 075° T., 35m.                9·1′ N.       45·9′ E.
                                 ─────────────────────────
    D.R. at time of meridian alt. 42° 39·1′ N.   31° 22·1′ W.

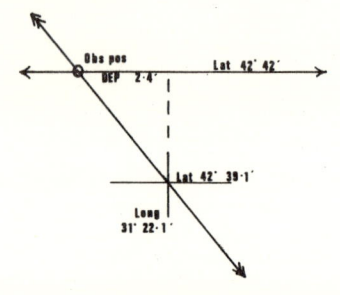

FIG. 9.9

PLOTTING POSITION LINES 151

**By measurement departure between observed position and D.R. = 2·4′ W.**

| | | |
|---|---|---|
| | D.R. longitude | 31° 22·1′ W. |
| | D. long. | 3·3′ W. |
| | Observed long. | 31° 25·4′ W. |
| Observed position | 42° 42·0′ N. | 31° 25·4′ W. |

**Example 7**

Position line from observation of a celestial body combined with the position line from observation of a shore object.

An observation of a celestial body gave bearing 220° T. and long. 115° 02′ E., by using D.R. lat. 32° 00′ S. Later, a point of land (lat. 32° 00′ S., long. 115° 31′ E.) bore 070° T. Between the observations the vessel steamed 145° T. for 17 nautical miles, and then 063° T. for 12 nautical miles. Find the vessel's position.

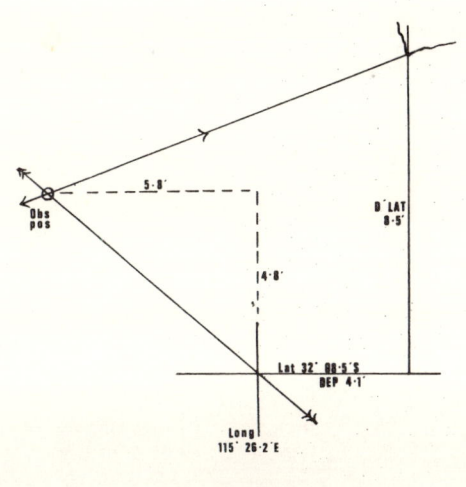

FIG. 9.10

| | | | |
|---|---|---|---|
| 'Co. 145° dist. 17′ d. lat. 13·9′ S. | Dep. | | 9·8′ E. |
| Co. 063° dist. 12′ | 5·4′ N. | | 10·7′ E. |
| D. lat. | 8·5′ S. | Dep. | 20·5′ E. |

## 152  PRACTICAL NAVIGATION FOR SECOND MATES

| | | | |
|---|---|---|---|
| D.R. lat. | 32° 00′ S. | Obs. long. | 115° 02′ E. |
| D. lat. | 8·5′ S. | D. long. | 24·2′ E. |
| Run up pos. lat. | 32° 08·5′ S. | Long. | 115° 26·2′ E. |
| Point of land lat. | 32° 00·0′ S. | Long. | 115° 31·0′ E. |
| D. lat. | 8·5′ N. | D. long. | 4·8 E. Dep. |
| | | | = 4·1′ E. |
| From plot  D. lat. 4·8′ N. | Dep. 5·8′ W. | D. long. 6·8′ W. | |
| Run up pos. lat. | 32° 08·5′ S. | Long. | 115° 26·2′ E. |
| D. lat. | 4·8′ N. | D. long. | 6·8′ W. |
| Ship's pos. lat. | 32° 03·7′ S. | Long. | 115° 19·4′ E. |

*Procedure*

1. Calculate the run up position from the D.R. lat. and the obs. long.

2. Plot the position of the point of land in relation to the run up position, converting the d. long. into departure, and draw in the line of bearing.

3. Measure the d. lat. and dep. from the ship's position to the run up position, convert the dep. into d. long. and thence find the lat. and long. of the ship's position.

### Example 10

On a vessel at anchor, an observation of the sun, during the afternoon, gave longitude 05° 05′ W. by using lat. 50° 04′ N. Vertical sextant angle observations taken later, put the ship 4 M. south and 6 M. east of this position. What was the sun's true bearing?

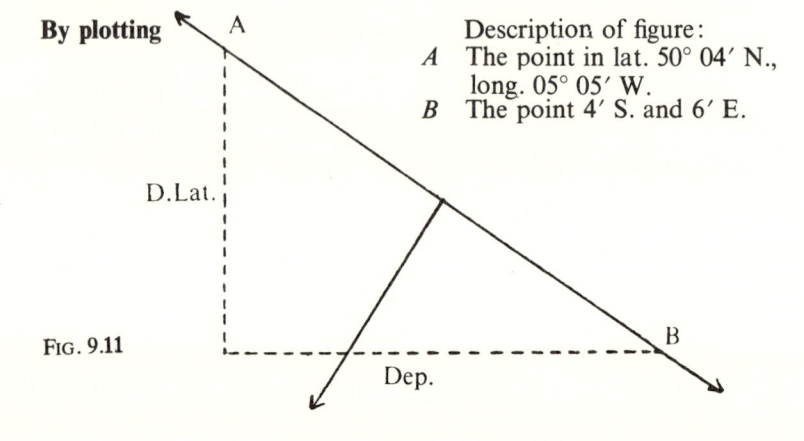

**By plotting**

Description of figure:
*A*  The point in lat. 50° 04′ N., long. 05° 05′ W.
*B*  The point 4′ S. and 6′ E.

FIG. 9.11

# PLOTTING POSITION LINES

The position line must pass through position *A* and it must also pass through *B* since that was the ship's actual position. Therefore a line at right angles to the line joining *A* and *B* will give the sun's true bearing.

Sun's bearing = 214° T.

## By calculation

From a figure similar to the plot,

$$\text{Tan} \angle A = \frac{6}{4} = 1\cdot5$$

$A = 56° \ 19'$    ∴ direction of $AB =$ S. 56° 19′ E.

Hence Sun's bearing = 213° 41′ T.

## Note

There are two possibilities for the bearing, i.e. either N. 34° E. or S. 34° W. As the body observed was the sun during the afternoon, it must therefore be west of the meridian. So S. 34° W. is the correct answer.

## Example 11

A morning observation of the sun worked with lat. 42° 10′ N. gave long. 35° 20′ W. and when worked with lat. 42° 20′ N. gave long. 35° 05·1′ W. What was the sun's bearing?

## By plotting

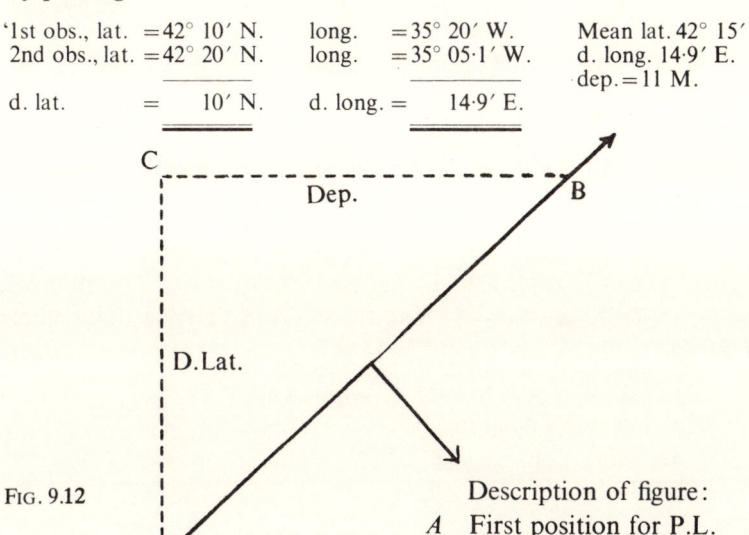

| | | | |
|---|---|---|---|
| 1st obs., lat. = 42° 10′ N. | long. = 35° 20′ W. | | Mean lat. 42° 15′ |
| 2nd obs., lat. = 42° 20′ N. | long. = 35° 05·1′ W. | | d. long. 14·9′ E. |
| | | | dep. = 11 M. |
| d. lat. = 10′ N. | d. long. = 14·9′ E. | | |

Fig. 9.12

Description of figure:
*A*  First position for P.L.
*B*  Second position for P.L.

The position line must pass through *A* and through *B*. Therefore by joining these two points, the position line is obtained, and the sun's bearing will be at right angles to this direction, and as the observation of the sun was taken during the morning then sun's bearing = 138° T.

**By calculation**

From a figure similar to the plot, and using the traverse table:
*AC* (d. lat.) 10, and *CB* (dep.) 11, give angle equal to 47° 44'
∴ PL. trends 047° 44' T.—227° 44' T., so that sun's bearing = 137° 44'

## EXERCISE 9A

1. Given chosen position lat. 40° 20' N., long. 18° 30' W., T.$ZX$ 38° 10·0', C.$ZX$ 38° 20·0', azimuth 120° T. Plot the position line, using scale of 1 cm. to 1 nautical mile, and state the position of the intercept terminal point.

2. D.R. position lat. 20° 20' S., long. 27° 30' W., true altitude 55° 28', C.$ZX$ 34° 26', azimuth 235° T. Plot the position line. State the position of the intercept terminal point.

3. In D.R. position lat. 40° 00' N., long. 30° 00' W., an observation of the sun gave true altitude 45° 02'. The calculated zenith distance was 45° 04', and the azimuth was 140° T. Plot the position line and state the position of the intercept terminal point.

4. From the following simultaneous observations, find the ship's position:
   Sun —bearing 130° T. intercept 6·0' towards
   Venus—bearing 210° T. intercept 8·0' away
The selected position was lat. 50° 10' N., long. 44° 20' W.

5. In estimated position lat. 40° 20' N., long. 34° 20' W., simultaneous observations gave:
   Sirius —bearing 136° T. intercept 10·0' away ⎫ Find ship's
   Venus—bearing 286° T. intercept  8·0' towards ⎭ position

6. In D.R. position lat. 48° 10' N., long. 50° 14' W., simultaneous observations of two stars gave:
   1. longitude 50° 08' W.  azimuth 070° T.
   2. longitude 50° 20' W.  azimuth 330° T.
Find the ship's position.

7. By using D.R. lat. 25° 20' N., simultaneous stellar observations gave:
   1. longitude 36° 15' W.  bearing 060° T.
   2. longitude 35° 50' W.  bearing 300° T.
Find the ship's position.

## PLOTTING POSITION LINES

8. From a vessel steering 035° T. a point of land bore 330° T. After the vessel had steamed 30 nautical miles, the point bore 250° T. Find the distance off the point at the second observation.

9. In D.R. position lat. 23° 40′ N., long. 52° 30′ W., a stellar observation gave intercept 4′ towards, and bearing 040° T. The vessel steamed 090° T. at 12 knots through a current setting 000° T. at 2·5 knots. Two hours later, another observation gave the intercept 5′ towards and bearing 120° T. Find the ship's position at the 2nd observation.

10. By using D.R. lat. 34° 11′ N. the longitude by observation was 42° 25′ W., bearing of the sun being 121° T., and log reading 40. The vessel steered 042° T. until noon, when the latitude by meridian altitude of the sun was 34° 11′ N., and the log read 72. Find the position at noon.

11. An observation of the sun gave longitude 36° 58′ W. and bearing 130° T., by using D.R. lat. 29° 32′ S. The ship then steamed 300° T. for 27 M. in a current setting 090° T. 5 M., when the latitude by meridian altitude of the sun was 29° 06′ S. Find the ship's position at noon.

12. In D.R. lat. 34° 20′ N. an observation of a star gave longitude 47° 58′ W., and bearing of the star as 222° T. At the same time an observation of another star gave longitude 47° 46′ W., and bearing 142° T. Find the ship's position.

13. By observation in D.R. position lat. 53° 47′ S., long. 178° 37′ W., the bearing of the sun was 076° T. intercept 11′ away. The ship then ran 284° T. for 47 M. through a current setting 256° T. for 7 M., when a second observation of the sun gave bearing 284° T., intercept 5′ towards. Find the ship's position at the second observation.

14. By using selected position lat. 16° 41′ S., long. 163° 29′ E., an observation of the sun gave intercept 18′ away, bearing 055° T. The ship then steamed 208° T. 33 M., when a second observation gave intercept 12′ towards, bearing 332° T. Find the ship's position at each observation.

15. An observation of the sun worked with lat. 42° 17′ S. gave longitude 76° 43′ E., bearing 123° T. The ship then steamed 237° T. 29 M. until noon, when the latitude by meridian altitude of the sun was 42° 27′ S. Find the ship's position at noon.

16. In D.R. position lat. 39° 39′ N., long. 130° 47′ E., an observation of the sun gave intercept 4′ towards, bearing 160° T. Later, a second observation, using lat. 39° 09′ N. gave longitude 130° 47′ E., bearing 200° T. Find the ship's position at the second observation, if during the interval the ship ran 196° T. 20 M. and 186° T. 18 M.

17. During the forenoon, the longitude was worked out on a vessel at anchor. Fog then set in. Later the fog cleared, and vertical

angle observations put the ship 6 M. north and 5 M. east of the observed position. What was the true bearing of the sun at sights?

18. An a.m. sight of the sun when worked with lat. 51° 55′ N. gave longitude 20° 04′ W., and when worked with lat. 52° 05′ N. gave longitude 19° 54·5′ W. What was the true bearing of the sun?

19. An observation worked with D.R. lat. 48° 20′ N. gave long. 35° 17′ W., and bearing 127° C. The vessel then steamed for 4 hours at 11 knots and a current set 090° T. at 3 knots. The course steered was 154° C., dev. 5° E., var. 12° W., wind N.E., and leeway 5°. A second observation then gave a star's bearing 252° C., intercept 10′ towards. Find the ship's position.

# CHAPTER 10

# CORRECTION OF ALTITUDES

## Definitions

*Visible Horizon*

The circle which bounds the observer's view of the earth's surface in a clear atmosphere.

*Sensible Horizon*

A plane which passes through the observer's eye and is at right angles to the vertical of the observer.

*Rational Horizon*

A plane which passes through the centre of the earth, and is at right angles to the observer's vertical. The rational horizon is therefore parallel to the sensible horizon.

The observed altitude of a heavenly body is the angle at the observer's eye between a line from the observer's eye to the point on the body observed, and a line from the observer's eye to the visible horizon. In figure 10.1 angle ∠ LEV.

The altitude required for navigational computations is the true altitude. This is the angle at the centre of the earth between a line joining the earth's centre and the centre of the heavenly body, and the plane of the observer's rational horizon.

The following corrections must be applied to the observed altitude to obtain the true altitude.

## Dip

This is defined as the angle at the observer's eye between the plane of the sensible horizon and a line joining the observer's eye and the visible horizon. In figure 10.1 angle ∠ SEV.

The application of dip to the observed altitude corrects the altitude above the visible horizon to an altitude above the sensible horizon. The altitude so corrected is called the Apparent altitude.

In the figure:
∠ LEV = Observed altitude
∠ LES = Apparent altitude
∠ SEV = Dip

Thus Apparent altitude = Observed altitude − Dip.

Fig. 10.1.

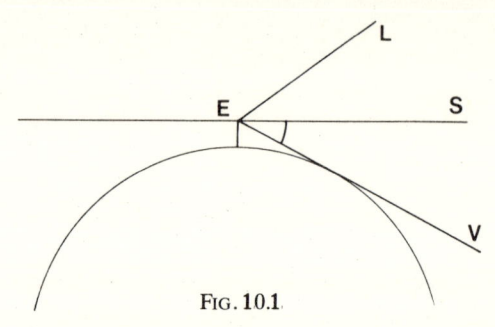

Thus dip is always negative. Values of dip are tabulated against height of eye in nautical tables and in the *Nautical Almanac*.

### Refraction

A ray of light entering a medium of greater density than that in which it has been travelling is bent or refracted towards the normal. (The normal is the perpendicular to the surface of the interface between the two mediums at the point of entry of the ray of light.)

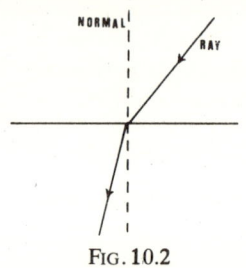

Fig. 10.2

Light entering the earth's atmosphere from space therefore is so refracted, but as there is no definite interface between the atmosphere and space, but a gradual increase in density of the air, the refraction is not sudden as in figure 10.2, but the ray is bent gradually as it approaches the earth's surface. The true path of a ray therefore is as shown as a solid line in figure 10.3, but the apparent direction of the body to an observer will be as shown as a dotted line. The altitude will always appear to be greater than it really is, and therefore a correction must be subtracted to allow for it.

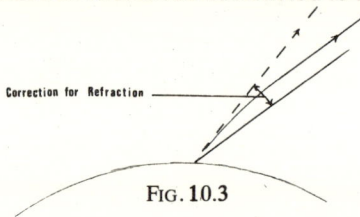

Fig. 10.3

# CORRECTION OF ALTITUDES

Refraction is greatest at low altitudes and is zero when the ray of light enters the atmosphere at right angles to the earth's surface, i.e. when the altitude is 90°. Uncertainty in the value of refraction at low altitudes makes it advisable to avoid taking sights of bodies near the horizon, if possible.

Values of refraction for standard conditions of atmosphere are tabulated against altitude in nautical tables.

**Semi-diameter**

In the case of the sun and the moon, it is easier and more accurate to take the altitude of the upper or the lower edge or limb of the disc rather than estimate the position of the centre of the disc. A correction must in this case be applied to obtain the altitude of the centre of the body.

The amount of the correction will be the angle subtended at the observer's eye by the radius or semi-diameter of the body. From figure 10.4 it will be positive to an observation of the lower limb, and negative to an observation of the upper limb.

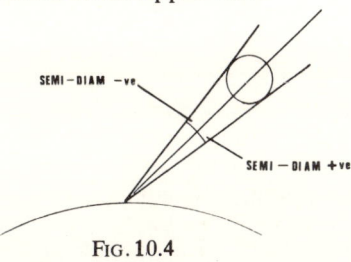

FIG. 10.4

Semi-diameter varies with the distance of the body from the earth, and values are given in nautical tables. They are also given for each day in the daily pages of the almanac at the foot of the 'sun' and 'moon' columns.

**Parallax**

This is defined as the angle at the centre of the body, subtended by the line from the centre of the earth to the observer's eye. The application of this correction changes the altitude observed at the observer's position on the surface of the earth, to the altitude as it would be observed from the centre of the earth.

In the figure ∠ ELC is the angle of parallax.
∠ LES is the altitude uncorrected for parallax.
∠ LCR is the altitude after correction for parallax, i.e. the true altitude.

The external angle of a triangle is equal to the sum of the two internal and opposite angles.

## 160 PRACTICAL NAVIGATION FOR SECOND MATES

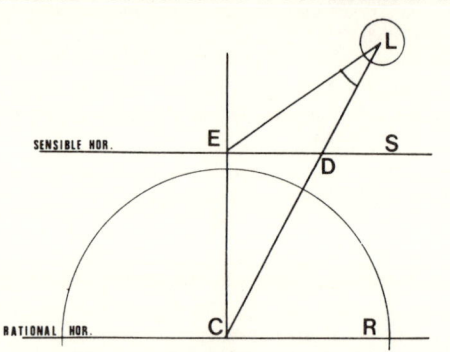

Fig. 10.5

Thus from figure 10.5 ∠LDS = ∠LCR = True altitude
and ∠LDS = ∠ELD + ∠LES
thus T. alt. = App. alt. + parallax

Thus the parallax correction is always positive to apparent altitude to give true altitude. The value of parallax will vary with the altitude, and will be maximum when the body is on the horizon, and zero when the altitude is 90°. Values for the sun are tabulated against altitude in nautical tables. For parallax of the moon, see 'Correction of Moon's Altitude'. The value of parallax for the stars, because of the vast distances of the stars, is negligible.

**Note**

The index error of the sextant should be applied before any correction of altitude is done, according to the rule:

Index error off the arc, error positive,
Index error on the arc, error negative.

**Summary**

To correct an altitude of the sun we need to apply the following corrections.

| Index error | −ve, or +ve |
| dip | −ve |
| refraction | −ve |
| semi-diameter | −ve (upper limb), +ve (lower limb) |
| parallax | +ve |

To correct an altitude of a star or planet, we need to apply the following corrections.

| Index error | −ve, or +ve |
| dip | −ve |
| refraction | −ve |

## CORRECTION OF THE SUN'S ALTITUDE

**Example**

The sextant altitude of the sun's lower limb was 45° 20′, index

# CORRECTION OF ALTITUDES

error 1·2′ on the arc; height of eye 15·4 metres; sun's semi-diameter 15·9′. Find the true altitude of the sun's centre.

|       |       |
|---|---|
| Sext. alt. | 45° 20·0′ |
| I.E. | −     1·2′ |
| Obs. alt. | 45° 18·8′ |
| Dip | −     6·93′ |
| App. alt. | 45° 11·87′ |
| ref. | −     0·94′ |
|  | 45° 10·93′ |
| S.D. | +    15·90′ |
|  | 45° 26·83′ |
| par. | +     0·11′ |
| True alt. | 45° 26·94′ |

## Notes

1. The corrections for dip. refraction, and parallax-in-altitude are obtained from the appropriate tables in *Norie's*, *Burton's*, etc.

2. The sun's semi-diameter is obtained from the daily page for the given date in the *Nautical Almanac*.

3. The corrections should be made in the order shown.

## Example 2

The sextant altitude of the sun's upper limb on 7th January, 1980, at a certain instant was 53° 14·4′; index error 1·4′ off the arc, height of eye 18 metres. Find the true altitude.

|       |       |
|---|---|
| Sext. alt. | 53° 14·4′ |
| I.E. | +     1·4′ |
| Observed alt. | 53° 15·8′ |
| dip | −     7·5′ |
| Apparent alt. | 53° 08·3′ |
| refraction | −     0·7′ |
|  | 53° 07·6′ |
| S.D. | −    16·3′ |
|  | 52° 51·3′ |
| parallax | +     0·1′ |
| True altitude | 52° 51·4′ |

## EXERCISE 10A

### CORRECTION OF THE SUN'S ALTITUDE

These examples are to be worked *fully* (as shown), i.e. using individual corrections.

**Find the true altitude of the sun's centre, given:**

1. The sextant altitude of the sun's lower limb was 52° 31·2′; index error 2·2′ on the arc; height of eye 8·3 metres; sun's semi-diameter 16·1′.
2. The sextant altitude of the sun's L.L. 33° 10′ 50″; I.E. 1·0′ off the arc; H.E. 12·0 metres; S.D. 15·9′.
3. Sextant altitude U.L. 71° 53′ 30″; index error 1′ 50″ off the arc; H.E. 11·0 metres; S.D. 16·0′.
4. The observed altitude of the sun's upper limb was 27° 46′ 40″; height of eye 7·7 metres; semi-diameter 15·8′.
5. Sextant alt. L.L. 62° 34·3′; I.E. 2·2′ off the arc; H.E. 9 metres; S.D. 16·1′.
6. Sextant altitude of the sun's upper limb was 55° 55′ 50″, index error 1′ 00″ on the arc; height of eye 7·4 metres; semi-diameter 16·3′.
7. The sextant altitude of the sun by back angle, using the limb nearest the clear horizon, was 110° 51·6′; index error 2·2′ off the arc; H.E. 11·5 metres; semi-diameter 16·2′.
8. The sextant altitude of the sun by reverse horizon, using the limb farthest from the clear horizon, was 98° 24·4′; index error 1·2′ off the arc; height of eye 10·5 metres; semi-diameter 16·1′.

### CORRECTION OF THE ALTITUDE OF A STAR OR A PLANET

**Example**

Find the true altitude of the star *Rigel* 11, the sextant altitude of the star being 29° 17·2′, index error 1·8′ off the arc, and height of eye 14·0 metres.

|  |  |  |
|---|---|---|
| Sext. alt. |   | 29° 17·2′ |
| I.E. | + | 1·8′ |
| Obs. alt. |   | 29° 19·0′ |
| dip | − | 6·6′ |
|   |   | 29° 12·4′ |
| ref. | − | 1·7′ |
| True alt. |   | 29° 10·7′ |

# CORRECTION OF ALTITUDES

## EXERCISE 10B

## FIND THE TRUE ALTITUDE OF THE FOLLOWING BODIES

|     | Sext. Alt. | Ind. Error | Ht. of eye (metres) | Body |
|-----|------------|------------|---------------------|---------|
| 1.  | 47° 29·6′  | 1·0′ on the arc  | 11·2 | Altair |
| 2.  | 32° 24·4′  | 0·8′ on the arc  | 7·3  | Canopus |
| 3.  | 21° 13·6′  | 0·4′ off the arc | 11·6 | Arcturus |
| 4.  | 47° 15·8′  | 1·4′ on the arc  | 15·3 | Polaris |
| 5.  | 37° 10·4′  | 1·8′ on the arc  | 8·4  | Dubhe |
| 6.  | 12° 17·0′  | 2·0′ off the arc | 14·0 | Saturn |
| 7.  | 53° 20·2′  | 0·6′ on the arc  | 7·7  | Venus |
| 8.  | 23° 14·0′  | 2·2′ off the arc | 11·0 | Jupiter |
| 9.  | 51° 56·0′  | 0·4′ on the arc  | 17·0 | Mars |
| 10. | 14° 38·2′  | 2·8′ on the arc  | 9·9  | Venus |

### Correction of the moon's altitude

Because of the moon's proximity to the earth, the correction of its altitude presents special problems.

### Horizontal parallax

This is the angle of parallax of a body when it is on the sensible horizon, i.e. its maximum value of parallax (see definition of parallax). But even the value of horizontal parallax varies with latitude of the observer, being maximum when the observer is on the equator, and minimum when the observer is at the poles. The reason for this can be understood by considering the definition of parallax, i.e. the angle at the centre of the body subtended by a line from the observer's eye to the earth's centre. When the observer is at the equator this line will be the equatorial radius of the earth. When the observer is at the poles it will be the polar radius. As the equatorial radius is greater then it will subtend a greater angle at the body's centre than the polar radius.

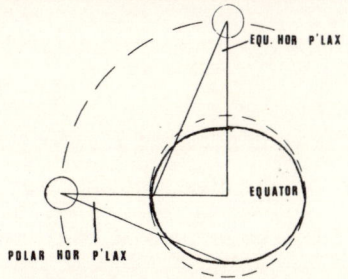

FIG. 10.6

The parallax given in the almanac against each hourly G.M.T. is the equatorial horizontal parallax, and it must be reduced to find the horizontal parallax for the latitude. This correction is given in nautical tables under the name of 'Reduction to the Moon's Horizontal Parallax for Latitude'.

Once the horizontal parallax is found thus, the parallax in altitude is obtained by:

Parallax in alt = Hor. Pax. × Cosine altitude.

This figure is then added to the apparent altitude to obtain true altitude.

## Augmentation of the moon's semi-diameter

Considering the distance of the moon from the earth's centre to be constant, the distance of the moon from an observer is greatest when the moon is on the rational horizon, i.e. when the altitude is zero. It is least when the moon's altitude is 90°. The difference will be the earth's radius.

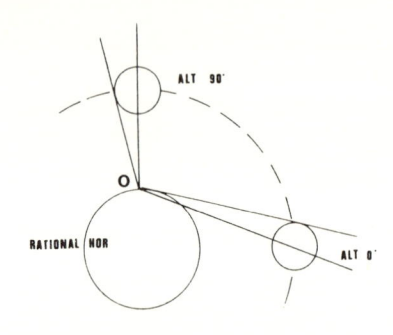

FIG.10.7

Hence the angular semi-diameter is greatest when the altitude is 90° and least when the altitude is zero.

The value given in the almanac is the least value when the altitude is zero, and must be increased or augmented for altitude. The value of the augmentation is given against altitude in nautical tables.

Hence to correct the moon's altitude the following corrections must be applied.

| | |
|---|---|
| Index error | −ve or +ve. |
| dip | −ve. |
| refraction | −ve. |
| semi-diameter | −ve or +ve after augmentation. |
| parallax in alt. | +ve obtained by reducing the equatorial parallax and multiplying by cosine altitude |

# CORRECTION OF THE MOON'S ALTITUDE

**Example**

The sextant altitude of the moon's lower limb was 16° 58·2′, index error 0·8′ off the arc, height of eye 5·4 m., semi-diameter 15·2′, horizontal parallax 55·7′, and latitude 12° 50′ N. Find the true altitude of the moon's centre.

| | | | | |
|---|---|---|---|---|
| Sextant altitude | | 16° 58·2′ | Semi-diameter | 15·2′ |
| Index error | + | 0·8′ | augmentation | 0·07′ |
| Observed alt. | | 16° 59·0′ | Augmented S.D. | 15·27′ |
| dip | − | 4·16′ | | |
| | | 16° 54·84′ | Hor. par. | 55·7′ |
| Semi-diameter | + | 15·27′ | reduction | Nil |
| Apparent alt. | | 17° 10·11′ | Reduced H.P. | 55·7′ |
| ref. | − | 3·05′ | | |
| App. alt. | | 17° 07·06′ | par.-in-alt. | |
| par.-in-alt. | + | 53·23′ | = H.P. × cos. app. alt. | |
| | | | = 55·7′ × cos. 17° 07·06′ | |
| True altitude | | 18° 00·29′ | = 53·23′ | |

**Notes**

1. The corrections for augmentation of the moon's semi-diameter, and the reduction for latitude to apply to the equatorial horizontal parallax, are taken from tables given in *Burton's*, *Norie's*, etc.

2. Working to the second place of decimals is not necessary—one place is quite sufficient. It is here shown solely for illustration.

## EXERCISE 10C

### CORRECTION OF THE MOON'S ALTITUDE

From the following information, find the true altitude of the moon's centre:

|   | Obs. limb | Sext Alt. | Index Error | Height of eye (metres) | S.D. | H.P. | Lat. |
|---|---|---|---|---|---|---|---|
| 1. | L.L. | 63° 12·8′ | 1·6′ off the arc | 7·3 | 15·3′ | 56·0′ | 50° N. |
| 2. | L.L. | 34° 14·8′ | 2·2′ on the arc | 13·0 | 15·1′ | 55·4′ | 39° S. |
| 3. | U.L. | 58° 16·2′ | 1·0′ on the arc | 10·4 | 16·1′ | 59·2′ | 44° N. |
| 4. | U.L. | 77° 51·6′ | 1·2′ off the arc | 9·0 | 14·8′ | 54·5′ | 22° N. |
| 5. | L.L. | 21° 38·8′ | 3·4′ on the arc | 11·5 | 15·8′ | 58·1′ | 00 |
| 6. | L.L. | 38° 21·8′ | 2·4′ off the arc | 9·0 | 16·3′ | 59·7′ | 41° 10′ S. |
| 7. | U.L. | 51° 17·0′ | 1·6′ on the arc | 16·0 | 14·9′ | 54·6′ | 37° 20′ N. |
| 8. | L.L. | 43° 18·4′ | Nil | 13·7 | 16·6′ | 61·0′ | 25° 15′ S. |

## Total correction tables

In practice correction of altitudes is simplified by the use of total correction tables. The most commonly used, and described here, are the convenient correction tables included in the *Nautical Almanac*. These are in three tables, for the sun, for stars and planets, and for the moon respectively. Each table is compiled with the apparent altitude as the argument so that the dip correction must first be applied to the observed altitude. A dip table is included with the total correction tables.

## The dip table

The dip is tabulated against height of eye in metres or in feet. The table is based upon the formula:

$$1 \cdot 76 \sqrt{\text{ht. of eye in metres.}}$$

The table is arranged as a critical entry table which means that one value of the correction is given for an interval of the argument, height of eye. This means that no interpolation is necessary, but it should be remembered that if the required height of eye corresponds to a tabulated value, then the upper of the two possible values of correction should be used. Thus the correction for a height of eye of **13·0** metres is −6·3 **(**see extracts of *Nautical Almanac*, correction tables**)**.

# CORRECTION OF ALTITUDES

## Sun total correction table

The sun correction table, found on the first page of the almanac, corrects for mean refraction, semi-diameter and parallax. The argument is the apparent altitude, that is the observed altitude corrected for dip. Two separate tables are used, one for the spring and summer months from April to September, and one for the autumn and winter months, from October to March. This allows annual variations in the semi-diameter to be allowed for. Each table contains corrections for lower limb observations in bold type and corrections for upper limb observations in feint type. The tables are arranged as critical tables which means that one value of the correction is given for an interval of the argument, apparent altitude. No interpolation is required but it should be remembered that if the required value of apparent altitude is a tabulated value then the correct correction is the upper of the two possible corrections. For example for an apparent altitude of 50° 46′ in the October to March table a correction of +15·4 (lower limb) should be used (see extracts from the *Nautical Almanac*).

## Example

The sextant altitude of the sun's lower limb was 48° 56·3′. Index error 1·2 on the arc. Height of eye 7·2 metres. Date June 16th. Find the true altitude.

| | |
|---|---|
| Sextant altitude | 48° 56·3′ |
| Index error | −1·2′ |
| Observed altitude | 48° 55·1′ |
| Dip | − 4·7′ |
| Apparent altitude | 48° 50·4′ |
| Correction | +15·1′ |
| True altitude | 49° 05·5′ |

## Stars and planets

The correction table for stars and planets found on the first page of the almanac, corrects for a mean refraction only. The corrections are tabulated against apparent altitude (observed altitude corrected for dip), and are arranged as critical entry tables.

For Mars and Venus an additional correction may be applicable, depending upon the date. These are given down the right hand side of the refraction correction table. The additional correction for planets corrects for the effect of parallax and phase, but the correction for Venus is only applicable when the sun is below the horizon. The correction for daylight observations may be calculated from data given in the explanation in the back of the almanac, but the magnitude of the corrections is such that this is unnecessary and may be ignored.

## Example

The sextant altitude of the star *Procyon* was 57° 18·9′ Index error 1·0′ off the arc. Height of eye 6·5 metres. Find the true altitude.

| Sextant altitude | 57° 18·9′ |
| Index error | +1·0′ |
| Observed altitude | 57° 19·9′ |
| Dip | −4·5′ |
| Apparent altitude | 57° 15·4′ |
| Correction | −0·6′ |
| True altitude | 57° 14·8′ |

## Example

The sextant altitude of Mars on 30th March 1980 was observed to be 38° 06·5′. Index error 0·5′ off the arc. Height of eye 5·0 metres. Find the true altitude.

| Sextant altitude | 38° 06·5′ |
| Index error | +0·5′ |
| Observed altitude | 38° 07·0′ |
| Dip | −3·9′ |
| Apparent altitude | 38° 03·1′ |
| Correction | −1·2′ |
|  | 38° 01·9′ |
| Additional corr. | +0·2′ |
| True altitude | 38° 02·1′ |

## Moon

The moon's total correction table, found on the last pages of the almanac, is in two parts. The main correction, in the upper part of the table corrects for refraction, semi-diameter and parallax, using mean values. It is tabulated against apparent altitude, and some interpolation is necessary to obtain the accuracy to within one decimal place.

The second correction allows for variations in the semi-diameter and parallax, both of which depend upon the horizontal parallax. The arguments are therefore, apparent altitude and horizontal parallax. Two values are given one for lower limb, and one for upper limb observations. These are arranged in columns, the correction being taken from the same column as that from which the main correction was extracted, and against H.P.

All corrections for the moon are additive to the apparent altitude, but those for upper limb observations have 30′ added to maintain them positive. This 30′ must therefore be subtracted from the final altitude.

# CORRECTION OF ALTITUDES

**Example**

The sextant altitude of the moon's lower limb was 16° 58·2'. Index error 0·8' off the arc. Height of eye 5·4 metres. The G.M.T. was 1400 on 27th June 1980. Find the true altitude.

From almanac H.P. =56·9

| | |
|---|---|
| Sextant altitude | 16° 58·2' |
| Index error | +0·8' |
| Observed altitude | 16° 59·0' |
| Dip | −4·1' |
| Apparent altitude | 16° 54·9' |
| Main correction | +62·7' |
| Second corr. | +4·0' |
| True altitude | 18° 01·6' |

## CHAPTER 11

## LATITUDE BY MERIDIAN ALTITUDE

An observation of any body whilst on the meridian of the observer is of particular value to the navigator as it provides a quick and easy method of finding a position line which will be coincident with the observer's parallel of latitude. The latitude obtained from the sight is therefore the observer's latitude.

Let the following diagram represent the earth and the celestial sphere on the plane of the observer's meridian.

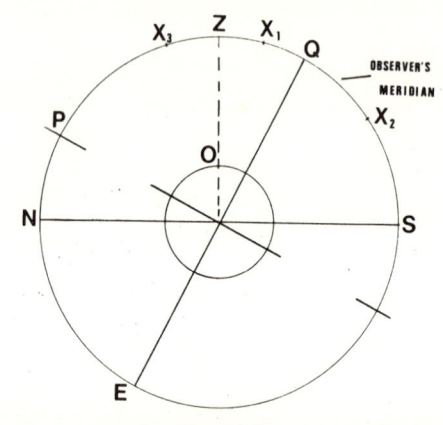

Fig. 11.1

Explanation

$O$ is the position of the observer in northerly latitude.
$Z$ is the observer's zenith on the celestial sphere.
$NS$ is the plane of the rational horizon.

If $EQ$ represents the plane of the equinoctial then $P$ will be the north celestial pole and arc $ZQ$ will be the latitude of the observer.

Consider a body $X_1$ of declination same name as the latitude and less than the latitude, while on the observer's meridian. Then $ZX_1$ is the angular distance of the body from the zenith, i.e. the zenith distance, and $X_1Q$ is the angular distance of the body from the equinoctial, i.e. the declination.

170

## LATITUDE BY MERIDIAN ALTITUDE

From the diagram:
$$ZQ = ZX_1 + X_1Q$$
Latitude = zenith distance + declination

Consider a body $X_2$ of declination opposite name to latitude. Then similarly:
$$ZQ = ZX_2 - QX_2$$
Latitude = zenith distance − declination.

Consider a body of $X_3$ of declination the same name as latitude and greater than latitude.
Then similarly:
$$ZQ = QX_3 - ZX_3$$
Latitude = declination − zenith distance

These results can be memorised but preferably the appropriate one can be simply derived in each problem as is shown in the first example to follow.

### Latitude by meridian altitude of a star

*Procedure*

1. Extract the declination of the star from the daily pages of the *Nautical Almanac* at the appropriate date. One value is given for each three day page in the list of stars. (If the star is not listed in the daily pages refer to the complete list of selected stars at the end of the almanac.)

2. Correct the sextant altitude for
    (i) Index error,
    (ii) Dip,
    (iii) Refraction (main correction from the table on the inside cover of almanac).

3. Subtract the true altitude from 90° to obtain the zenith distance.

4. Draw a rough sketch on the plane of the rational horizon to determine the appropriate rule, thus:
Insert the position of Z (the central point of the diagram). Mark on X the body, either to the north or to the south of Z according to the bearing of the body at meridian passage, and at a distance from Z to represent the zenith distance. Mark on Q, the point where the equinoctial cuts the observer's meridian, either to the north or to the south of X according to the name of the declination, and at a distance from X to represent the declination.

The relationship between ZX and QX should now be evident in order to find ZQ.
(See figure in example 1 for illustration.)

5. Apply the declination to the zenith distance according to the rule derived, to give latitude.

172  PRACTICAL NAVIGATION FOR SECOND MATES

**Notes**

The bearing of the body must either be 000° or 180°. The position line must therefore run along a parallel of latitude upon which the observer must lie.

If the bearing of the body is not given in the question it can be inferred by inspection of the latitude of the D.R. and the declination. If declination is greater north than a northerly latitude then the body must pass to the north of the observer at meridian passage. If they are of opposite names then the bearing must be the same as the name of the declination.

Note that the time of meridian passage is not required for the calculation. This is because the declination of a star is constant over relatively large periods of time, and the G.M.T. is not required therefore for extracting it. In practice the time will be required, however, in order to know when to take the sight.

**Example 1**

18th December, 1980, the sextant altitude of the star *Diphda* 4 on the meridian, bearing 180°T., was 46° 15·4′, index error 1·4′ on the arc, height of eye 12 metres, D.R. position lat. 25° 33′ N., long. 33° 52′ W. Find the latitude and P.L.

| | | |
|---|---|---|
| Sext. alt. | 46° 15·4′ S. | Dec. 18° 05·7 S. |
| ind. err. | − 1·4′ | |
| Obs. alt. | 46° 14·0′ | |
| dip | − 6·1′ | |
| | 46° 07·9′ | |
| Main corr. − | 0·9′ | |
| True alt. | 46° 07·0′ S. | |
| | 90° | |
| zen. dist. | 43° 53·0′ N. | |
| dec. | 18° 05·7′ S. | |
| lat. | 25° 47·3′ N. | |

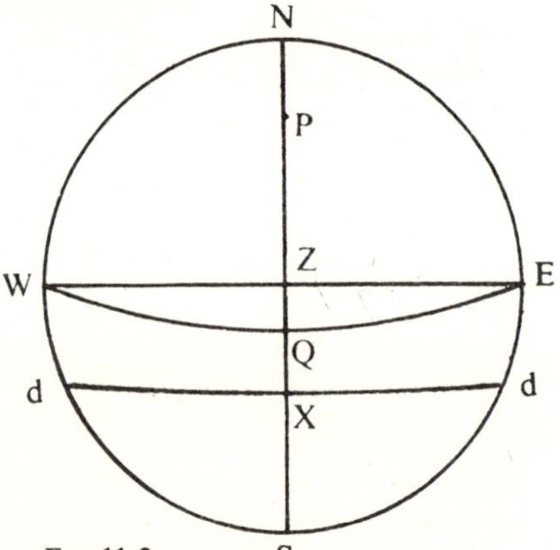

FIG. 11.2

*NESW* Represents the plane of the observer's rational horizon.

| | | | |
|---|---|---|---|
| Z | The observer's zenith. | dd | The parallel of declination. |
| P | The elevated pole. | X | The body on the observer's meridian. |
| PZS | The observer's meridian. | | |
| WQE | The equinoctial. | ZX | The true zenith distance. |
| WZE | The prime vertical. | QZ | The observer's latitude. |

P.L. trends 090° T.−270° T. through lat. 25° 47·3′ N., long. 33° 52′ W.

## LATITUDE BY MERIDIAN ALTITUDE

**Example 2**

On 5th January, 1980, the sextant altitude of the star *Fomalhaut* when on the meridian south of the observer was 77° 52·4'. Index error 3·0' off the arc. Height of eye 11·0 metres. Find the latitude and state the direction of the position line.

| Sext. alt. | 77° 52·4' | Declination 29° 43·9' S. |
|---|---|---|
| ind. err. + | 3·0' | |
| Obs. alt. | 77° 55·4' | |
| dip − | 5·8' | |
| | 77° 49·6' | |
| Main corr. − | 0·2' | |
| True alt. | 77° 49·4' | |
| Zen. dist. | 12° 10·6' | |
| Decl. | 29° 43·9' S. | |
| Latitude | 17° 33·3' S. | |

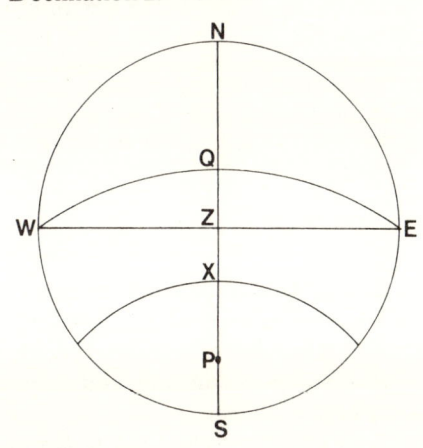

FIG. 11.3

### EXERCISE 11A

1. 19th September, 1980, to an observer in long. 42° 10' W., the sextant altitude of *Aldebaran* 10, on the meridian, was 71° 22·8', index error 1·4' off the arc, height of eye 14·5 metres, the star bearing 180° T. Find the latitude and P.L.

2. 19th December, 1980, the sextant altitude of *Dubhe* 27 on the meridian, and bearing 000° T. to an observer in long. 18° 30' W., was 28° 06·2', index error 1·2' off the arc, height of eye 10·5 metres. Find the P.L. and latitude of the point through which it is drawn.

3. 5th January, 1980, in D.R. position, lat. 49° 50' S., long. 42° 10' W. the sextant altitude of the star *Regulus* 26, on the meridian, was 28° 14·4', index error 0·6' on the arc, height of eye 15·3 metres. Find the latitude and the P.L.

4. 18th September, 1980, *Rigel* 11 was observed on the meridian bearing 000° (T.), sextant altitude 68° 10·9', index error 0·4' on the arc, height of eye 14·5 metres. Find the latitude and P.L.

5. 27th June, 1980. Find the latitude of an observer, given: sextant altitude of *Alioth* 32, on the meridian, was 34° 03·5', bearing 000° T., index error 1·8' off the arc, height of eye 12·0 metres.

### Latitude by meridian altitude of the sun

The true or apparent sun is on the observer's meridian at apparent noon or 1200 Local Apparent Time each day. However, we require the mean time when this occurs, in order to extract the declination.

The L.M.T. of apparent noon may be earlier or later than 1200 hours by the value of the equation of time, and is given for each day at the foot of each right hand daily page in the *Nautical Almanac*, in the box labelled SUN under the heading 'mer. pass'. The longitude in time can then be applied to this figure to obtain the G.M.T. (See chapter 7 on finding times of meridian passages.) It is sufficient in this case to obtain the G.M.T. to the nearest minute.

*Procedure*

1. Take out the L.M.T. of meridian passage from the almanac.

2. Apply the longitude in time to obtain G.M.T. (longitude WEST, Greenwich BEST, longitude EAST, Greenwich LEAST).

3. Extract the declination for this G.M.T.

4. Correct the altitude and subtract from 90° to obtain zenith distance.

5. Apply the declination to the zenith distance as explained for the problem with a star.

6. State the direction of the position line, which will always be east/west.

### Note

If it is preferred to remember rules of thumb to obtain the latitude, given the zenith distance and the declination, then the following may be helpful.

Put the bearing of the sun, i.e. N. or S., after the sextant altitude and the true altitude and apply the reverse name to the zenith distance.

Then:
lat. = zen. dist. + decl. (if the names are the same).
lat. = zen. dist. ~ decl. (if the names are different, and name the lat. the same name as the greater).

### Example

18th December, 1980, in D.R. position 22° 05′ N., 154° 20′ W., the sextant altitude of the sun's L.L. on the meridian was 44° 20·8′, index error 0·4′ off the arc, height of eye 15·3 metres. Find the latitude and P.L.

## LATITUDE BY MERIDIAN ALTITUDE

|   |   |   |
|---|---|---|
| L.M.T. mer. pass. 18th | 11h 57m |
| Long. W. | 10h 17m |
| G.M.T. 18th | 22h 14m |

| | | |
|---|---|---|
| Sext. alt. | 44° 20·8′ S. | Dec. 23° 24·6′ S. |
| I.E. | + 0·4′ | |
| obs. alt. | 44° 21·2′ | |
| dip | − 6·9′ | |
| App. alt. | 44° 14·3′ | |
| main corr. + | 15·3′ | |
| True alt. | 44° 29·6′ S. | |
| | 90° | |
| zen. dist. | 45° 30·4′ N. | |
| dec. | 23° 24·6′ S. | |
| latitude | 22° 05·8′ N. | |

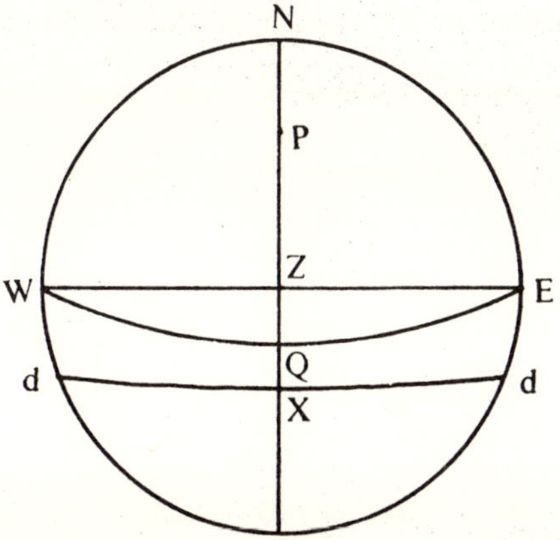

FIG. 11.4

P.L. trends 090°T.−270°T. through lat. 22° 05·8′ N., long. 154° 20′ W.

### EXERCISE 11B

1. 18th December, 1980, in D.R. position lat. 00° 20′ N., long. 162° 20′ W., the sextant altitude of the sun's lower limb on the meridian was 66° 10·4′ bearing south, index error 1·2′ on the arc, height of eye 13·2 metres. Find the latitude and P.L.

2. 26th June, 1980, the sextant altitude of the sun's lower limb when on the meridian was 41° 26·4′, index error 2·4′ off the arc, height of eye 7·3 metres. The D.R. position of the observer was lat. 25° 10′ S., long. 40° 20′ W. Find the latitude and P.L.

3. 6th January, 1980, an observation of the sun on the meridian by an observer in E.P. 51° 30′ S., 96° 35′ W., gave the sextant altitude of the sun's upper limb 61° 25′, index error was 1·4′ on the arc, height of eye 11·5 metres. Find the latitude and P.L.

4. From the following data, find the latitude and P.L.
   Date at ship, 30th September, 1980.
   Observer's E.P. lat. 36° 55′ N., long. 165° 30′ E.
   Body observed: the sun on the meridian, bearing 180° T., sextant altitude of the lower limb 50° 11·8′, index error 1·6′ off the arc, height of eye 14·0 metres.

5. 19th September, 1980, an observation of the sun on the meridian bearing 000° T. gave the sext. alt. of the sun's lower limb as 37° 37·6′, index error 1·6′ off the arc, height of eye 13·0 metres. The D.R. long. was 141° 10·8′ E.
Find the latitude and position line.

**Latitude by meridian altitude of the moon**

It is particularly important in the case of the moon to obtain an accurate G.M.T. for the time of meridian passage as the declination is usually changing rapidly. (See chapter 7 on finding time of meridian passage of the moon.)

*Procedure*

1. Extract the L.M.T. of meridian passage for the day in question. These are given for each day at the foot of the right hand of each daily page in a box labelled MOON under the heading of 'Mer. Pass. Upper'.

2. Extract the time for the following day if in westerly longitude *or* the preceding day if in easterly longitude, and take the difference between the two. Entering table II with this difference and the longitude, extract the correction for longitude.

3. Apply this correction for longitude to the time of meridian passage for the required day, adding if in west longitude, or subtracting if in east longitude.

4. Apply the longitude in time to obtain the G.M.T. of meridian passage, and extract the declination from the almanac.

5. Correct the sextant altitude to true altitude and subtract from 90° to obtain the zenith distance.

6. Apply the declination to obtain latitude and state the direction of the position line.

**Note**

Particular care should be taken over the correction of the moon's altitude. Study chapter 10 on correction of altitudes.

**Example**

On 27th June in longitude 58° 45′ W. the sextant meridian altitude of the moon's lower limb was 67° 48·6′ south of the observer. Index error 2·0′ off the arc. Height of eye 9·5 metres.

L.M.T. mer. pass. 27th, long. 0°   23h 42m
L.M.T. mer. pass. 28th, long. 0°   24h 37m
difference                          55m

# LATITUDE BY MERIDIAN ALTITUDE

correction for longitude = 9 m (from Table II)

| | |
|---|---|
| L.M.T. mer. pass. 27th, long. 0° | 23h 42m |
| long. corr. | 9m |
| L.M.T. mer. pass. long. 58° 45′ W. | 23h 51m |
| longitude in time | 3h 55m |
| G.M.T. mer. pass. long. 58° 45′ W. | 03h 46m  28th |

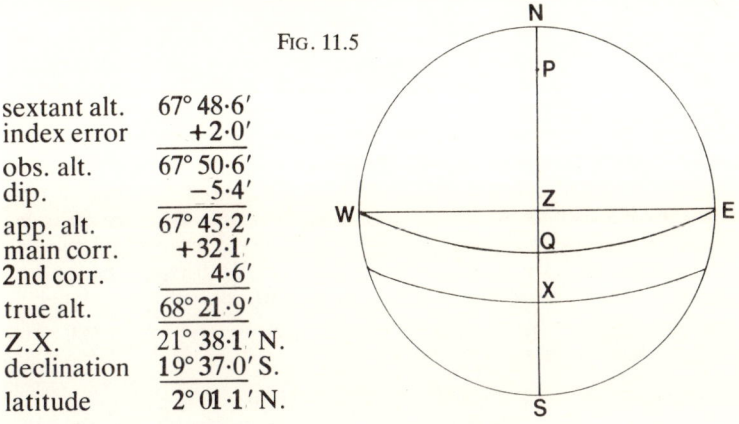

| | | |
|---|---|---|
| declination | 19° 36·1′ S. | H.P. 57·2 |
| 'd' corr. | +0·9′ | |
| declination | 19° 37·0′ S. | |

Fig. 11.5

| | |
|---|---|
| sextant alt. | 67° 48·6′ |
| index error | +2·0′ |
| obs. alt. | 67° 50·6′ |
| dip. | −5·4′ |
| app. alt. | 67° 45·2′ |
| main corr. | +32·1′ |
| 2nd corr. | 4·6′ |
| true alt. | 68° 21·9′ |
| Z.X. | 21° 38·1′ N. |
| declination | 19° 37·0′ S. |
| latitude | 2° 01·1′ N. |

P/L runs 090°/270° through 2° 01·1′ N. 58° 45′ W.

## EXERCISE 11C

1. On 5th January, 1980, in longitude 45° 20′ E. the observed altitude of the moon's lower limb when on the meridian north of the observer was 40° 18·5′. Index error nil. Height of eye 5·5 metres. Find the latitude.

2. On 19th September, 1980, in longitude 167° 15′ W., the sextant meridian altitude of the moon's upper limb was 30° 30·5′ south of the observer. Index error 1·5′ on the arc. Height of eye 10 metres. Find the latitude.

3. On 19th December, 1980, in longitude 130° E. the observed altitude of the moon's upper limb when bearing south was 70° 30·0′. Height of eye 9·0 metres. Find the latitude.

4. On 30th September, 1980, in longitude 0° the observed altitude of the moon's lower limb when bearing 000° T. was 88° 18·6′. Height of eye 8·6 metres. Find the latitude.

## To compute the altitude of a star on the meridian and find the time of the star's meridian passage

A practical problem arises when selecting suitable stars to observe in order to obtain a position. It is advantageous if a star can be found at its meridian passage at a time suitable for observation. The altitude can be computed, and this angle clamped on the sextant and the star found in the sextant telescope, and the accurate meridian altitude observed.

To enable this to be done we must first find the time when the star will be on our meridian, and this time must be at a time which is suitable for the observation of stars. In other words it must be during twilight.

In practice any stars which have their meridian passages during twilight can be found by extracting the time of nautical twilight from the almanac, and converting it to G.M.T. The S.H.A. of an imaginary star which has a G.H.A. equal to the longitude at this time can be computed, and the list of stars inspected to find stars which have S.H.A.s similar to this one.

Once a star is selected its exact time of meridian passage can be computed and this will give the navigator the time to observe. The D.R. latitude can then be used with the declination to find the zenith distance, from which the true altitude and hence the sextant altitude can be worked. The meridian altitude problem is worked in reverse to do this, all corrections being applied with the opposite sign to that in the normal way.

### Example

19th September, 1980, compute the sextant altitude and find the L.M.T. when the star *Aldebaran* 10 is on the meridian to an observer in D.R. position lat. 55° 18′ N., long. 142° 10′ W. Height of eye 13·3 metres, index error 0·6′ off the arc.

When a body is on the observer's meridian,

G.H.A. body = W. long. of observer.

Thus:

| | | |
|---|---|---|
| G.H.A. Aldebaran | 142° 10′ | |
| S.H.A. Aldebaran | 291° 17·9′ | |
| G.H.A. Aries | 210° 52·1′ | (G.H.A. + 360° − S.H.A.) |
| G.H.A. 14h 19th | 208° 37·5′ | |
| increment | 2° 14·6′ | = 8m 57s |

| | | |
|---|---|---|
| G.M.T. mer. pass Aldebaran 19th | 14h 8m 57s | 19th |
| longitude | 9h 28m 40s | |
| L.M.T. | 04h 40m 17s | 19th |

(see chapter 7 for full explanation of this method).

# LATITUDE BY MERIDIAN ALTITUDE

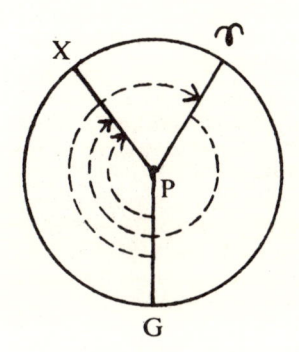

FIG. 11.6

It will be noted that this time occurs during a.m. twilight for the observer's latitude, see *Nautical Almanac*.

| | |
|---|---|
| latitude | 55° 18·0′ N. |
| declination | 16° 28·2′ N. |
| Z.X. | 38° 49·8′ |
| True alt. | 51° 10·2′ |
| correction | +0·8′ |
| apparent alt. | 51° 11·0′ |
| dip. | +6·4′ |
| obs. alt. | 51° 17·4′ |
| index error | −0·6′ |
| sextant alt. | 51° 16·8′ |

Computed altitude 51° 16·8′ T.

## EXERCISE 11D

1. Compute the sextant altitude and find the L.M.T. of the star *Vega* 49 on the meridian to an observer in E.P. lat. 5° 50′ N., long. 22° 30′ W., index error 0·4′ off the arc, height of eye 8·4 metres. Date at ship 19th September, 1980.

180    PRACTICAL NAVIGATION FOR SECOND MATES

2. 6th January, 1980. Compute the altitude to set on a sextant and find the L.M.T. for observation of *Menkar* 8 on the meridian to an observer in lat. 35° 10′ S., long. 32° 10′ E., index error 1·2′ off the arc, height of eye 12·0 metres.

3. Compute the altitude of *Gienah* 29 for setting on the sextant and find the L.M.T. for observation when on the meridian, observer's D.R. position 39° 20′ N., 35° 30′ W., height of eye 13·2 metres, index error 0·6′ off the arc, date at ship 19th December, 1980.

4. 27th June, 1980. Find the L.M.T. of meridian passage and compute the altitude of *Spica* 33 for seting on the sextant, index error 1·8 on the arc, height of eye 16·2 metres, D.R. lat. 12° 18′ N., long. 60° 35′ E.

5. 20th September, 1980, E.P. lat. 36° 15′ N., long. 142° 04′ W., compute the sextant altitude of *Betelgeuse* 16 and find the L.M.T. when on the meridian, index error 2·2′ off·the arc, height of eye 17·0 metres.

**Lower meridian passage**

The daily apparent motion of all heavenly bodies is to describe a circle around the pole, once in a sidereal day. Thus during this period as well as crossing the observer's meridian it must also cross the observer's antimeridian, i.e. the meridian 180° removed from the observer's meridian. Under certain circumstances the body will remain visible to an observer during the whole period, and will never set below the horizon. Such a body is called a circumpolar body. The conditions for circumpolarity are:

Latitude greater than polar distance
or lat. > (90° − declination)

A circumpolar body will, of course, be visible at the time when it crosses the observer's antimeridian. This occurrence is called the 'Lower Meridian Passage', 'Lower Meridian Transit', or 'on the meridian below the pole'.

The latitude can just as easily be found from an observation at lower meridian passage as at upper meridian passage.

At lower meridian passage, always:

Latitude = True altitude + (90° − declination)
or lat.    = T.A. + polar distance

**Latitude by a star on the meridian below the pole**
*Procedure*

1. Take out the star's declination from the *Nautical Almanac*.
2. Subtract the declination from 90° to obtain the polar distance.
3. Correct the altitude of the star.
4. Add the polar distance to the true altitude to obtain the latitude.
5. Name the latitude the same as the declination.

## LATITUDE BY MERIDIAN ALTITUDE

**Example**

18th September, 1980, the sextant altitude of *Atria* 43 on the meridian below the pole, was 19° 41·8′, index error 0·8′ on the arc, height of eye 9·7 metres. Find the latitude.

Sext. alt.  19° 41·8′         Dec.     68° 59·8′ S.
I.E.   −    0·8′                       90°

Obs. alt.  19° 41·0′         Polar dist. 21° 00·2′
dip    −    5·5′

App. alt.  19° 35·5′
Corr.  −    2·7′

T. alt.  19° 32·8′
Polar dist. 21° 00·2′

Latitude  40° 33·0′ S.

True bearing 180°
Position line 090° − 270°

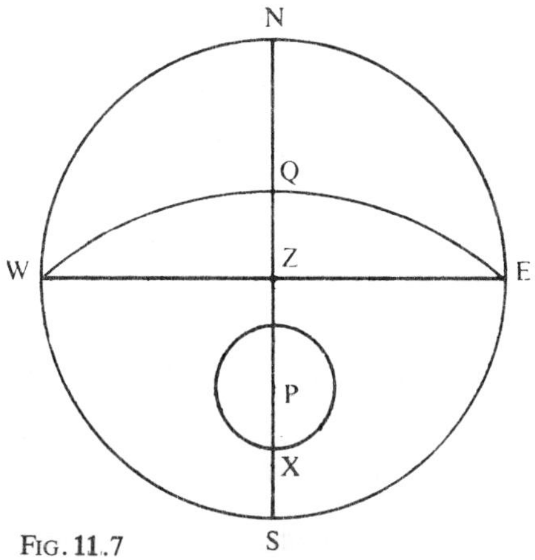

FIG. 11.7

## EXERCISE 11E

1. The sextant altitude of *Dubhe* 27 on the meridian below the pole on 18th December, 1980, was 22° 19·5′, index error 2·2′ on the arc, height of eye 12·8 metres. Find the latitude.

2. 19th December, 1980, find the latitude by *Alkaid* 34, on the meridian below the pole, sextant altitude 12° 27·9′, index error 2·4′ on the arc, height of eye 12·8 metres.

3. 7th January, 1980, the sextant altitude of *Schedar* 3 on the meridian below the pole was 21° 48·0′, index error 0·8′ off the arc, height of eye 13·2 metres. Find the latitude.

4. 20th September, 1980, the star *Avior* 22 was observed at its lower transit, sextant altitude 19° 32·4′, index error 1·2′ off the arc, height of eye 14 metres. Find the latitude.

5. 26th June, 1980, the sextant altitude of *Achernar* 5, on the meridian below the pole, was 13° 00·4′, index error 1·4′ on the arc, height of eye 12·5 metres. Find the latitude.

## CHAPTER 12

## THE CALCULATION OF A POSITION LINE BY OBSERVATION OF A BODY OUT OF THE MERIDIAN

A knowledge of the use of the spherical haversine formula is assumed. If necessary a text on spherical trigonometry should be consulted for its derivation and its use.

The solution of any navigational problem is basically the solution of a spherical triangle on the celestial sphere, the three points of which are: the elevated pole (P), the position of the body (X), and the position of the zenith (Z).

The three sides of the triangle will therefore be:

PX   the angular distance of the body from the pole, i.e. the polar distance, i.e. $90° -$ declination.

PZ   the angular distance of the observer from the pole, i.e. the co-lat., i.e. $90° -$ latitude.

ZX   the angular distance of the observer's zenith from the body, i.e. the zenith distance, i.e. $90° -$ altitude.

The angles of the triangle are:

∠P   the angle between the observer's meridian and the meridian of the body (see definition of L.H.A.). Angle P is equal to the L.H.A. when the body is setting, and is equal to $360° -$ L.H.A. when the body is rising.

∠Z   the angle between the direction of the meridian and that of the body, i.e. the azimuth.

∠X   the parallactic angle. Is not used in the normal reduction of sights.

The triangle is usually represented by a figure on the plane of the rational horizon, i.e. looking down from above the observer's zenith.

# CALCULATION OF POSITION LINES

FIG. 12.1

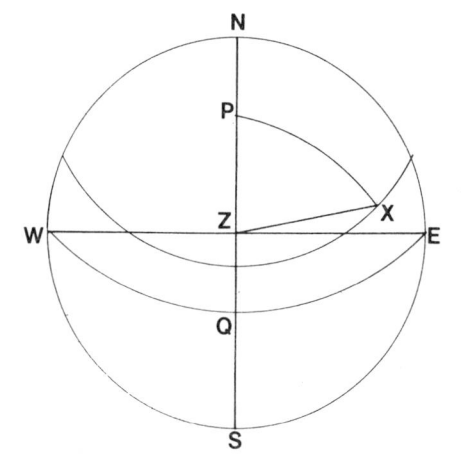

To solve the triangle we need to know three of its elements.

## The Marcq St. Hilaire (Intercept) Method

This is a popular method as any sight may be reduced by its use. The three elements used are:

1. An assumed latitude (D.R. lat.) to give a value for PZ.
2. Polar distance (PX) (90° − declination).
3. An assumed longitude (D.R. long.), which is combined with the G.H.A. to give the L.H.A. and thus angle P.

With these arguments we solve the triangle for the side ZX, the zenith distance, by use of the haversine formula, thus:

Hav ZX = (Hav P. sin PZ. sin PX) + Hav(PZ ∼ PX)
and as PZ = complement of latitude
  and PX = complement of declination
Hav ZX = (Hav P. cos lat. cos dec.) + Hav (lat. ∼ dec.)

Having found this calculated zenith distance it can be compared with the true zenith distance, which is found by correcting the sextant altitude to a true altitude and subtracting it from 90°. The difference is the intercept (see chapter 9).

The true bearing can be calculated by the use of the ABC tables as described in chapter 8, and we are then in a position to plot a position line as described in chapter 9.

*Procedure*

1. From the chronometer reading, deduce the G.M.T. This is done by taking the approximate L.M.T. (ship's time indicated by clock is quite accurate enough), and applying the longitude in time to obtain the approximate G.M.T. From this can be decided:

(a) Whether to add 12 hours to the chronometer time or not, i.e. 02h indicated on the chronometer may either be 02h or 14h.
(b) The correct date at Greenwich. (Date given in the problem is the date at the ship. The date at Greenwich may be the day before or the day after, depending on the longitude.)

2. With the G.M.T. extract the G.H.A. and the declination of the body.

3. Apply the longitude to the G.H.A. and obtain the L.H.A. and thus deduce the angle P. (body setting $\angle P = L.H.A.$, body rising $\angle P = 360° - L.H.A.$).

4. Combine the latitude and the declination to obtain (PZ~PX). If lat. and dec. are of the same name take (L~D). If of opposite name take (L+D).

5. Use the haversine formula to calculate the zenith distance.

6. Correct the altitude and subtract from 90° to obtain the zenith distance.

7. Apply the C.Z.X. to the T.Z.X. and obtain the intercept.

8. Using ABC tables find the true bearing.

We now have information enough to plot a position line as explained in chapter 9.

### Example 1. By an observation of the sun

On 30th September, 1980, at about 0900 at ship in D.R. position, latitude 41° 15′ N., longitude 175° 30′ W., when the chronometer, which was correct on G.M.T., showed 8h 30m 15s, the sextant altitude of the sun's lower limb was 29° 24·6′, index error 0·4′ off the arc, height of eye 15·8 metres. Required the position line and a point through which it passes.

```
Approx. L.M.T. 30th  09h 00m
Long. W.             11h 42m
                     ─────────
Approx. G.M.T. 30th  20h 42m          G.M.T. 30th 20h 30m 15s
                     ─────────        ─────────────────────────

From N.A. G.H.A. 20h  122° 33·1′       decl.  3° 05·7′ S.
          Increment     7° 33·8′       'd'       0·5′
          G.H.A.      130° 06·9′       decl.  3° 06·2′ S.
                      360°             lat.  41° 15·0′ N.
                      ─────────        L±D   44° 21·2′
                      490° 06·9′
          Longitude   175° 30·0′ W.
                      ─────────

          L.H.A.      314° 36·9′
          ∠p           45° 23·1′
```

## CALCULATION OF POSITION LINES

Hav. ZX = Hav. P. cos lat. cos dec. + hav (lat. ⚹ dec.).

| | | | |
|---|---|---|---|
| P. = 45° 23·1' | log hav. | $\bar{1}$·17269 | Sext. alt. 29° 24·6' |
| lat. = 41° 15' | log. cos. | $\bar{1}$·87613 | I.E. +0·4' |
| dec. = 3° 06·2' | log cos | $\bar{1}$·99936 | Obs. alt. 29° 25·0' |
| | | $\bar{1}$·04818 | dip. −7·0' |
| | | 0·11173 | App. alt. 29° 18·0' |
| lat ⚹ dec. | nat. hav. | 0·14248 | Corr. +14·3' |
| CZX = 60° 33·3' | | 0·25421 | True alt. 29° 32·3' |
| | | | T.Z.X. 60° 27·7' |
| | | | C.Z.X. 60° 33·3' |

5·6' Towards

A. 0·865 S.
B. 0·076 S.
─────────
C. 0·941 S.
Az. S. 54·7 E. or 125·3°

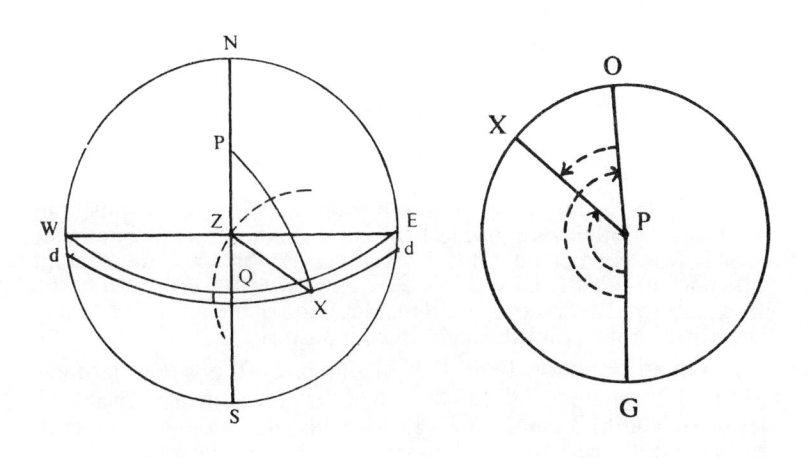

FIG. 12.2 (a)   FIG. 12.2 (b)

FIG. 12.3

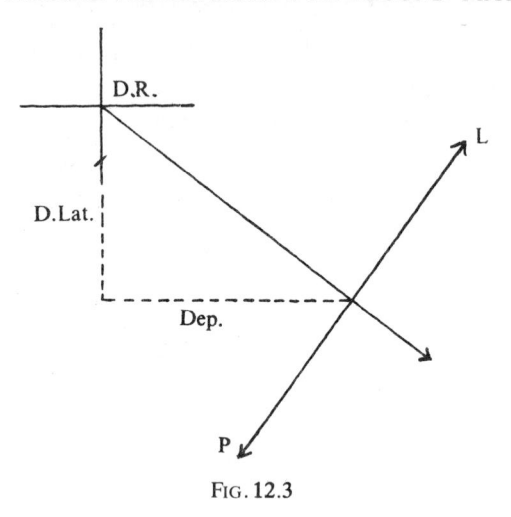

**To calculate the position of the I.T.P.**

| | | | |
|---|---|---|---|
| D.R. lat. | 41°15·0 N. | Long. | 175°30·0' W. |
| Co. 125·3° dist. 5·6' | 3·2'S. | | 6·0' E. (dep.=4·5') |
| I.T.P. | 41°11·8' N. | | 175°24·0' W. |

*Answer:* Position line runs 035·3°/215·3° through position 41° 11·8' N. 175° 24·0' W.

## EXERCISE 12A

### BY OBSERVATION OF THE SUN

1. On 26th June, 1980, at about 0930, at ship in D.R. position, latitude 29° 30' S., longitude 121° 20' W., when the chronometer which was correct on G.M.T. showed 5h 45m 20s, the sextant altitude of sun's L.L. was 26° 52·2', index error 1·6' on the arc, height of eye 12·0 metres. Find the direction of the position line and the position of a point through which it passes.

2. On 8th January, 1980, 1530 at ship in D.R. position, latitude 32° 15' S., longitude 48° 16' W., when the chronometer which was correct on G.M.T. indicated 18h 31m 24s, the sextant altitude of the sun's U.L. was 46° 58·0', index error 0·4' on the arc, height of eye 11·0 m. Find the P.L. and the position of a point through which it passes.

# CALCULATION OF POSITION LINES

3. On 19th September, 1980, at about 4 p.m., at ship in an estimated position, latitude 0°00·0′, longitude 160° 55′ W., when the chronometer which was correct on G.M.T. showed 2h 30m 15s, the sextant altitude of the sun's U.L. was 32°12·9′, index error 0·′6 off the arc, height of eye 12·5 m. Find the position line and the position of a point through which it passes.

4. On 18th December, 1980, at about 0900, at ship in D.R. position, latitude 43° 12′ N., longitude 38° 25′ W., when the chronometer which was 2m 21s fast on G.M.T. showed 11h 51m 52s, the sextant altitude of the sun's L.L. was 13° 33·3′, index error 1·6′ off the arc, height of eye 11·5 m. Find the position line and the position of a point through which it passes.

5. On 30th September, 1980, at ship in D.R. position, latitude 44° 05′ N., longitude 27° 41′ W. at 09h 41m 02s G.M.T., the sextant altitude of the sun's L.L. was 18° 57·5 , index error 1·4′ on the arc, height of eye 9·0 m. Find the direction of the position line and the position of a point through which it passes.

## Example 2. By an observation of a star

On 9th January 1980 at approximately 1900 at ship in D.R. position 35°10′ S. 127° 50 E., the sextant altitude of the star *Sirius* was observed to be 36° 58·1′. Index error 0·4′ on the arc. Height of eye 15 metres. A chronometer which was correct on G.M.T. showed 11h 15m 10s. Find the direction of the position line and the I.T.P.

| | | | |
|---|---|---|---|
| Approx. L.M.T. 1900 | | Chron. 11h 15m 10s | |
| long. E. 0831 | | G.M.T. 11h 15m 10s 9th Jan. | |

Approx. G.M.T. 1029 9th

| | | | | |
|---|---|---|---|---|
| From *N.A.* | G.H.A. ♈ 11h | 273° 08·9′ | Dec. | 16° 41·5′ S. |
| | incr. | 3° 48·1′ | Lat. | 35°10·0′ S. |
| | G.H.A. ♈ | 276° 57·0′ | Lat. ~ Dec. | 18° 28·5′ |
| | S.H.A.* | 258° 55·8′ | | |
| | G.H.A.* | 535° 52·8′ | | |
| | | 360° | | |
| | G.H.A.* | 175° 52·8′ | | |
| | Long. E. | 127° 50·0′ | | |
| | L.H.A.* | 303° 42·8′ | | |
| | ∠ P. | 56°17·2′ | | |

## 188 PRACTICAL NAVIGATION FOR SECOND MATES

Hav. $ZX$ = Hav. P. cos lat. cos dec. + hav (lat. ~ dec.)

| | | | | | |
|---|---|---|---|---|---|
| P. | 56° 17·2′ | log hav. | $\bar{1}$·34729 | Sext. alt. | 36° 58·1′ |
| Lat. | 35° 10·0′ | log cos | $\bar{1}$·91248 | I.E. | −0·4′ |
| Dec. | 16° 41·5′ | log cos | $\bar{1}$·98130 | | |
| | | | | Obs. alt. | 36° 57·7′ |
| | | | $\bar{1}$·24107 | Dip. | −6·8′ |
| | | | 0·17421 | App. alt. | 36° 50·9′ |
| Lat. ~ Dec. | | nat. hav. | 0·02577 | corr. | −1·3′ |
| C.Z.X. = 53° 07·6′ | | | 0·19998 | T. alt. | 36° 49·6′ |
| | A. ·470 N. | | | T.Z.X. | 53° 10·4′ |
| | B. ·360 N. | | | C.Z.X. | 53° 07·6′ |
| | C. ·110 N. | | | Intercept | 2·8′ Away |
| | Az. N. 84·9′ E. | | | | |

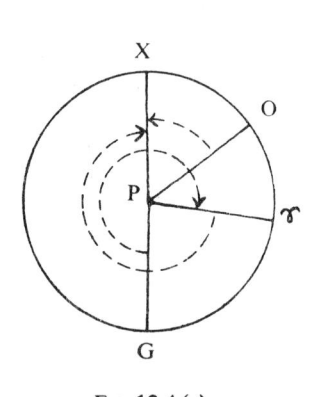

FIG. 12.4 (a) 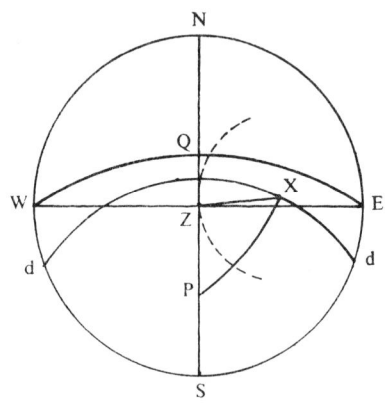     FIG. 12.4 (b)

### To calculate the position of I.T.P.

| | |
|---|---|
| D.R. lat. | 35° 10·0′ S. |
| Co. 84·9° dist. 2·8′ d. lat. | 0·3′ S. |
| I.T.P. | 35° 10·3′ S. |
| Long. | 127° 50·0′ E. |
| d. long. | 3·4 W. (dep. 2·8′) |
| | 127° 46·6′ E. |

FIG. 12.5

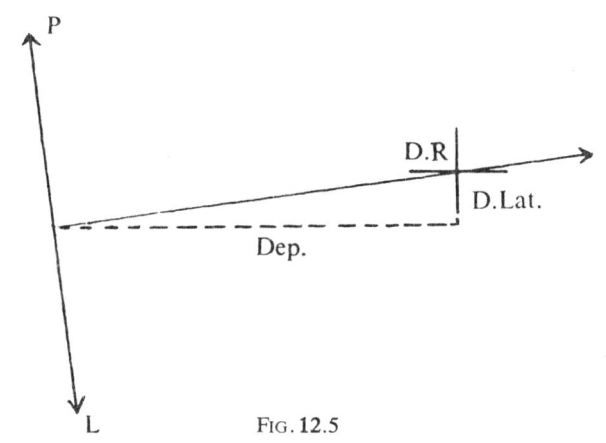

*Answer*

P.L. trends 354·9T–174·9T through position latitude 35° 10·3' S., longitude 127° 46·6' E.

## EXERCISE 12B

### BY OBSERVATION OF A STAR

1. On 18th September, 1980, at ship in D.R. position, latitude 24° 50' N., longitude 145° 10' E., at 08h 59m 50s G.M.T. the sextant altitude of the star *Arcturus* 37 was 31° 30·5', index error 0·8' on the arc, height of eye 12 m. Find the direction of the position line and the position of a point through which it passes.

2. On 30th September, 1980, at ship in D.R. position, latitude 43° 05' N., longitude 177° 16' W., at 17h 01m 44s G.M.T. the sextant altitude of the star *Schedar* 3 was 41° 04·2', index error 0·2' off the arc, height of eye 13·2 m. Find the direction of the position line and the position of a point through which it passes.

3. On 19th September, 1980, at ship in estimated position, latitude 17° 53·6' N., longitude 47° 30' W., the sextant altitude of the star *Alphard* 25 during morning twilight was 18° 59·2', index error 0·5' on the arc, height of eye 18·6 m. The chronometer, which was 04m 53s slow on G.M.T., showed 8h 10m 23s. Find the direction of the position line and the position of a point through which it passes.

4. On 18th December, 1980, in estimated position latitude 42° 40' N., longitude 172° 10' W., at 17h 59m 30s G.M.T. the sextant altitude of the star *Alphecca* 41 was 48° 05·9', index error 1·3' on the arc, height of eye 17·5 m. Find the direction of the P.L. and the position of a point through which it passes.

5. On 26th June, 1980, at ship in D.R. position latitude 40° 59·5′ S., longitude 56° 57′ W., at 21h 26m 00s G.M.T. the sextant altitude of the star *Procyon* 20 was 15° 23·5′, index error 0·6′ off the arc, height of eye 9·0 m. Find the direction of the P.L. and the position of a point through which it passes.

### Example 3. By observation of the moon

An observer in D.R. position 14° 38′ S. 154° 14′ W. observes the altitude of the moon's lower limb to be 52° 07·5′. Index error nil, height of eye 12·0 metres. The chronometer showed 04h 45m 14s at the time and was correct on G.M.T. The approximate ship's time was 0639 on 30th September, 1980.

Find the direction of the position line and a position through which to draw it.

| | | | | | | |
|---|---|---|---|---|---|---|
| App. L.M.T. | 0639 30th | | | Chron. | 04h 45m 14s | |
| Long. W. | 10 17 | | | G.M.T. | 16h 45m 14s 30th | |
| App. G.M.T. | 16 56 30th | | | Dec. | 19° 38·7′ N. | |
| | | | | 'd' | +1·1′ | |
| G.H.A. 16h | 157° 37·7′ | | | | | |
| Incr. | 10° 47·6′ | | | Decl. | 19° 39·8′ N. | |
| 'v' (7·8) | 5·9′ | | | Lat. | 14° 38·0′ S. | |
| G.H.A. | 168° 31·2′ | | | L.~ D. | 34° 17·8′ | |
| Long. | 154° 14·0′ W. | | | | | |
| L.H.A. | 14° 17·2′ = ∠P. | | | | | |

| | | | | | |
|---|---|---|---|---|---|
| ∠ P. 14° 17·2′ | log hav. | $\overline{2}$·18931 | Sext. alt | 52° 07·5′ | |
| Lat. 14° 38′ | log cos | $\overline{1}$·98568 | I.E. | – | |
| Dec. 19° 39·8′ | log cos | $\overline{1}$·97391 | | | |
| | | | Obs. alt. | 52° 07·5′ | |
| | | $\overline{2}$·14890 | Dip. | −6·1′ | |
| | | 0·01409 | App. alt. | 52° 01·4′ | |
| L. ~ D. | nat. hav. | 0·08693 | M. corr. | 45·4′ | |
| | | | 2nd corr. | 5·1′ | |
| CZX = 37° 03·9′ | | 0·10102 | T. alt. | 52° 51·9′ | |
| | A 1·025+ | | T.Z.X. | 37° 08·1′ | |
| | B 1·448+ | | C.Z.X. | 37° 03·9′ | |
| | C 2·473+ | | Intercept | 4·2′ away | |
| | Az. N. 22·7° W. | | | | |

## CALCULATION OF POSITION LINES

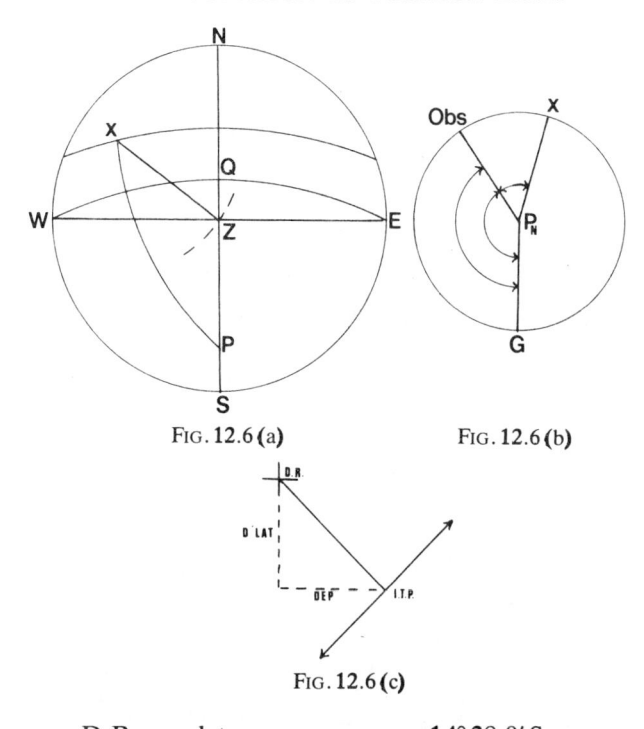

FIG. 12.6(a)

FIG. 12.6(b)

FIG. 12.6(c)

| D.R. pos. lat. | 14° 38·0′ S. |
| Course 22·7° dist. 4·2′ d. lat. | 3·9′ S |
| I.T.P. | 14° 41·9′ S. |
| Long. | 154° 14·0′ W. |
| D. long. | 1·7′ E. (dep. 1·6′) |
| | 154° 12·3′ W. |

*Answer*

Position line runs 067·3°/247·3° through 14° 41·9′ S. 154° 12·3′ W.

### EXERCISE 12C

**1.** At approximately 18.15 on 26th June at ship, in D.R. position 42° 50′ S. 41° 30′ W., the sextant altitude of the moon's lower limb was 29° 10·8′ Index error 2·0′ off the arc. Height of eye 10 metres. A chronometer which was correct on G.M.T. showed 09h 10m 02s at the time. Find the direction of the position line and a position through which it passes.

2. On 9th January 1980 at approximate L.M.T. 0900, in D.R. position 25° 30′ N. 175° 00′ E., the sextant altitude of the moon's upper limb was 27° 21·5′. Index error 2·0′ on the arc. Height of eye 12·0 metres. A chronometer which was slow on G.M.T. by 1m 24s showed 09h 14m 21s at the time. Find the direction of the position line and the I.T.P.

### The longitude by chronometer method

In this method only a D.R. latitude is assumed. This gives the side PZ in the triangle, and this is used with the polar distance and the observed zenith distance (ZX), in the haversine formula to calculate the angle P. From this the L.H.A. is deduced, and the G.H.A. applied to it to obtain the longitude.

Note that this longitude will only be the correct longitude, if the assumed latitude is correct. Thus the D.R. latitude and the longitude by calculation give a position through which to draw the position line. The true bearing must also be calculated as in other methods to find the direction of the position line.

### Note

For one particular observation there can only be one position line. Whether the observation is worked by Marcq St. Hilaire or by longitude by chronometer, the same position line will result. The positions calculated through which to draw the position line will, however, differ.

Figure 12.7 shows one position line, and the information obtained from each method.

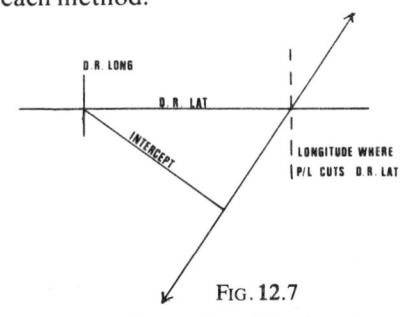

Fig. 12.7

Thus the arguments used to solve the triangle are:
1. PZ, obtained from the D.R. latitude.
2. PX, obtained from the declination.
3. ZX, the true zenith distance obtained from the sextant altitude.

and by haversine formula:

$$\text{Hav. P} = (\text{hav. ZX} - \text{hav. (PZ} \sim \text{PX)}) \operatorname{cosec} \text{PZ} \operatorname{cosec} \text{PX}$$
$$= (\text{hav. ZX} - \text{hav. (lat.} \sim \text{dec.)}) \sec \text{lat.} \sec \text{dec.}$$

# CALCULATION OF POSITION LINES

*Procedure*
1. From the chronometer time deduce the G.M.T. as in the intercept method.
2. Using the G.M.T. extract the G.H.A. and the declination from the almanac.
3. Correct the sextant altitude and find the zenith distance.
4. By haversine formula, using lat. dec. and ZX find the angle P and hence L.H.A.
5. Apply L.H.A. to G.H.A. to obtain the longitude.
6. Calculate the true bearing.

We now have information for plotting a position line.

**Note**

This method of determining a position through which the position line passes is suitable provided the body is not too close to the observer's meridian. In this case there is a considerable change in longitude for a small change in azimuth, and in general it may be said that the longitude method can be used if the observed body is more than 2 hours from meridian passage. It should be noted that there is no such limitation for the Marcq St. Hilaire method.

**Example 4. By an observation of the sun**

(Using Example 1 worked by the longitude method).

On 30th September, 1980, at about 0900, at ship in D.R. latitude 41° 15′ N., when the G.M.T. was 20h 30m 15s, the sextant altitude of the sun's L.L. was 29° 24·6′, index error 0·4′ off the arc, height of eye 15·8 m. Required the P.L. and the longitude in the D.R. latitude through which it passes.

G.M.T. 30th 20h 30m 15s

| | | | |
|---|---|---|---|
| G.H.A. 20h | 122° 33·1′ | Dec. | 3° 05·7′ |
| incr. | 7° 33·8′ | corr.   + | 0·5′ |
| G.H.A. | 130° 06·9′ | Dec. | 3° 06·2′ S. |
| Sextant alt. | 29° 24·6′ | Lat. | 41° 15′ N. |
| I.E.    + | 0·4′ | Dec. | 3° 06·2′ S. |
| Obs. alt. | 29° 25·0′ | | |
| Dip    − | 7·0′ | (Lat.+dec.) | 44° 21·2′ |
| App. alt. | 29° 18·0′ | | |
| Tot. corr. + | 14·3′ | | |
| T. alt. | 29° 32·3′ | | |
| | 90° | | |
| *TZX* | 60° 27·7′ | | |

Hav. $P.= \{(\text{hav. } ZX - \text{hav. (lat. } \sim \text{dec.)}\} \sec \text{lat. } \sec \text{dec.}$

| | | |
|---|---|---|
| $ZX$ 60° 27·7′ | nat. hav. | 0·25350 |
| l $\sim$ d 44° 21·2′ | nat. hav. | 0·14248 |
| | | 0·11102 |

|  |  |  |  |
|---|---|---|---|
| | | $\overline{1}$·04540 | A ·870 S. |
| lat. 41° 15′ | log sec | 0·12388 | B ·076 S. |
| dec. 3° 06·2′ | log sec | 0·00064 | |
| | | | C ·946 S. |
| $\angle P$ = 45° 13·9′ | | $\overline{1}$·16992 | Az. S. 54·6° E. |

L.H.A. = 314° 46·1′
G.H.A. = 130° 06·9′ (Long. = G.H.A. + 360 − L.H.A.)

Long. 175° 20·8′ (see figure 12.2 (b))

*Answer*

Position line runs 035·4°/215·4° through position 41° 15′ N. 175° 20·8′ W.

**Note**

The position line is the same as that calculated in Example 1 and the only difference is in the position given through which it passes. Using the intercept and azimuth it is possible to find the longitude in the D.R. latitude through which the P.L. can be drawn.

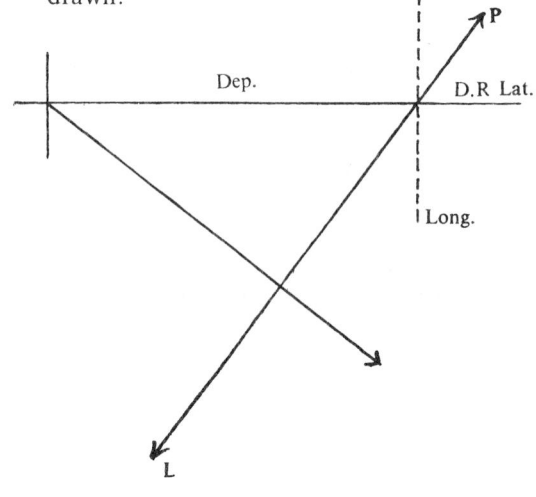

Fig. 12.8

CALCULATION OF POSITION LINES 195

In figure 12.8:
Intercept=5·6' T. Az.=S54·7E. (from example 1).

Longitude calculated by long. by Chron. is that shown by a dashed line. Thus by measurement departure=6·9.

D. long.=9·2'E.   D.R. long 175° 30·0' W.
                 D. long.      9·2' E.
                              _____
Long. by chron.               175° 20·8' W.

This was the longitude calculated in example 4 which shows that the same position line is obtained, whatever method of reduction is used.

## EXERCISE 12D

### BY OBSERVATION OF THE SUN

1. On 26th June, 1980, at about 1600, at ship in D.R. latitude 10° 25' N., when the chronometer, which was 4m 27s fast on G.M.T., indicated 11h 59m 53s, the sextant altitude of the sun's lower limb was 31° 33·3', index error 1·2' on the arc, height of eye 17·0 m. Find the direction of the position line and the longitude in the D.R. latitude through which it passes.

2. On 19th September, 1980, at ship in D.R. position, latitude 18° 44' N., longitude 117° 12' W., the sextant altitude of the sun's upper limb was 24° 34·5', index error 0·6' off the arc, height of eye 18·0 m., at 00h 01m 42s G.M.T. Find the direction of the position line and the longitude in the D.R. latitude through which it passes.

3. On 20th December, 1980, a.m., at ship in D.R. position latitude 35° 24' S., longitude 171° 15' E., the sextant altitude of the sun's lower limb was 43° 09·7', index error 0·4' on the arc, height of eye 14·5 m., when the chronometer, which was 01m 17s slow on G.M.T., showed 9h 00m 35s. Find the direction of the position line and the longitude in which it cuts the D.R. latitude.

4. On 5th January, 1980, a.m., at ship in D.R. latitude 0° 30' S., the sextant altitude of the sun's upper limb was 30° 27·1', index error 1·4' on the arc, height of eye 19·5 m., at 08h 15m 35s G.M.T. Find the direction of the position line and the longitude in which it cuts the D.R. latitude.

5. On 30th September, 1980, at ship in D.R. position, latitude 44° 05' N., longitude 27° 41' W., at 09h 41m 02s G.M.T., the sextant altitude of the sun's lower limb was 18° 57·5', index error 1·4' on the arc, height of eye 9·0 m. Find the direction of the position line and the longitude in which it cuts the D.R. latitude.

(The answer to this problem may be verified from Exercise 12A, No. 5.)

### Example 5. By an observation of a star

(Using example 2 worked by the longitude method.)

On 9th January, 1980, p.m., at ship in D.R. position latitude 35° 10′ S., longitude 127° 50′ E., at 11h 15m 10s G.M.T., the sextant altitude of the star *Sirius* to the east of the meridian was 36° 58·1′, index error 0·4′ on the arc, height of eye 15·0 m. Find the direction of the position line and the longitude in which it cuts the D.R. latitude.

G.M.T. 9th 11h 15m 10s

| | | | | |
|---|---|---|---|---|
| From *N.A.* G.H.A. ♈ | 11h | 273° 08·9′ | Dec. | 16° 41·5′ S. |
| | incr. | 3° 48·1′ | | |
| | G.H.A. ♈ | 276° 57·0′ | | |
| | S.H.A.* | 258° 55·8′ | | |
| | G.H.A.* | 535° 52·8′ | | |
| | | 360 | | |
| | G.H.A.* | 175° 52·8′ | | |

| | | | |
|---|---|---|---|
| Sextant alt. | 36° 58·1′ | Lat. | 35° 10·0′ S. |
| I.E. | −0·4′ | Dec. | 16° 41·5′ S |
| Obs. alt. | 36° 57·7′ | (Lat. ∼ Dec.) | 18° 28·5′ |
| Dip. | −6·8′ | | |
| App. alt. | 36° 50·9′ | | |
| Corr. | −1·3′ | | |
| T. alt. | 36° 49·6′ | | |
| T.Z.X. | 53° 10·4′ | | |

Hav. $P = \{$hav. $ZX −$ hav. (lat. ∼ dec.)$\}$sec lat. sec dec.

| | | | |
|---|---|---|---|
| $ZX$ 53° 10·4′ | | nat. hav. | 0·20030 |
| (lat. ∼ dec.) 18° 28·5′ | | nat. hav. | 0·02577 |
| | | | 0·17453 |
| | | | $\overline{1}$·24187 |
| lat. 35° 10′ | | log sec | 0·08752 |
| dec. 16° 41·5′ | | log sec | 0·01870 |
| ∠ P = 56° 20·6′ | | | $\overline{1}$·34809 |

# CALCULATION OF POSITION LINES

L.H.A. = 303° 39·4'  　　　　A  ·469+
G.H.A. = 175° 52·8'  　　　　B  ·360−

Long. = 127° 46·6' E.  　　　C  ·109+

Az. N. 84·9° E.

(longitude = L.H.A. − G.H.A. See figure 12.4 (a)).

*Answer*

Position line runs 354·9°/174·9° through position 35° 10·0' S 127° 46·6' E.

From Example 2

Using intercept 2·8' away. Az. N. 84·9° E. to verify the above answer.

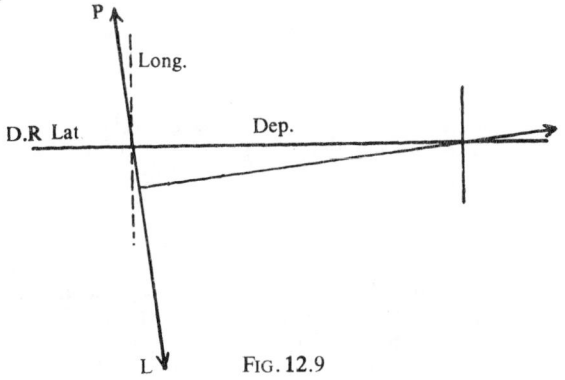

Fig. 12.9

From figure dep. = 2·8'
D.R. lat. 35° 10' S.　long.　　127° 50' E.
　　　　　　　　　　d. long.　　　3·4' W.

D.R. lat. 35° 10' S.　long.　　127° 46·6' E.

## EXERCISE 12E

### BY OBSERVATION OF A STAR

**1.** On 5th January, 1980, at ship in D.R. position, latitude 30°30' N., longitude 44° 40' W. at 09h 15m 07s G.M.T. the sextant altitude of the star *Rasalhague* 46 east of the meridian was 27° 56·5', index error 1·3' off the arc, height of eye 16·8 m. Find the direction of the position line and the longitude in which it cuts the D.R. latitude.

2. On 27th June, 1980, at ship in D.R. latitude 29° 40′ S., the observed altitude of the star *Procyon* 20 at p.m. twilight was 14° 49·8′, height of eye 13·2 m., west of the meridian, when the chronometer, which was 3m 47s slow on G.M.T., indicated 02h 47m 24s. Find the direction of the position line and the longitude in which it cuts the D.R. latitude. D.R. long. 134° 55′ W.

3. On 19th September, 1980, p.m., at ship in D.R. position, latitude 27° 30′ N., longitude 178° 10′ E., at 06h 40m 12s G.M.T. the sextant altitude of the star *Arcturus* 37 was 32° 21·4′, index error 2·4′ off the arc, height of eye 15·8 m. Find the direction of the position line and the longitude in which it cuts the D.R. latitude.

4. On 1st October, 1980, at ship during morning twilight in D.R. position, latitude 32° 15′ S., longitude 78° 33′ E., the sextant altitude of the star *Regulus* 26 was 13° 24·6′, index error 0·8′ on the arc, height of eye 11·5 m. The chronometer, which was 2m 16s fast on G.M.T., showed 11h 53m 04s. Find the direction of the position line and the longitude in which it cuts the D.R. latitude.

5. On 26th June, 1980, at ship in D.R. position, latitude 40° 59·5′ S., longitude 56° 57′ W., at 21h 23m 42s G.M.T., the sextant altitude of the star *Procyon* 20 was 15° 23·5′, index error 0·6′ off the arc, height of eye 9·0 m. Find the direction of the position line and the longitude in which it cuts the D.R. latitude.

## Noon position by longitude by chronometer and meridian altitude

The most popular method of obtaining a noon position at sea is to transfer a position line obtained in the forenoon by observation of the sun, up to the time of noon, i.e. the time of meridian passage of the sun. It can then be crossed with a position line obtained from the meridian altitude which will run east–west (see chapter 9 for transferred position lines).

This problem can, however, be solved without resort to any plotting.

Let the figure represent a position line obtained by observation

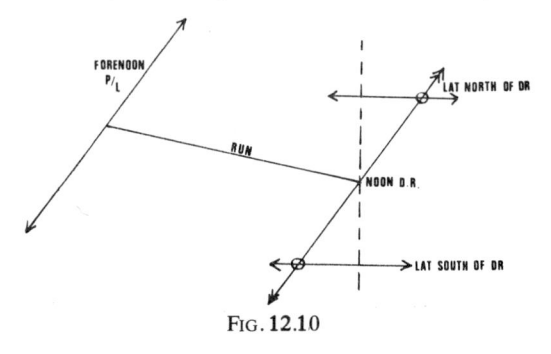

FIG. 12.10

# CALCULATION OF POSITION LINES

of the sun and worked by longitude by chronometer, during the forenoon. This position line is then transferred to the time of noon by application of the course and distance steamed. The transferred position line is marked with double arrows.

We can say that at noon, if our D.R. latitude used in the forenoon sight was correct then our noon longitude is our D.R. longitude. However, the latitude obtained at noon will probably indicate that our true latitude is to the north or the south of our D.R. latitude, and therefore our longitude will be in error. The amount of the error in longitude can be found by taking the difference in minutes of d. lat. between the D.R. latitude and the observed latitude, and multiplying this by the value of 'C' from the ABC table calculation when finding the azimuth for the forenoon sight.

The value of 'C' in this respect can be taken as the error in longitude caused by an error of 1 minute in the latitude when working the sight.

The direction of the longitude error must be found by inspection of the direction of the position line and the direction of the error in latitude. Thus in the figure 12·10, where the position line runs SW/NE if the observed latitude is south of the D.R. latitude the true longitude must be to the west of the D.R. longitude. If the observed latitude is to the north, the observed longitude must be to the east.

If the position line runs NW/SE then the opposite will apply. The appropriate case must be found from a rough sketch of the position line and the observed latitude.

## Example

On 19th December, 1980, at 0810 ship's time in D.R. position 25° 50′ N. 57° 37′ W. an observation of the sun's lower limb gave a sextant altitude of 15° 47·5′. Index error was 3·0′ on the arc. Height of eye 13·6 metres. The chronometer showed 11h 58m 04s and was 1m 03s slow on G.M.T. The ship then steamed 210° T. for 55 miles, when the meridian altitude of the sun's lower limb was 41° 19·8′ south of the observer. Find the ship's position at the time of the meridian altitude.

| Approx. ship's time | 0810 | Chronometer | 11h 58m 04s |
| Longitude | 0349 | Error | 1m 03s |
| Approx. G.M.T. | 1159 | G.M.T. | 11h 59m 07s |

200 PRACTICAL NAVIGATION FOR SECOND MATES

| | | | |
|---|---|---|---|
| G.H.A. 11h | 345° 41·2' | Dec. | 23° 25·3' S. |
| incr. | 14° 46·8' | Lat. | 25° 50·0' N. |
| G.H.A. | 360° 28·0' | Lat. ~ Dec. | 49° 15·3' |
| | = 0° 28·0' | | |

| | | | |
|---|---|---|---|
| Sext. alt | 15° 47·5' | ZX 74° 09·1' nat. hav. | 0·36345 |
| I.E. | −3·0' | Lat. Dec. nat. hav. | 0·17365 |
| Obs. alt. | 15° 44·5' | | 0·18980 |
| Dip. | −6·5' | | |
| | | | 1·27830 |
| App. alt. | 15° 38·0' | Lat. 25° 50·0' log sec | 0·04573 |
| Corr. | +12·9' | Dec. 23° 25·3' log sec | 0·03734 |
| True alt. | 15° 50·9' | ∠P = 57° 17·4' | 1·36137 |
| T.Z.X. | 74° 09·1' | L.H.A. = 302° 42·6' | (long. = G.H.A. − |
| | | G.H.A. = 360° 28·0' | L.H.A.) |
| A ·311+ | | | |
| B ·515+ | | Long. = 57° 45·4' W. | |
| C ·826+ | Az. S. 53·4° E. | | |

| | | |
|---|---|---|
| D.R. pos. at sights | 25° 50·0' N. | 57° 45·4' W. (using |
| Run 210° T. dist. 55 miles | 47·6' S. | 30·5' W. longitude |
| D.R. at noon | 25° 02·4' N. | 58° 15·9' W. calculated) |

*Meridian alt.*

| | | | |
|---|---|---|---|
| Sext. alt. | 41° 19·8' | Mer. pass. | 1157 |
| I.E. | −3·0' | Long. | 0353 |
| Obs. alt. | 41° 16·8' | G.M.T. | 1550 |
| Dip. | −6·5' | Decl. | 23° 25·4' S. |
| App. alt. | 41° 10·3' | D.R. lat. | 25° 02·4' N. |
| Corr. | +15·2' | Obs. lat. | 25° 09·1' N. |
| True alt. | 41° 25·5' | Difference | 6·7' N. |
| T.Z.X. | 48° 34·5' | 'c' | × ·826 |
| Decl. | 23° 25·4' S. | D. long. | 5·5' E. |
| Lat. | 25° 09·1' N | | |

# CALCULATION OF POSITION LINES

Noon D.R. long.   58°15·9′ W.
D. long.                5·5′ E.
                    ──────
Noon longitude   58°10·4′ W.

Noon position 25° 09·1′ N. 58° 10·4′ W.

## EXERCISE 12F

1. On 30th September, 1980, in D.R. latitude 46° 17′ S., the sextant altitude of the sun's lower limb was observed to be 32° 15′, during the forenoon when the G.M.T. was 19h 34m 51s on the 30th. The index error was 3·0′ off the arc and height of eye 11·0 m.
The ship then steamed 300° T. for 45 miles when the sextant meridian altitude of the sun's lower limb was 46° 47·9′ north of the observer. Find the ship's position at the time of the meridian altitude.

2. On 27th June, 1980, in D.R. position 38° 15′ S. 169° 15′ E., at approximate ship's time 0919 hrs., the sextant altitude of the sun's lower limb was 17° 18·2′. Index error was 1·0′ on the arc. Height of eye 8.0 m. The chronometer showed 10h 05m 17s at the time. The ship then steamed 045° T. for 40 miles until noon when the sextant meridian altitude of the sun's lower limb was 28° 39·4′. Find the ship's position at the time of the meridian altitude.

## CHAPTER 13

## LATITUDE BY EX-MERIDIAN PROBLEM

This is another method of finding a point-through which to draw the position line.

In this case if the longitude is assumed, the latitude in that longitude through which the position line passes can be calculated.

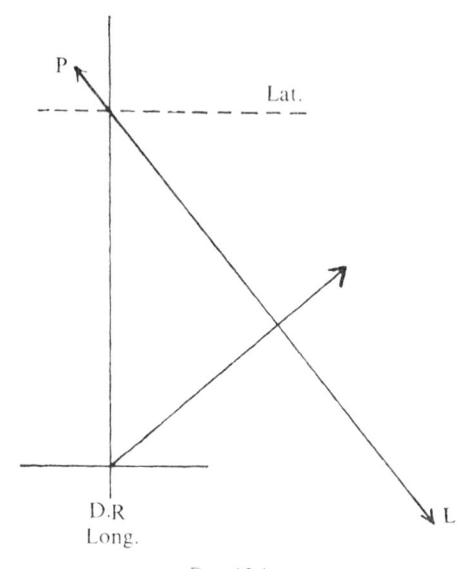

Fig. 13.1

This method is limited to cases where the body observed is near the meridian, i.e., where the hour angle is small. The actual limits of hour angle before or after meridian passage will depend upon the rate of change of altitude.

If the declination of the body and the observer's latitude are the same name, the rate of change of altitude will be greater if the latitude and declination are of opposite names. This means that the limits of hour angle within which this method can be used will be less when the latitude and declination are the same name.

# THE EX-MERIDIAN PROBLEM

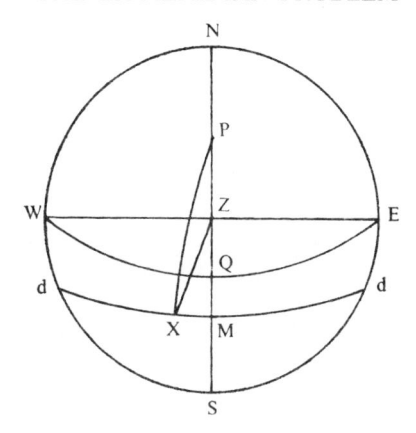

FIG. 13.2

Description of figure:

*NESW* represents the plane of the observer's rational horizon.
*NZS* the observer's meridian.
*P* the elevated celestial pole.
*Z* the observer's zenith.
*X* the body.
*WZE* the prime vertical.
*WQE* the equinoctial.
*M* the position of the body when on the observer's meridian.
*dMXd* the parallel of declination of the body.

From the figure $QZ = ZM - QM$:
   i.e. Latitude = meridional zenith distance − declination.
   and $ZM = PZ \sim PM$
and assuming that the declination remains constant between the time of sight and the time when the body is on the observer's meridian:
$$PM = PX = 90° \pm \text{declination}$$
Then $ZM = PZ \sim PX$
Thus meridional zenith distance $= (PZ \sim PX)$
From the haversine formula:

$$\text{Hav. } P = \frac{\text{hav } ZX - \text{hav } (PZ \sim PX)}{\text{Sin } PZ \text{ Sin } PX}$$

∴ Hav. $P$ sin $PZ$ sin $PX$ = hav. $ZX$ − hav. $(PZ \sim PX)$
∴ Hav. $(PZ \sim PX)$ = hav. $ZX$ − hav. $P$ sin $PZ$ sin $PX$
   i.e. Hav. $MZD$ = hav. $ZX$ − hav. $P$ sin $PZ$ in $PX$
This can be further simplified, so that:
   Hav. $MZD$ = hav. $ZX$ − hav. $P$ cos lat. cos dec.

204   PRACTICAL NAVIGATION FOR SECOND MATES

The procedure is as follows:

1. Using the G.M.T., find the G.H.A. of the body observed from the *Nautical Almanac*, and thence the L.H.A.

2. Correct the sextant altitude to obtain the true altitude and thence the zenith distance.

3. From the formula find the *MZD*.

4. Apply the declination to the *MZD* to obtain the latitude of the point in the D.R. longitude through which the position line passes.

5. Find the azimuth of the body by any convenient method, and thence the position line.

6. Draw the position line on the chart, or state the position.

**Note**

Before deciding on the method to use, if the hour angle is small, it is advisable to verify that the ex-meridian method is appropriate. This can be found from a table in *Norie's* or *Burton's*, which gives the limits of time before and after meridian passage.

**Example 1. By an observation of the sun**

On 19th September, 1980, in D.R. position 45° 40′ S. 52° 35′ W., the sextant altitude of the sun's lower limb near the meridian was observed to be 41° 57·6′. Index error 2·2′ off the arc. Height of eye 12·0 metres. A chronometer showed 04h 01m 20s at the time. Find the direction of the position line and a position through which it passes.

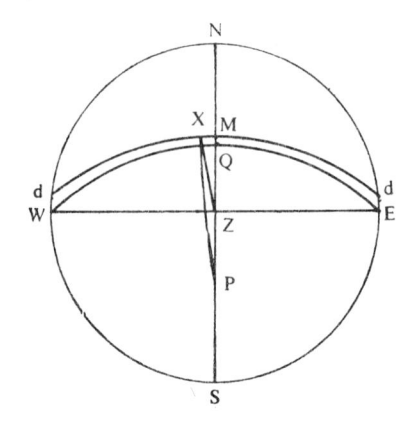

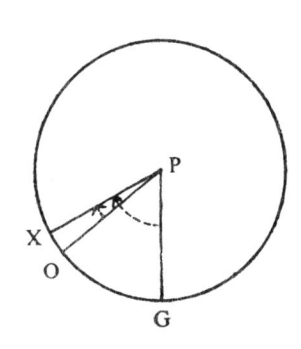

FIG. 13.3

FIG. 13.3(*a*)

## Note
The fact that the sun is near the meridian means that the approximate local time can be taken as 1200.

| | | | |
|---|---|---|---|
| Approx. L.M.T. 19th | 1200 | Chron. | 04h 01m 20s |
| Long. W. | 0330 | | |
| Approx. G.M.T. 19th | 1530 | G.M.T. 19th | 16h 01m 20s |

| | | | |
|---|---|---|---|
| G.H.A. 16h | 61° 35·6′ | Dec. | 1° 15·0′ N. |
| Increment | 0° 20·0′ | 'd' | 0·0′ |
| G.H.A. | 61° 55·6′ | Dec. | 1° 15·0′ N. |
| Long. W. | 52° 35·0′ | | |
| L.H.A. | 9° 20·6′ | | |

| | | | |
|---|---|---|---|
| Sext. alt. | 41° 57·6′ | L.H.A. 9° 20·6′ log hav. | 3̄·821 74 |
| I.E. | +2·2′ | Lat. 45° 40′ log cos | 1̄·844 37 |
| Obs. alt. | 41° 59·8′ | Dec. 1° 15·0′ log cos | 1̄·999 90 |
| Dip. | −6·1′ | | |
| | | | 3·666 01 |
| App. alt. | 41° 53·7′ | | 0·004 63 |
| Corr. | +14·9′ | T.Z.X. 47° 51·4 nat. hav. | 0·164 51 |
| True alt. | 42° 08·6′ | M.Z.X. 47° 08·0′ | 0·159 88 |
| T.Z.X. | 47° 51·4′ | | |

| | | | |
|---|---|---|---|
| Mer. zen. dist. | 47° 08·3′ | A | 6·22+ |
| Dec. | 1° 15·0′ N. | B | ·13+ |
| Lat. | 45° 53·3′ S. | C | 6·35+ |
| | | Az. N. 12·7° W. | |

*Answer*

Position line runs 257·3°/077·3° through position 45° 53·3′ S. 52° 35′ W.

## Ex-meridian tables

There are certain approximations inherent in the ex-meridian method which may be avoided if the sight were worked by the Marcq St. Hilaire method. The ex-meridian method by haversine formula is therefore rarely used in practice, but may be encountered in Department of Trade examinations. The ex-meridian method however is still of practical importance as it may be used to reduce a sight much more rapidly than by Marcq St. Hilaire if ex-meridian tables are used. Examples of these may be found in nautical tables such as Nories' or Burton's.

In both these commonly used tables the tabulation is in two parts, Table I and Table II. Table I gives a factor (called A in Nories' and F in Burton's), which depends upon the latitude and the declination. The factor is extracted from Table I and used as an argument in Table II with the hour angle at the time of sight, to extract the 'reduction'. This is the amount by which the zenith distance at the time of the sight should be reduced to obtain the zenith distance at the time of meridian passage, assuming a stationary observer and a constant declination. The latitude is then found by the usual meridian observation formulae.

*Procedure*

1. Extract the G.H.A. from the almanac, apply longitude and hence find the L.H.A.

2. Correct the sextant altitude and obtain the observed zenith distance.

3. Enter Table I of the ex-meridian tables with D.R. latitude and declination and extract the factor (A or F). Take care to note whether latitude and declination are same name or opposite name and use the appropriate table.

4. With the factor and the hour angle (L.H.A.), enter Table II and extract the reduction.

5. Enter Table III of the ex-meridian tables with the reduction and the altitude and extract a second correction, which is a small correction to be subtracted from the reduction. In most cases this second correction is negligible.

6. Subtract the reduction (with second correction applied if necessary), from the zenith distance at the time of observation to obtain the meridional zenith distance (M.Z.X.).

7. Apply declination to the MZX to obtain latitude as for a meridian observation.

## Example (using Example 1 as worked by haversine method)

From Example 1 L.H.A.=9° 20·6′. Declination=1° 15·0′ N. D.R. Lat.=45° 40′ S. Zenith distance=47° 51·4.

From Table I (lat. and dec. contrary names)
    Factor=1·88 (interpolating to second decimal place).

From Table II for hour angle 9° 20·6′
| | |
|---|---|
| For factor of 1·0 reduction | =23·2 |
| For factor of 0·8 reduction | =18·6 |
| For factor of 0·08 reduction | = 1·9 |
| reduction= | 43·7 |
| From Table III second correction = | ·2 |
| reduction | 43·5 |

THE EX-MERIDIAN PROBLEM 207

       Zenith distance  47° 51·4'
       Reduction           43·5'

       M.Z.X.             47° 07·9'
       Declination       1° 15·0' N

       Latitude          45° 52·9' S.

**Note**

The azimuth must be calculated as in the haversine method. Ex-meridian tables do not give the latitude, but only the latitude in which the position line cuts the D.R. longitude. The answer is therefore:

position line runs 257·3°/077·3° through 45° 52·9' S. 52° 35' W.

## EXERCISE 13A

1. On 18th September, 1980, in D.R. position 49° 00' N. 35° 20' W., the sextant altitude of the sun's lower limb near the meridian was 42° 19·5'. Index error 1·2' off the arc. Height of eye 10·0 metres. A chronometer showed 02h 40m 56s at the time and was correct on G.M.T. Find the direction of the position line and the latitude in which it cuts the D.R. longitude.

2. On 19th September, 1980, in D.R. position 41° 28' N. 28° 40' W., the sextant altitude of the sun's upper limb was observed to be 49° 28' when near the meridian. Index error 0·6' off the arc. Height of eye 12·6 metres. A chronometer showed 01h 15m 59s and was correct on G.M.T. Find the direction of the position line and the latitude in which it cuts the D.R. longitude.

3. On 19th December, 1980, in D.R. position 41° 04' N. 179° 30' E., the sextant altitude of the sun's lower limb when near the meridian was 24° 39·0'. Index error 1·2' on the arc. Height of eye 11·0 metres. A chronometer which was 3m 20s slow on G.M.T. showed 11h 10m 41s at the time. Find the direction of the position line and the latitude in which it cuts the D.R. longitude.

**Example 2. By an observation of a star**

On 26th June, 1980, at ship, in D.R. position, latitude 34° 40' N., longitude 40° 20' W., the sextant altitude of the star *Arcturus* 37 near the meridian during evening twilight was 74° 14·1', index error 0·8' off the arc, height of eye 12·0 m. The chronometer, which was 1m 20s fast on G.M.T., showed 10h 55m 39s. Find the direction of the position line and the latitude in which it cuts the D.R. longitude.

208   PRACTICAL NAVIGATION FOR SECOND MATES

| Approx. L.M.T. 26th | 20m 00s | Chron. | 22h 55m 39s |
| Long. W. | 02m 41s | Error | — 1m 20s |
| Approx. G.M.T. 26th | 22m 41s | G.M.T. 26th | 22h 54m 19s |

From *N.A.* 26th 22h G.H.A. ♈ 245° 10·4'   *Arcturus*
 Incr.  13° 37·0'   S.H.A.* 146° 18·5'
                    Dec.     19° 17·3' N.

G.H.A. ♈ 258° 47·4'
S.H.A.*  146° 18·5'

G.H.A.*  405° 05·9'
          360

G.H.A.*   45° 05·9'
Long. W.  40° 20·0'

L.H.A.    4° 45·9'

| Sext. alt. | 74° 14·1' | P 4° 45·9' log hav. | 3̄·23754 |
| I.E. | +0·8' | Lat. 34° 40·0' log cos | 1̄·91512 |
| Obs. alt. | 74° 14·9' | Dec. 19° 17·3' log cos | 1̄·97491 |
| Dip. | −6·1' | | 3̄·12757 |
| App. alt. | 74° 08·8' | | 0·00134 |
| Corr. | −0·3' | T.Z.X. 15° 51·5' nat. | 0·01903 |
| True alt. | 74° 08·5' | M.Z.X. 15° 17·2' | 0·01769 |
| T.Z.X. | 15° 51·5' | | |

M.Z.X. 15° 17·2'         A 8·30 S.
Dec.   19° 17·3'         B 4·21 N.

Lat.   34° 34·5' N.      C 4·09 S.    Az. S. 16·6° W.

*Answer*
  Position line runs 286·6°/106·6° through 34° 34·5' N. 40° 20' W.
By ex-meridian tables.
  As above L.H.A.=4° 45·9', dec.=19° 17·3' N. D.R. lat.=34° 40' N. ZX=15° 51·5'.

THE EX-MERIDIAN PROBLEM

From Table I (lat. and dec. same names)
$$\text{Factor} = 5 \cdot 75$$

From Table II for hour angle 4° 45·9'
| | |
|---|---|
| For factor of 5·0 reduction | =30·3 |
| For factor of 0·7 reduction | = 4·2 |
| For factor of 0·05 reduction | = 0·3 |
| reduction | =34·8 |
| From Table III second correction | =−0·6 |
| reduction | =34·2 |

| | |
|---|---|
| Zenith distance | 15° 51·5' |
| Reduction | 34·2' |
| M.Z.X. | 15° 17·3' |
| Declination | 19° 17·3' |
| Latitude | 34° 34·6' |

*Answer*

Position line runs 286·6°/106·6° through 34° 34·6' N. 40° 20' W.

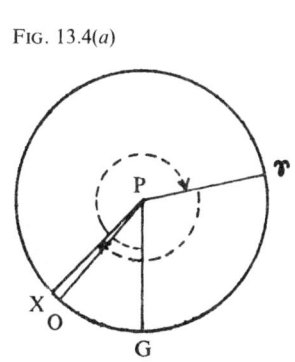

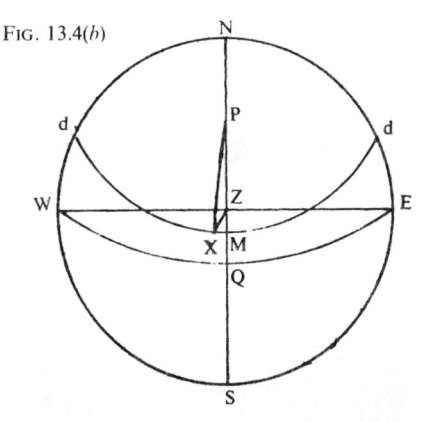

FIG. 13.4(*a*)   FIG. 13.4(*b*)

EXERCISE 13B

**By an observation of a star**

1. On 18th December, 1980, at ship, at about 0625 in D.R. position, latitude 45° 10' N., longitude 136° 02' W., the sextant altitude of the star *Denebola* 28 was 59° 02·5', index error 0·8' on the arc, height of eye 11·0 m., when the chronometer, which was correct on G.M.T., showed 3h 31m 16s. Find the direction of the position line and the latitude in which it cuts the D.R. longitude.

2. In D.R. position, latitude 36° 10' N., longitude 40° 15' W., on 26th June, 1980, an observation of the star *Fomalhaut* 56 near the meridian gave a sextant altitude of 23° 26·4', index error 0·4' off the arc, height of eye 12·0 m. G.M.T. 06h 41m 03s. Find the direction of the position line and the latitude in which it cuts the D.R. longitude.

3. On 19th September, 1980, at ship, in D.R. latitude 25° 44' N., longitude 144° 25' E., the observed altitude of the star *Rigel* 11 near the meridian was 55° 28·3', height of eye 10·5 m., at 19h 26m 02s G.M.T. Find the direction of the position line and the latitude in which it cuts the D.R. longitude.

4. At ship, on 18th December, 1980, in D.R. position, latitude 30° 10' S., longitude 137° 50' W., the sextant altitude of the star *Alphard* 25 near the meridian was 67° 49·1', index error 0·8' on the arc, height of eye 15·4 m. when the chronometer, which was 1m 10s fast on G.M.T., indicated 13h 12m 45s. Find the direction of the position line and the latitude in which it cuts the D.R. longitude.

5. On 18th September, 1980, at ship, in estimated position, latitude 18° 40' S., longitude 120° 25' W., at 13h 15m 28s G.M.T., the sextant altitude of the star *Capella* 12 near the meridian was 25° 29·1', index error 1·4' on the arc, height of eye 17·7 m. Find the direction of the position line and the latitude in which it cuts the D.R. longitude.

# CHAPTER 14

## LATITUDE BY THE POLE STAR

Reference has already been made to the fact that the altitude of the Celestial Pole is equal to the latitude of the observer. If the Celestial Pole could be marked in some way, the latitude of an observer could be obtained at any time simply by finding the altitude.

The star *Polaris* has a declination in excess of 89° N., so that it moves around the Celestial Pole, describing a small circle with an angular radius of less than 1°. As it is so near, it is called the Pole Star, and the altitude can be adjusted by small corrections so that the latitude of the observer can be derived from it.

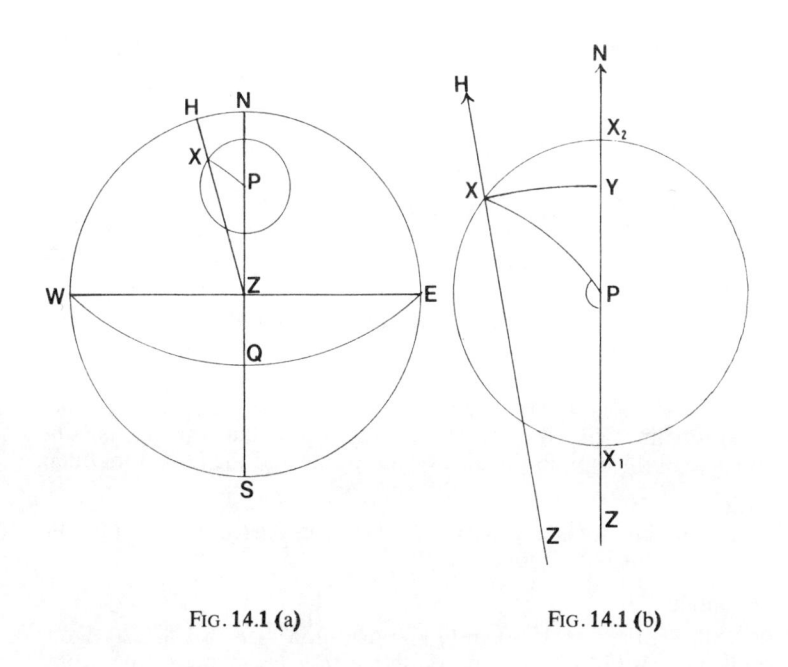

Fig. 14.1 (a)          Fig. 14.1 (b)

It is apparent from figure **14.1**, which represents the daily path of the star about the Celestial Pole, that if the star is at position

$X_1$, the angular distance $PX_1$ must be subtracted from the altitude $NX_1$ to obtain the latitude. Similarly if the star is at position $X_2$, then the polar distance must be added to the altitude. There will be two instants during the star's daily motion around the pole when the altitude of the pole star will be the same as that of the pole. At all other times the correction to apply will be the arc $PY$, in figure 14.1(*b*), and this may be additive as shown, or negative. The solution to the triangle $XPY$ for $PY$ is tabulated in the 'Pole Star Tables' in the *Nautical Almanac*.

The solution is arranged in three separate quantities, $a^0$, $a^1$, and $a^2$. To each is added a constant. The sum of the three constants is 1 degree. This is done to ensure that all values of the three quantities are positive. The 1 degree is subtracted from the final result.

Because the correction depends upon the L.H.A., and because the S.H.A. can be considered constant, the separate corrections are tabulated in the *Nautical Almanac* for values of L.H.A. ♈.

i.e.: L.H.A.* = L.H.A. ♈ + a constant.

The procedure is as follows:

1. Obtain the G.M.T. and then find the L.H.A. ♈ for the time of observation.

i.e. L.H.A. ♈ = G.H.A. ♈ + E. Long.
− W. Long.

2. Correct the sextant altitude for index error, dip and star's total correction, to obtain the true altitude.

3. Using the L.H.A. ♈, from the Pole Star Tables, find the column appropriate to its value. The three corrections and the azimuth will be found in the same column reading in sections down the page.

4. Find the corrections $a_0$, $a_1$ and $a_2$ and add these to the true altitude and subtract 1° from the total to obtain the observer's latitude.

5. From the tabulated azimuths find the bearing of the star. The position line will then lie at right angles to the bearing, passing through a position in the observed latitude and the D.R. longitude.

**Note**

It will be necessary to interpolate for correction $a^0$, but unnecessary for corrections $a^1$ and $a^2$.

**Example**

18th September, 1980, in D.R. position 37° 58′ N., 52° 30′ E., at 01h 30m 24s G.M.T., an observation of *Polaris* gave sextant altitude 38° 40·4′, *i.e.* 2·2′ off the arc, height of eye 11·7 m. Find the latitude and the direction of the position line.

# LATITUDE BY POLE STAR

G.M.T. 18th 01h 30m 24s

| | |
|---|---|
| G.H.A. ♈ | 12° 06·3' |
| Incr. | 7° 37·2' |
| G.H.A. ♈ | 19° 43·5' |
| Long. E. | 52° 30·0' |
| L.H.A. ♈ | 72° 13·5' |
| Sext. alt. | 38° 40·4' |
| I.E. + | 2·2' |
| Obs. alt. | 38° 42·6' |
| Dip − | 6·0' |
| App. alt. | 38° 36·6' |
| Tot. corr. − | 1·2' |
| T. alt. | 38° 35·4' |
| $a_0$ | 0° 20·5' |
| $a_1$ | 0·5' |
| $a_2$ | 0·3' |
| Total | 38° 56·7' |
| | − 1° |
| Latitude | 37° 56·7' N. |

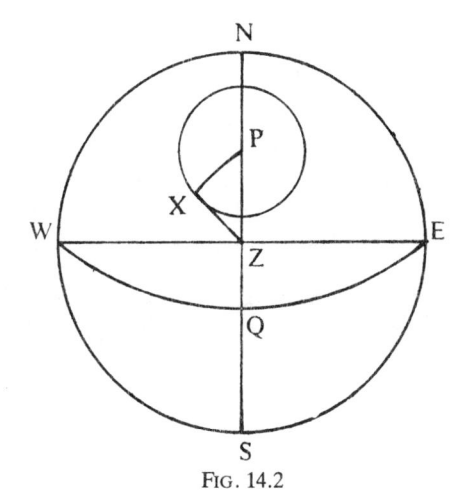

FIG. 14.2

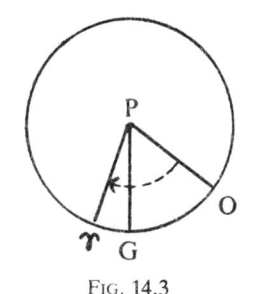

FIG. 14.3

T. Az. 359·3
P.L. 269·3−089·3
P.L. trends 269·3 T.−089·3° T. through latitude 37° 56·7' N., long. 52° 30' E.

## EXERCISE 14A

1. 8th January, 1980, at 19h 45m 22s G.M.T. in D.R. position 49° 10' N., 36° 20·4' W., the sextant altitude of *Polaris* was 50° 09·4', index error 1·6' off the arc, height of eye 12·8 m. Find the latitude and the direction of the position line.

2. 20th September, 1980, in D.R. lat. 35° 25' N., long. 36° 25' W., at 21h 15m 40s G.M.T., the sextant altitude of *Polaris* was 35° 15·8', index error 0·8' on the arc, height of eye 11·5 m. Find the latitude and position line.

3. 26th June, 1980, at ship in D.R. lat. 47° 15′ N., long. 158° 40′ W., the sextant altitude of *Polaris* was 47° 42′, index error 1·4′ off the arc, height of eye 6·0 m., at 13h 26m 44s G.M.T. Find the latitude and position line.

4. 4th January, 1980, p.m., at ship in D.R. position, lat. 22° 40′ N., long. 163° 20′ W. at 04h 58m 20s G.M.T., the sextant altitude of *Polaris* was 23° 40·4′, index error 0·8′ on the arc, height of eye 13·2 m. Find the latitude and position line.

5. On 30th September, 1980, at about 0520, at ship in D.R. lat. 50° 40′ N., long. 162° 10·8′ E. when the chronometer, which was 2m 08s slow on G.M.T., showed 6h 13m 17s, the sextant altitude of *Polaris* was 51° 10·8′, index error 1·2′ off the arc, height of eye 14·0m. Find the latitude and position line.

6. 19th September, 1980, in D.R. position, lat. 32° 05′ N., long. 31° 20′ E., at 03h 00m 21s G.M.T., the sextant altitude of *Polaris* was 32° 44·2′, index error 1·6′ off the arc, height of eye 13·2 m. Find the latitude and position line.

7. 26th June, 1980, in long. 57° 02′ W. at 23h 51m 14s G.M.T., the sextant altitude of the star *Polaris* was 40° 35·4′ index error 0·6′ on the arc, height of eye 10·5 m. Find the latitude and the position line.

## CHAPTER 15
## GREAT CIRCLE SAILING

This method of sailing between two positions on the earth's surface is used over long ocean passages. Its use involves a knowledge of spherical trigonometry, and this knowledge is assumed. If necessary a text on this subject should be consulted, for the use of the spherical haversine formula, and Napier's rules for the solution of right-angled and quadrantal spherical triangles.

Between any two positions on the surface of a sphere, unless the two positions are at opposite ends of a diameter, there is one only great circle that can be drawn through the two positions. The track along the shorter arc of this great circle is the shortest distance along the surface of the sphere between those two positions.

The main disadvantage in steaming such a track along the surface of the earth is that the course is constantly changing, and to attempt to make good a great circle a ship must steer a series of short mercator courses which correspond approximately to the curve of the great circle. The course must be altered at frequent intervals. The problem becomes initially to find the distance over a great circle track and then to find the course at the departure point, and the course at a series of positions along the track. These positions become the alter course positions.

To solve this problem a spherical triangle is formed by the intersection of the three great circles:
  (i) The great circle track,
  (ii) The meridian through the departure point,
  (iii) The meridian through the arrival point.

Thus the three points of the triangle are the two positions sailed between and one of the poles of the earth, usually the nearest pole.

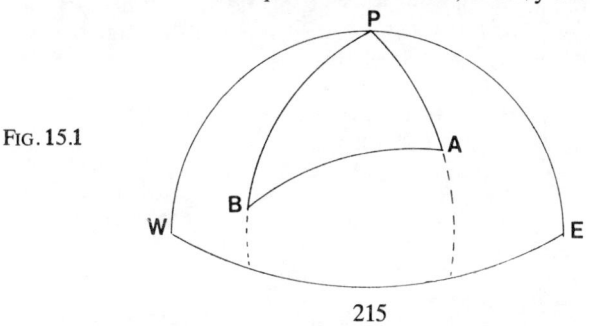

FIG. 15.1

In the figure $PA$ = Colat. of $A = 90° -$ lat. $A$
$PB$ = Colat. of $B = 90° -$ lat. $B$
$\angle P$ = D. long. between the two positions
$WE$ = Equator

Thus using the haversine formula:

Hav. $p =$ (hav. $P$. sin. $PA$. sin. $PB$) + hav. $(PA \sim PB)$

Thus hav. dist. = (hav. d. long. cos. lat. $A$. cos. lat. $B$) + hav. d. lat.

The initial course is then found from angle $A$:

Hav. $A =$ {hav. $PB -$ hav. $(AB \sim PA)$} × cosec. $AB$ cosec $AP$

The final course may be found by calculating angle $B$.

## The vertex of a great circle

This is the point on the great circle which is closest to the pole. Thus every great circle will have a northerly vertex and a southerly vertex.

## To find the position of the vertex

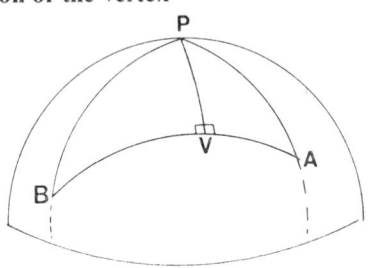

Fig. 15.2

In figure 15.2 let $V$ be the vertex of the great circle through $A$ and $B$. At the vertex the great circle runs in a direction 090°/270°. Thus it cuts the meridian through the vertex at right angles.

Thus $\angle PVA = \angle PVB = 90°$.

Solving the right-angled triangle by Napier's rules for $PV$ and $\angle P$ will give the latitude of the vertex and the d. long. between the vertex and position $A$ respectively.

Given  $\angle A$ = initial course
$PA$ = colat. of $A$
Sin. $PV$ = sin. $PA$. sin. $\angle A$
and  Cot. $\angle P$ = cos. $PA$. tan. $\angle A$

Having found the position of the vertex, a series of positions along the track can be found and the course of the great circle at each of these positions calculated, thus:

Assume longitudes at regular intervals of d. long. and solve a triangle $PVX$ in figure 15.3.

## GREAT CIRCLE SAILING

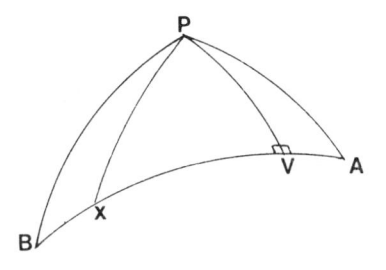

FIG. 15.3

Where $X$ is the position where the meridian of the assumed longitude cuts the great circle track.

Thus given angle $\angle P$ = d. long. between assumed longitude and the vertex

$PV$ = colat. of the vertex
Cot. $PX$ = cos. $<P$. cot. $PV$
and Cos. $\angle X$ = cos. $PV$. sin. $\angle P$

Solving for $PX$ and $\angle X$ will give the latitude corresponding to the assumed longitude, and the great circle course at that point respectively.

This can be done for a number of assumed longitudes as required, the working being tabulated as in the following example.

**Note**

The vertex of the great circle need not be between the two given positions. If it is not, then either the angle at $A$ or that at $B$ will be greater than 90°, and the course will lie in the same quadrant between the two positions. If the vertex lies between the two positions then the course will change quadrants at the vertex. After solving the angles at the positions then the course must be found by inspecting the triangle.

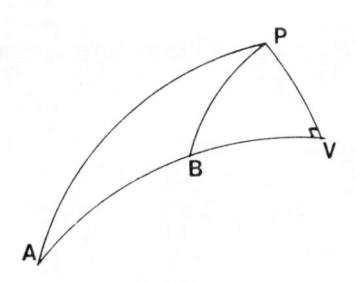

FIG. 15.4

## Example

Find the total distance, the initial course, of the great circle track between:

$$A \quad 41° \ 00' \ S. \quad 175° \ 00' \ E.$$
$$B \quad 33° \ 00' \ S. \quad \ \ 71° \ 30' \ W.$$

Find also the latitudes at which the meridians of 90° W., 110° W., 130° W., 150° W., and 170° W. cut the great circle track, and the course at these points.

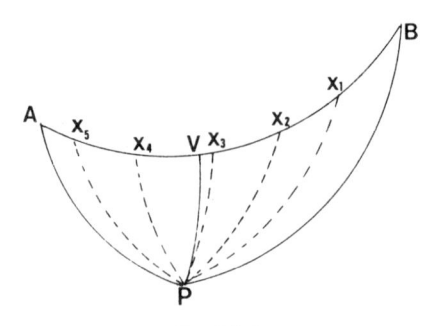

Fig. 15.5

| | | | |
|---|---|---|---|
| $A$ | 41° 00′ S. | 175° 00′ E. | |
| $B$ | 33° 00′ S. | 71° 30′ W. | |
| d. lat. | 8° 00′ | 113° 30′ | = d. long. = $\angle P$ |

Hav. dist. = (hav. d. long. cos. lat. $A$. cos. lat. $B$) + hav. d. lat.
Hav. dist. = (hav. 113° 30′. cos. 41°. cos. 33°) + hav. 8°

|  | Number | Log |
|---|---|---|
|  | hav. d. long. | $\bar{1}$·84471 |
|  | cos. 33° | $\bar{1}$·92359 |
|  | cos. 41° | $\bar{1}$·87778 |
| Dist. = 83° 58·6′ |  |  |
| = 5038·6 miles |  | $\bar{1}$·64608 |
|  |  | 0·44267 |
|  | nat. hav. 8° | 0·00487 |
|  |  | 0·44754 |

# GREAT CIRCLE SAILING

## To find initial course

Hav. $A = \{\text{hav. } PB - \text{hav. } (AB \sim PA)\}$ cosec. $AB$ cosec. $PA$

Hav. $A = (\text{hav. } 57° - \text{hav. } 34° \; 58 \cdot 6') $ cosec. $83° \; 58 \cdot 6'$. cosec. $49°$

|  | Number | o |
|---|---|---|
| nat. hav. 57° | 0·22768 | |
| 34° 58·6' | 0·09031 | |
| | 0·13737 | |
| | $\overline{1}$·13790 | |
| cosec. 49° | 0·12222 | |
| cosec. 83° 58·6' | 0·00240 | |
| | $\overline{1}$·26252 | |

$A = 50° \; 39 \cdot 5'$
Course = S. 50·5° E.

## To find the position of the vertex

In triangle $APV$
Sin. $PV$ = sin. $\angle A$. sin. $PA$
= sin. 50° 39·5'. sin. 49°
$PV$ = 35° 42·6'
Lat. of vertex = 54° 17·4' S.

| Number | Log |
|---|---|
| Sin. 50° 39·5' | $\overline{1}$·88839 |
| Sin. 49° | $\overline{1}$·87778 |
| | $\overline{1}$·76617 |

And cot. $\angle P$ = cos. $AP$. Tan $\angle A$
= cos. 49°. tan. 50° 36·5'
$P$ = 51° 22·7'
Long. of vertex = 133° 37·3' W.

| Number | Log |
|---|---|
| Cos. 49° | $\overline{1}$·81694 |
| Tan. 50° 36·5' | 0·08557 |
| | $\overline{1}$·90251 |

## To find the latitude where the track cuts the given longitudes and the course at those points

In $VPX$ Cot. $PX$ = cos. $\angle P$ cot. $PV$

220  PRACTICAL NAVIGATION FOR SECOND MATES

|  | $X_1$ | $X_2$ | $X_3$ | $X_4$ | $X_5$ |
|---|---|---|---|---|---|
| Longitude | 90°W | 110°W | 130°W | 150°W | 170°W |
| ∠P | 43° 37·3′ | 23° 37·3′ | 3° 37·3′ | 16° 22·7′ | 36° 22·7′ |
| Cos ∠P | 1̄·85968 | 1̄·96199 | 1̄·99913 | 1̄·98201 | 1̄·90586 |
| Cot PV | 0·14337 | 0·14337 | 0·14337 | 0·14337 | 0·14337 |
| Cot PX | 0·00305 | 0·10536 | 0·14250 | 0·12538 | 0·04923 |
| PX = | 44° 48′ | 38° 07′ | 35° 45·8′ | 36° 50·5′ | 41° 45·5′ |
| Latitude | 45° 12′ S | 51° 53′ S | 54° 14·2′ S | 53° 09·5′ | 48° 14·5′ |
| and Cos X = Cos PV Sin ∠P | | | | | |
| Sin ∠P | 1̄·83878 | 1̄·60282 | 2̄·80050 | 1̄·45022 | 1̄·77314 |
| Cos PV | 1̄·90954 | 1̄·90954 | 1̄·90954 | 1̄·90954 | 1̄·90954 |
| Cos ∠X | 1̄·74832 | 1̄·51236 | 2̄·71004 | 1̄·35976 | 1̄·68268 |
| ∠X | 55° 56′ | 71° 01′ | 87° 04′ | 76° 46′ | 61° 13′ |
|  | N 55° 56′ E | N 71° 01′ E | N 87° 04′ E | S 76° 46′ E | S 61° 13′ E |

*Answer*

Position on g.c.   Course
48° 14·5′ S.  170° W.   S. 61·25° E.
53° 09·5′ S.  150° W.   S. 76·75° E.
54° 14·2′ S.  130° W.   N. 87° E.
51° 53·0′ S.  110° W.   N. 71° E.
45° 12·0′ S.   90° W.   N. 56° E.

**Example 2**

Find the distance, and the initial course and the position of the vertex between the following positions.

$A$   48° 30·0′ N.    5° 10·0′ W.
$B$   22° 00·0′ S.   40° 40·0′ W.

d. lat.  70° 30·0′ S.   35° 30′ W. = d. long.

Hav. dist. = (hav. 35° 30′. cos. 48° 30′. cos. 22°) + hav. 70° 30′

| | Number | Log |
|---|---|---|
| Dist. = 77° 18·8′ = 4638·8′ = 4638·8 miles | hav. 35° 30′<br>cos. 48° 30′<br>cos. 22° | 2̄·96821<br>1̄·82126<br>1̄·96717 |
| | | 2̄·75664 |
| Figure will be similar to fig. **15.4** | | 0·05710<br>0·33310 |
| | | 0·39020 |

# GREAT CIRCLE SAILING

Hav. $\angle A =$ {hav. $PB -$ hav. $(AB \sim AP)$}. cosec. $AB$. cosec. $AP$

$\phantom{\text{Hav. } \angle A} = ($hav. $112° - $hav. $35° \; 48 \cdot 8')$ cosec. $77° \; 18 \cdot 8'$

$\phantom{\text{Hav. } \angle A = ($hav. $112° - $hav. $35° \; 48 \cdot 8')} $ cosec $41° \; 30'$

$\angle A \quad = 146° \; 31'$
Course $= $ S. $33° \; 29'$ W.

| Number | Log |
|---|---|
| hav. $112°$ | 0·68730 |
| hav. $35° \; 48 \cdot 8'$ | 0·09453 |
| | 0·59277 |
| | $\overline{1}$·77289 |
| cosec. $77° \; 18 \cdot 8'$ | 0·01074 |
| cosec. $41° \; 30'$ | 0·17874 |
| | $\overline{1}$·96237 |

**To find the position of the vertex**

In triangle $PVA$
$\angle A \quad = 33° \; 29'$
$PA \quad = 41° \; 30'$
and by Napier's rules
Sin. $PV = $ sin. $\angle A$. sin. $PA$
$\phantom{\text{Sin. } PV} = $ sin. $33° \; 29'$. sin. $41° \; 30'$
$PV \quad = 21° \; 26 \cdot 6' = $ co-lat. of vertex
Latitude $= 68° \; 33 \cdot 4'$ N.

| Number | Log |
|---|---|
| Sin. $33° \; 29'$ | $\overline{1}$·74170 |
| Sin. $41° \; 30'$ | $\overline{1}$·82126 |
| | $\overline{1}$·56296 |

and Cos. $PA = $ cot. $\angle P$. cot. $\angle A$
Cot. $\angle P = $ cos. $PA$. tan. $\angle A$
$\phantom{\text{Cot. } \angle P} = $ cos. $41° \; 30'$. tan. $33° \; 29'$
$\angle P \quad = 63° \; 38 \cdot 7'$

| Number | Log |
|---|---|
| Cos. $41° \; 30'$ | $\overline{1}$·87446 |
| Tan. $33° \; 29'$ | $\overline{1}$·82051 |
| | $\overline{1}$·69497 |

Longitude of vertex $= \phantom{0}5° \; 10 \cdot 0'$ W.
$\phantom{\text{Longitude of vertex} = \;} 63° \; 38 \cdot 7'$ E.

$\phantom{\text{Longitude of vertex} = \;} \overline{58° \; 28 \cdot 7' \; \text{E.}}$

Position of vertex $68° \; 33 \cdot 4'$ N.   $58° \; 28 \cdot 7'$ E.

## EXERCISE 15A

1. Find the initial course, the final course and the distance by great circle sailing from:

$$\begin{array}{ccc} A & 55°\ 25'\ \text{N.} & 7°\ 12'\ \text{W.} \\ \text{to } B & 51°\ 12'\ \text{N.} & 56°\ 10'\ \text{W.} \end{array}$$

2. Find the great circle distance and the initial course, and the position of the vertex of the great circle from:

$$\begin{array}{ccc} A & 34°\ 55'\ \text{S.} & 56°\ 10'\ \text{W.} \\ \text{to } B & 33°\ 55'\ \text{S.} & 18°\ 25'\ \text{E.} \end{array}$$

3. Find the saving in distance by steaming a great circle track as opposed to a mercator track from:

$$\begin{array}{ccc} A & 43°\ 36'\ \text{S.} & 146°\ 02'\ \text{E.} \\ \text{to } B & 26°\ 12'\ \text{S.} & 34°\ 00'\ \text{E.} \end{array}$$

4. Find the distance, the initial course, and the positions where the meridians of 140° W., 160° W., 180°, 160° E. cut the track, and the courses at these positions, on a great circle from:

$$\begin{array}{ccc} A & 48°\ 24'\ \text{N.} & 124°\ 44'\ \text{W.} \\ \text{to } B & 34°\ 50'\ \text{N.} & 139°\ 50'\ \text{E.} \end{array}$$

## Composite great circle sailing

If the vertex of a great circle lies between the two positions involved, then the great circle track will take the vessel into a higher latitude than either of the two positions. In some circumstances this may not be desirable. Due regard must be had to the weather conditions likely to be encountered.

In a composite great circle sailing a limit is put on the latitude to which a vessel goes, the limit being decided by the navigator. An example of the circumstances in which such a track might be employed is the voyage between Cape of Good Hope and Australia, in which case the great circle would reach very high southerly latitudes.

The track now becomes a great circle track from the departure position to the parallel of the limiting latitude, a parallel sailing along this parallel, and then another great circle track from the limiting latitude to the destination position.

The great circle to be taken from the departure position is that great circle which has its vertex in the limiting latitude. It will therefore form a right angle with the meridian through the vertex, and all triangles can be solved with Napier's rules. The same applies to the track from the limiting latitude to the arrival position.

# GREAT CIRCLE SAILING

Figure 15.6 represents such a track.

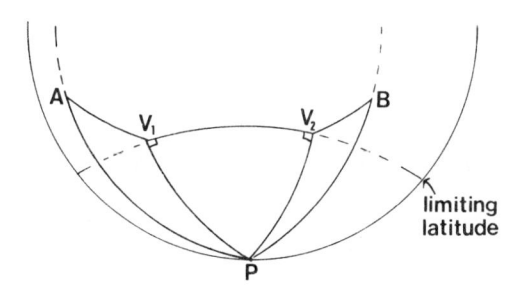

FIG. 15.6

Description of figure:

| | |
|---|---|
| $A$ | = Departure position |
| $B$ | = Arrival position |
| $AP$ | = Co-lat. of $A$ |
| $BP$ | = Co-lat. of $B$ |
| $V_1$ | = Vertex of great circle $AV_1$ |
| $V_2$ | = Vertex of great circle $BV_2$ |
| $V_1 V_2$ | = Parallel of the limiting latitude |
| $V_1 P$ | = $V_2 P$ = Co-lat. of limiting latitude |

## To find total distance and initial course

The right-angled triangle $AV_1 P$ can be solved using $AP$ and $V_1 P$, to find $AV_1$:

$$\text{Cos. } AV_1 = \text{cos. } AP . \text{ sec. } V_1 P$$

to find $\angle A$:

$$\text{Sin. } \angle A = \text{sin. } V_1 P . \text{ cosec. } AP$$

to find $\angle P$:

$$\text{Cos. } \angle P = \text{tan. } V_1 P . \text{ cot. } AP$$

The right-angled triangle $BV_2 P$ can be solved using $V_2 P$, and $BP$, with the same formulae, to find $V_2 B$, and $\angle P$.

$V_1 V_2$ can then be found by the parallel sailing formula;

$$\text{dep. (dist.)} = \text{d. long.} \times \text{cos. latitude}$$

The d. long. being that between $V_1$ and $V_2$ which is found by

d. long. between $A$ and $B - \angle APV_1 - \angle BPV_2$

The problem of finding a series of positions along the two great circles $AV_1$ and $V_2 B$, and the course of those positions is solved in the same way as in the normal great circle sailing problem.

224  PRACTICAL NAVIGATION FOR SECOND MATES

**Example**

Find the initial course and the total distance along a composite great circle track from:

$A$    35° 00′ S.    20° 00′ E.
to $B$    43° 40′ S.    146° 50′ E.

It is required not to go south of latitude 48°.
Give also the longitude of each vertex.

$A$    35° 00·0′ S.    20° 00·0′ E.
$B$    43° 40·0′ S.    146° 50·0′ E.

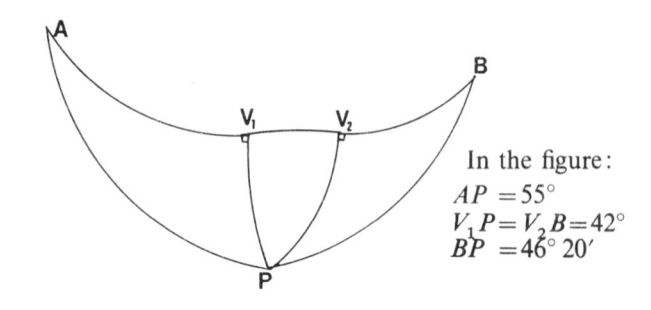

126° 50·0′ = d. long.

In the figure:
$AP = 55°$
$V_1P = V_2B = 42°$
$BP = 46° 20′$

FIG. 15.7

In triangle $APV_1$
By Napier's rules
  Cos. $AV_1$ = cos. $AP$. sec. $V_1P$
         = cos. 55° sec. 42°
$AV_1$     = 39° 29′
        = 2369′

| Number | Log |
|---|---|
| Cos. 55° | $\bar{1}$·75859 |
| Sec. 42° | 0·12893 |
| | $\bar{1}$·88752 |

Sin. ∠$A$ = sin. $V_1P$ cosec. $AP$
       = sin. 42° cosec. 55°
∠$A$    = 54° 46·3′

| | |
|---|---|
| Sin. 42° | $\bar{1}$·82551 |
| Cosec. 55° | 0·08664 |
| | $\bar{1}$·91215 |

Cos. $P$  = cot. $AP$ tan. $V_1P$
        = cot. 55° tan. 42°
∠$P$    = 50° 54·9′

| | |
|---|---|
| Cot. 55° | $\bar{1}$·84523 |
| Tan. 42° | $\bar{1}$·95444 |
| | $\bar{1}$·79967 |

## GREAT CIRCLE SAILING

In triangle $BPV_2$
Cos. $BV_2$ = cos. 46° 20′ sec. 42°
$BV_2$ = 21° 42′
 = 1302′

| Number | Log |
|---|---|
| Cos. 46° 20′ | $\bar{1}\cdot 83914$ |
| Sec. 42° | $0\cdot 12893$ |
| | $\bar{1}\cdot 96807$ |

Cos. $\angle P$ = cot. 46° 20′ tan. 42°
$\angle P$ = 30° 44·8′

| Number | Log |
|---|---|
| Cot. 46° 20′ | $\bar{1}\cdot 97978$ |
| Tan. 42° | $\bar{1}\cdot 95444$ |
| | $\bar{1}\cdot 93422$ |

Thus in triangle $V_1 V_2 P$
$P$ = 126° 50′ − 50° 54·9′ − 30° 44·8′
 = 45° 10·3′
 = 2710·3′

and $V_1 V_2$ = d. long. × cos. lat.
 = 2710·3 × cos. 48°
 = 1813·5

| Number | Log |
|---|---|
| 2710·3 | $3\cdot 43302$ |
| Cos. 48° | $\bar{1}\cdot 82551$ |
| | $3\cdot 25853$ |

Total distance = $AV_1 + V_1V_2 + V_2B$
 = 2369 + 1813·5 + 1302
 = 5484·5 miles
Initial course = $\angle V_1AP$ = S. 54° 46·3′ E.
Long. of $V_1$ = 20° 00′ E. + $\angle APV_1$
 = 70° 54·9′ E.
Long. of $V_2$ = 146° 50′ E. − $\angle V_2PB$
 = 116° 05·2′ E.

### EXERCISE 15B

1. A composite great circle track from Montevideo (34° 55′ S. 56° 10′ W.) to Cape Town (33° 55′ S. 18° 25′ E.) is required with a limiting latitude of 38° S. Find the total distance to steam and the initial course.

2. Find the total distance, the initial course, and the longitudes where the track reaches and leaves the limiting latitude, on the composite great circle from:
$A$   $A$   26° 12′ S.   34° 00′ E.
  to $B$   43° 36′ S.   146° 02′ E.   Limiting latitude 45° S.

3. Find the total distance, and initial course on the composite great circle, with limiting latitude 49° S., from:

$A$   45° 30′ S.   71° 37′ W.
to $B$   46° 40′ S.   168° 20′ E.

## SPECIMEN PAPERS

The following papers are typical of the practical navigation papers set for Department of Trade Class IV certificates. The time allowed is 2 hours and all questions must be attempted.

## PAPER 1

1. From the following information find the direction of the position line and a position through which the position line passes.

Time at ship 0805 on 20th September 1980
D.R. 5° 58′ S. 126° 03′ E.
Sextant altitude of Saturn 40° 49·5′
Index error 1·5′ off the arc
Height of eye 14·5 metres
Chronometer showed 0h 27m 38s.
Chron. error nil.

2. Find the G.M.T. and L.M.T. of meridian passage of the star *Vega*, and the setting to put on a sextant to observe this passage.

Date at ship 19th September 1980
D.R. 13° 00′ S. 138° 55′ E.
Index error 1·8′ off the arc
Height of eye 17·0 metres

3. Find by mercator sailing the true course and distance from 48° 11′ S. 169° 50′ E. to 23° 36′ S. 161° 42′ W.

4. From the following information find the compass error and the deviation for the ship's head.

Time at ship 1004 on 18th December 1980
D.R. 55° 08′ N. 5° 13′ E.
Sun bore 162° by compass
Chronometer showed 10h 2m 17s
  error 1m 40s fast on G.M.T.
Variation 7° W.

## PAPER 2

1. From the following information find the direction of the position line and a position through which it passes:

Time at ship 1930 on 26th June 1980   D.R. 33° 05' N. 131° 18' W.
Sextant altitude of *Regulus*
  35° 54·4'
Index error 1·0' on the arc   Height of eye 16·7 metres
Chronometer showed 4h 12m 13s
  and was fast on G.M.T. by 1m 3s

2. From the following information find the direction of the position line and the latitude in which it cuts the D.R. longitude.

Time at ship 1210 on 7th January   D.R. 26° 17' N. 48° 11' W.
  1980
Sextant altitude of sun's lower limb
  near the meridian 41° 16·9'
Index error 2·0' on the arc   Height of eye 13 metres
Chronometer showed 3h 18m 06s
  error 7m 14s fast on G.M.T.

3. From the following observation of *Polaris* during evening twilight find the latitude:

Date at ship 27th June 1980   D.R. 21° 03' N. 153° 16' W.
Sextant altitude 20° 15'
Index error 2·0' on the arc   Height of eye 11·0 metres
Chronometer showed 5h 27m 42s.
  Error 3m 29s slow on G.M.T.

4. From the following sights find the position of the ship at the time of the second observation:

E.P. 40° 12' S. 94° 30' E. Observed longitude 94° 33' E. Az. 079° T. Run 3·5 hours at 16 knots. Course 352° T. Current 260° T. at 2 knots throughout. Using E.P. run up intercept 3·0' towards Az. 012° T.


## PAPER 3

1. From the following information find the direction of the position line and a position through which it passes:

Time at ship 1429 on 2nd October   D.R. 47° 30' N. 45° 20' W.
  1980
Sextant altitude of sun's lower limb
  27° 41·3'
Index error 2·0' on the arc   Height of eye 6·0 metres
Chronometer showed 5h 31m 13s.
  Error 1m 03s fast on G.M.T.

2. From the following meridian observation find the latitude and state the direction of the position line:

Date at ship 9th January 1980     D.R. 46° 28' S. 136° 30' E.
Sextant altitude moon's lower limb
   40° 15·7'     Height of eye 16·5 metres
   Index error nil

3. Find by traverse table the vessel's position at the end of the fourth course.

|  |  |  |
|---|---|---|
| Initial position | 46° 45' N | 45° 00' W. |
| First course | 202° T. distance | 85 miles |
| Second course | 272° T distance | 63 miles |
| Third course | 337° T. distance | 40 miles |
| Fourth course | 050° T. distance | 36 miles |

4. From the following information find the error of the compass and the deviation for the ship's head:

Date at ship 1st October 1980     D.R. 10° 50' N.
Sun bore 273° by compass at setting     157° 17' W.
Variation 4° E.

## PAPER 4

1. From the following information find the direction of the position line and a position through which it passes:

Time at ship 0850 on 9th January 1980     D.R. 42° 10' N.
Sextant altitude moon's upper limb     50° 05' W.
   28° 09·6'
Index error 1·0' on the arc     Height of eye 8·2 metres
Chronometer showed 11h 59m 01s.
   Error 0m 48s slow on G.M.T.

2. From the following observation find the direction of the position line and the latitude in which it cuts the D.R. longitude:

Time at ship 0508 on 19th September     D.R. 38° 40' S.
   1980     138° 46' E.
Sextant altitude of *Bellatrix* near the
   meridian 44° 50'
Index error 3·0' off the arc     Height of eye 10 metres
Chronometer showed 8h 12m 19s.
   Error 2m 18s fast on G.M.T.

3. From the following observation find the latitude and state the direction of the position line:

Date at ship 20th September 1980     D.R. 26° 00' N. 116°
Sextant altitude of the sun's lower limb     37' W.
   64° 45'
Index error 1·5' on the arc     Ht. of eye 17·9 metres

230 PRACTICAL NAVIGATION FOR SECOND MATES

4. From the following sights find the position of the ship at the time of the second observation:

Time 1300; E.P. 23° 57' N. 92° 07' W.; Intercept 3·0' Towards, Az. 287°T.; Run 95 miles; Course 147°T.
Time 1830; Using E.P. run up, intercept 5·0' Away, Az. 030°T.

## PAPER 5

1. On 18th December, 1980, in D.R. position 29° 15' S. 134° 18' E. at approximately 1950 L.M.T., the sextant altitude of the star *Aldebaran* was observed to be 28° 41·0'. Index error 1·0' on the arc. Height of eye 13·8 metres. A chronometer which was 1m 09s slow on G.M.T. showed 10h 57m 40s at the time. Find the direction of the position line and a position through which it passes.

2. On 8th January, 1980, in D.R. position 30° 58' S. 151° 46' W., an observation of the moon's lower limb when on the meridian gave sextant altitude 55° 04·4'. Index error 0·9' off the arc. Height of eye 15·0 metres. Find the latitude and state the direction of the position line.

3. A vessel steams at 15·5 knots for 24 hours on a course of 064°T. If the departure position is 20° 18·4' S, 175° 50·0' W., find the position arrived at.

4. An observation in D.R. position 32° 30' N. 32° 15' W. gave a longitude of 32° 08' W. when worked by longitude by chronometer. Bearing of body observed was 060°T. The vessel then steamed 070°T. for 40 miles when a meridian observation gave a latitude of 32° 46·0' N. Find the vessel's position at the time of the meridian observation.

## PAPER 6

1. From the following observation find the direction of the position line and the longitude in which it cuts the D.R. latitude:

Approximate time at ship 0840 on 28th September, 1980.
D.R. position 30° 40' N. 175° 18' E.
Sextant altitude sun's lower limb, 33° 35·0.
Index error 1·0' off the arc. Height of eye 10·3 metres.
Chronometer 08h 50m 05s, correct on G.M.T.

2. From the following meridian observation find the latitude and state the direction of the position line.

Date at ship 6th January, 1980  D.R. long. 96° 35' W.
Sextant altitude sun's upper limb bearing north, 61° 25·0'.
Index error 1·4' on the arc  Height of eye 11·5 m

## SPECIMEN PAPERS

3. From the following observation of *Polaris* find the latitude and the direction of the position line:

Date at ship 27th June, 1980
D.R. position 47° 15′ N 125° 40′ W.
Sextant altitude 47° 52′
Height of eye 6·1 metres
Chronometer 01h 20m 44s
   error nil

Approx. L.M.T. 0500

I. E. 1·4′ off the arc

4. From the following data find the position of the ship at the time of the third observation;

Course 071°T., speed 20 knots. D.R. 42° 11′ S. 161° 17′ E.

Time 1731, star A using D.R. given, intercept 5·8′ Towards Az. 026°T.

Time 1737, star B using D.R. given, intercept 2·9′ Away, Az. 272°T.

Time 1746, star C using D.R. given, intercept 1·7′ Towards, Az. 319°T.

# ANSWERS

### Exercise 1A
1. 425' N.  709' W.
2. 910' N.  635' E.
3. 930' S.  741' W.
4. 2026' N.  522' E.
5. 741' N.  1278' W.
6. 1005' S.  300' E.
7. 995' N.  3712' W.
8. 2910' N.  4425' E.
9. 1508' N.  8226' W.
10. 2983' N.  3516' E.

### Exercise 1B
1. 2° 46' W.
2. 12° 24' N.  165° 34' W.
3. 43° 37' N.  17° 46' E.
4. 42° 08·2' N.  34° 14·4' W.
5. 17° 45·1' S.  170° 59·5' E.

### Exercise 2A
1. 15° E.
2. 19° E.
3. 33° W.
4. 30° W.
5. 14° E.
6. 15° W.
7. 17° W.
8. Nil.
9. 55° W.
10. 38° E.

### Exercise 2B
1. 207°
2. 351°
3. 345°
4. 283°
5. 022°
6. 318°
7. 106°
8. 204°
9. 096°
10. 195°

### Exercise 2C
1. 203°
2. 021°
3. 187°
4. 199°
5. 359°
6. 087°
7. 118°
8. 178°
9. 319°
10. 198°

### Exercise 2D
1. 24° E.
2. 9° E.
3. 3° E.
4. 4° W.
5. 2° W.
6. 5° E.
7. 10° E.
8. 16° E.
9. 6° E.
10. 5° W.

### Exercise 2E
1. 2° W.
2. 6° E.
3. 4° E.
4. 1° W.
5. 7° W.
6. 8° W.
7. 5° E.
8. 3° W.
9. 2° E.
10. 12° W.
11. 15° W.
12. 7° E.
13. 25° W.
14. 4° W.
15. 5° W.

### Exercise 2F
1. 6° E., 20° W.
2. 217°, 5° E.
3. 284°, 262°
4. 5° W., 15° E.
5. 245°, 230°
6. 172°, 12° E.
7. 346°, 348°
8. 280°, 275°
9. 3° E., 25° W.
10. 201°, 175°
11. Nil., 42° E.
12. 2° W., Nil.

### Exercise 2G
1. 118½° C.
2. 041¼° C.
3. 350° C.
4. 243¼° C.
5. 163¾° C.

### Exercise 2H
1. 049°
2. 121°
3. 259°
4. 322°
5. 105°
6. 107°
7. 013°
8. 178°
9. 240°
10. 250°

# ANSWERS

### Exercise 2I
1. 152°19·6 knots
2. 295½°13·36 knots
3. 291°2·1 knots
4. 343°T.
5. 177°14·2 knots

### Exercise 4A
1. Lat. 41° 24·6′ N. or S.
2. Lat. 70° 31·9′ N. or S.
3. D. long. 9° 22′ E. or W.
4. Dist. 348·5′
5. Lat. 56° 26·75′ N.
6. Lat. 31° 42′ N., long. 23° 07·8′ W.
7. Lat. 50° 20·1′ N. or S.
8. Angle at pole 6° 15·1′
9. Lat. 48° 11·3′ N.
10. Lat. 39° 00′ N., long. 50° 19·4′ W.

### Exercise 4B
1. Angle at pole 11° 08·9′
2. Dist. 44·06′
3. Lat. A 51° 19·1′ N.
   Lat. B 28° 57·3′ N.
4. Rate 574·5 knots
5. D. long. 6° 02·2′
6. Dist. 594·9′
7. Clocks advanced 20m 23s
8. Lats. 53° 07·8′ N., 25° 50·5′ S., d. lat. 78° 58·3′
9. Lat. 44° 25′ N. or S.
10. Speed 9·77 knots
11. D. lat. 29° 49·3′, dist. 1789·3′
12. Set 090° T., drift 32·3′

### Exercise 4C
1. Course N. 63°53′W. Distance 259 miles
2. 39°31·5′N 166°11·3′W.
3. Course S. 59°46·5′E. Distance 2620·2 miles
4. 35°04·75′S. 176°04·5′W
5. Course S. 63°19·3′W. Distance 6514·6 miles

### Exercise 4D
1. (a) 848·9, (b) 1862·0, (c) 2244·1, (d) 3962·8
2. D. lat. 1909′ S., D.M.P. 1927·1, d. long. 1128′ W., course S. 30° 20·6′ W., dist. 2212′
3. D. lat. 1795′ S., D.M.P. 2006·5, d. long. 506′ W., course S. 14° 09·2′ W. dist. 1851·2′
4. D. lat. 2943′ S., D.M.P. 3171·7, d. long. 5635′ E., course S. 60° 37·6′ E., dist. 6000′
5. D. lat. 649·5′ S., D.M.P. 708·9′, d. long. 409·3′ W., lat. 18° 40·5′ N., long. 155° 30·7′ E.
6. D. lat. 1296·3′ N., D.M.P. 1295·2′, d. long. 396′ E., lat. 11° 24·2′ N.. dist. 1355·3′
7. (1) D. lat. 565·7′ S., D.M.P. 581·89, d. long. 581·89′ W.
   (2) D. lat. 565·7′ S., D.M.P. 617·42, d. long. 617·42′ W.
   Lat. 28° 51·4′ S., long. 0° 35·5′ E.
8. D. lat. 781·8′ N., D.M.P. 873·3, d. long. 1872·9 E., lat. 33° 13·8′ N., long. 150° 07·1′ W.
9. D. lat. 1780′ S., D.M.P. 1814·6, d. long. 1978′ E., course S. 47° 28′ E., dist. 2633′
10. D. lat. 343′ S., D.M.P. 413·1, d. long. 995′ E., course S. 67° 27·1′ E., dist. 894·5′

### Exercise 4E
1. 41°24·0′N. 27°41·5′W.
2. N. 42°50·3′E. (043°), 2011·5 miles
3. S. 21°03·6′E. (159°), 375·1 miles
4. N. 29°16·2 W. 330¾°), 233·8 miles
5. 54°52·4′N. 5°15·7′E.
6. 45°56·8′N. 9°50·5′W.
7. N. 8°54·6′E. (009°), 404·9 miles
8. 50°03·1′N. 14°40·9′W.
9. 58°19·0′N. 1°11·0′E.
10. 57°50·0′N. 1°02·4′E.

### Exercise 5A

1. 215·7'  100·6'
2. 327·9'  57·8'
3. 386·9'  324·6'
4. 73·2'  201·1'
5. 103·5'  142·4'
6. 80·8'  191·2'
7. 241·6'  456·0'
8. 37°  348·2'
9. N. 24° E.  490·0'
10. S. 33° W.  421·0'
11. N. 18° E.  46·9'
12. S. 36·5° W.  388·5'
13. N. 50° W.  480·0'
14. S. 24° E.  936·0
15. 1230'  995·1'

### Exercise 5B

1. 552'
2. 319·6'
3. 333·7'
4. 250·7'
5. 391·0'
6. 1395·0'
7. 478·5'
8. 406·7'
9. 470·7'
10. 408·9'

### Exercise 5C

1. 199·5'
2. 234·0'
3. 32·49'
4. 34·08'
5. 450·6'
6. 36·26'
7. 416·8'
8. 59·98'
9. 204·6'
10. 314·2'

### Exercise 5D

| | D. lat. | D. long. | M. lat. | Dep. | Course | Dist. |
|---|---|---|---|---|---|---|
| 1. | 590·0' S. | 590·0' W. | 45·75° N. | 411·7 | S. 34·9° W. | 719·4' |
| 2. | 160·0' N. | 230·0' W. | 36·5° N. | 184·9 | N. 49·1° W. | 244·5' |
| 3. | 148·0' N. | 189·0' E. | 24° S. | 172·7 | N. 49·4' E. | 227·4' |
| 4. | 17·0' N. | 260·0' W. | 38° N. | 204·9 | N. 85·3° W. | 205·6' |
| 5. | 70·0' S. | 330 0' E. | 9·5° N. | 325·4 | S. 77·8° E. | 332·9' |
| 6. | 15·0' N. | 31·0' E. | 50·33° N. | 19·79 | Set N. 52·8° E., Drift 24·8' |
| 7. | 82·9' S. | 73·0' W. | 40° N. | 55·9 | Lat. 39° 17·1' N., Long. 5° 17·0' W. |
| 8. | 107·5' N. | 726·0' E. | 48° N. | 485·8 | N. 77·5° E. | 497·6' |
| 9. | 170·0' S. | 242·0' W. | 21·1° S. | 225·7 | S. 53° W. | 282·6' |
| 10. | 249·8' N. | 157·5' W. | 20·7° N. | 147·3 | N. 30·5° W. | 290·0' |

### Exercise 5E

1. 46° 36·9' N. 47° 34·6' W.
2. 60° 29·7' N. 17° 13·2' W.
3. 36° 13·5' N. 6° 22·0' W.
4. 13° 38·4' S. 48° 15·0' E.
5. Course N. 83·3' E., distance 162·8 miles. E.T.A. 0410 following day.
6. S. 63° 12·4' W. (243°), 157·5 miles
7. S. 51° 32·8' W. (231½°), 217·1 miles
8. 33° 50·3' S. 172° 56·5' E.

### Exercise 5F

1. 9·6 miles
2. 7·0 miles
3. 052° T., distant 6·8 miles
4. 49° 54·1' N. 5° 46·8' W.
5. 49° 57·0' N. 5° 45·3' W.
6. 49° 50·6' N. 5° 44·1' W.
7. 49° 58·5' N. 7° 35·7' W.

### Exercise 6A

1. H.W. 0449, 10·9 metres L.W. 1119, 2·3 metres.
   H.W. 1725, 11·3 metres. L.W. 2355, 1·9 metres.
2. 11·2 metres
3. Ht. above C.D. 7·2 metres. Clearance 3·4 metres
4. 9·1 metres. Will not dry.
5. Ht. above C.D. 6·0 m. Interval from H.W. −3h 16m. 0827 B.S.T.
6. Factor ·82, Interval from H.W. −1h 43m, 2138 G.M.T.
7. Factor ·24, Ht. above C.D. 5·1, Ht. of light 56·1 metres.

# ANSWERS 235

### Exercise 6B
1. L.W. 0202, 0·4 metres H.W. 0553, 7·3 metres.
   L.W. 1443, 0·4 metres. H.W. 1831, 7·8 metres.
2. 8·4 metres, 1532 G.M.T.
3. Factor ·39, ht. above C.D. 4·9 metres, clearance 2·6 metres.
4. Factor ·19, correction to soundings (ht. above C.D.), +3·0 metres.
5. Ht. above L.W. 7·6, factor ·62, interval from H.W. −2h 20m. 0858 B.S.T.
6. Ht. above L.W. 4·3, factor ·44, interval from H.W. (Sp.) +3h 30m, 2128 G.M.T.
7. Factor ·66, Ht. above C.D. 7·2 metres, sounding dries 7·2 metres.

### Exercise 8A
1. Az. S. 60·8°E., compass error 29·8°W., deviation 5·3°E.
2. Az. N. 56·2°W., compass error 17·8°E., deviation 5·2°W.
3. Az. S. 37·1°E., compass error 20·9°E., deviation 8·9°E.
4. Az. N. 51·2°E., compass error 5·8°W., deviation 3·8°W.
5. Az. S. 76·9°W., compass error 5·6°W., deviation 6·6°W.

### Exercise 8B
1. Az. S. 35·1°E., compass error 10·9°E., deviation 2·1°W.
2. Az. S. 73·3°W., compass error 6·3°E., deviation 0·2°W.
3. Az. S. 67·2°E., compass error 12·2°W., deviation 5·2°W.
4. Az. N. 49·0°W., compass error 45·0°W., deviation 18·0°W.
5. Az. S. 33·8°W., compass error 36·2°W., deviation 3·2°W.

### Exercise 8C
1. Ampl. E. 3·2°S., compass error 14·7°E., deviation 3·7°E.
2. Ampl. E. 2·4°N., compass error 2·6°E., deviation 2·4°W.
3. Ampl. W. 31·2°N., compass error 0·3°W., deviation 25·7°E.
4. Ampl. E. 30·0°S., compass error 18·0°W., deviation 3·0°E.
5. Ampl. W. 36·0°S., compass error 24·0°W., deviation 48·0°W.
6. Ampl. E. 32·5°N., compass error 4·5°W., deviation 24·5°W.

## CLASS V

### Revision Paper 1—Chartwork and Practical Navigation
1. Tide 247°T., 2·3 knots, compass course 266°C., E.T.A. 1h 40m after departure.
2. Compass error 4°E., True course 216°, position 49° 55·9′ N. 5° 08·1′ W.
3. Height of light 70m. Required distance off light 2·0 miles, vertical sextant angle 1° 05′.
4. Compass error 19°W., tide 300°T., drift 2·1 miles.

### Practical navigation
1. Total d. lat. 136·4′ s., total dep. 89·4′ W., final position 47° 13·6′ N. 10° 14·6′ W.
2. Course N. 28° 34·8′ W. (331½°), distance 131·0 miles.
3. Amplitude E. 38·6°N., compass error 1·6°W., deviation 6·2°E.

### Revision Paper 2—Chartwork and Practical Navigation
1. Compass course 267¾°.
2. Position 50° 16·2′ N. 4° 44·7′ W., compass course 110°.
3. Compass error 3°W., set and rate of current 033°T. at 3·3 knots, course and speed made good 340° 13·8 knots.
4. 20·7 miles.

### Practical navigation
1. 55°17·9′ N. 3° 56·1′ E.
2. Total d. lat. 43·0′ N., total dep. 153·3′ W., position 50° 13·0′ N. 11° 57·8′ W.
3. L.H.A. 70° 37·1′, Az. S. 73·6° W., error 8·4° W., deviation 3·6′ E.


## CLASS IV

### Revision Paper 1—Chartwork
1. Position 51° 07·7′ N. 4° 27·3′ W., compass error 2° W., deviation 7° E.
2. Position 51° 28·4′ N. 4° 59·6′ W., course 113° C., distance off Helwick abeam 5·8 miles, time 1302.
3. Tide 285° T. 4·1 knots, compass course 252½°, speed required 18 knots.
4. Brg. of Lundy 140° T., brg. of Breaksea 087½° T., course made good 082° T., tide 075° T. 3·4 knots, position 51° 18·0′ N. 4° 19·6′ W.
5. Predicted range 5·1 m., interval from H.W. −0213, factor 0·72 (using neap curve), height above C.D. 7·1 m., clearance 0·6 m.

### Revision Paper 2—Chartwork
1. 49° 58·5′ N. 5° 44·1′ W.
2. Course 121° T., 16·2 miles, compass course 158¼°, E.T.A. 2 hours after departure.
3. 50° 06·3′ N 5° 00·9′ W., compass error 11° W.
4. Position 49° 59·6′ N. 5° 37·8′ W., set and drift 091½° T. 1·1 miles.
5. L.W. 0009, 0·2 metres, H.W. 0422, 6·1 metres, L.W. 1252, 0·2 metres, H.W. 1652, 6·7 metres.


### Exercise 9A
1. I.T.P. lat. 40° 15′ N., long. 18° 18·6′ W.
2. I.T.P. lat. 20° 16·6′ S., long. 27° 24·8′ W.
3. I.T.P. lat. 39° 55·4′ N., long. 29° 55′ W.
4. Pos. lat. 50° 13·3′ N., long. 44° 03·7′ W.
5. Pos. lat. 40° 28·1′ N., long. 34° 27·9′ W.
6. Pos. lat. 48° 13·8′ N., long. 50° 10·1′ W.
7. Pos. lat. 25° 00·4′ N., long. 36° 02·5′ W.
8. Dist. off 27·6 miles
9. Pos. lat. 23° 44·4′ N., long. 51° 56·7′ W.
10. Noon pos. lat. 34° 11′ N., long. 42° 16·3′ W.
11. Noon pos. lat. 29° 06′ S., long. 37° 07·1′ W.
12. Pos. lat. 34° 15·9′ N., long. 47° 52·4′ W.
13. Pos. lat. 53° 29·8′ S., long. 179° 32·1′ E.
14. (a) Pos. lat. 16° 41·4′ S., long. 163° 07′ E.
    (b) Pos. lat. 17° 10·5′ S., long. 162° 50·2′ E.
15. Noon pos. lat. 42° 27′ S., long. 76° 15·2′ E.
16. Pos. lat. 39° 04·7′ N., long. 131° 02·3′ E.
17. True bearing 129° 48′
18. True bearing 120° 20′
19. Pos. lat. 47° 30·6′ N., long. 34° 37·2′ W.

# ANSWERS

### Exercise 10A

|   | Dip. | Ref. | S.D. | Par. | True Alt. |
|---|---|---|---|---|---|
| 1. | −5·2′ | −0·7′ | +16·1′ | +0·1′ | 52° 39·3′ |
| 2. | −6·2′ | −1·4′ | +15·9′ | +0·1′ | 33° 20·2′ |
| 3. | −5·9′ | −0·3′ | −16·0′ | Nil | 71° 33·2′ |
| 4. | −4·9′ | −1·8′ | −15·8′ | +0·1′ | 27° 24·3′ |
| 5. | −5·6′ | −0·5′ | +16·1′ | +0·1′ | 62° 46·8′ |
| 6. | −4·8′ | −0·6′ | −16·3′ | +0·1′ | 55° 33·2′ |
| 7. | −6·0′ | −0·4′ | −16·2′ | +0·1′ | 68° 55·7′ |
| 8. | −5·7′ | −0·1′ | +16·1′ | Nil | 81° 56·1′ |
| 9. | Nil | −0·8′ | +16·0′ | +0·1′ | 48° 33·2′ |
| 10. | Nil | −0·7′ | −15·8′ | +0·1′ | 51° 40·6′ |

### Exercise 10B

|   | Dip. | Ref. | True Alt. |
|---|---|---|---|
| 1. | −5·96′ | −0·87′ | 47° 21·77′ |
| 2. | −4·80′ | −1·5′ | 32° 17·3′ |
| 3. | −6·04′ | −2·44′ | 21° 05·52′ |
| 4. | −6·93′ | −0·88′ | 47° 06·59′ |
| 5. | −5·19′ | −1·27′ | 37° 02·14′ |
| 6. | −6·65′ | −4·32′ | 12° 08·03′ |
| 7. | −4·9′ | −0·7′ | 53° 14·0′ |
| 8. | −5·88′ | −2·22′ | 23° 08·1′ |
| 9. | −7·33′ | −0·74′ | 51° 47·53′ |
| 10. | −5·54′ | −3·63′ | 14° 26·23′ |

### Exercise 10C

|   | Dip. | Ref. | S.D. | Par. | True Alt. |
|---|---|---|---|---|---|
| 1. | −4·8′ | −0·48′ | +15·53′ | +25·02′ | 63° 49·67′ |
| 2. | −6·35′ | −1·38′ | +15·23′ | +45·70′ | 35° 05·80′ |
| 3. | −5·71′ | −0·58′ | −16·33′ | +31·43′ | 58° 24·01′ |
| 4. | −5·37′ | −0·21′ | −15·03′ | +11·76′ | 77° 43·95′ |
| 5. | −6·04′ | −2·36′ | +15·90′ | +53·98′ | 22° 36·88′ |
| 6. | −5·37′ | −1·18′ | +16·48′ | +46·61′ | 39° 20·74′ |
| 7. | −7·08′ | −0·77′ | −15·09′ | +34·41′ | 51° 26·87′ |
| 8. | −6·57′ | −1·01′ | +16·80′ | +44·26′ | 44° 11·88′ |

### Exercise 11A

1. Lat. 35° 11·0′ N., P/L 090°/270°.  2. Lat. 0° 09·1′ S., P/L 090°/270°.
3. Lat. 49° 51·3′ S., P/L 090°/270°.  4. Lat. 30° 09·9′ S., P/L 090°/270°.
5. Lat. 0° 02·1′ N., P/L 090°/270°.

### Exercise 11B

1. G.M.T. 2246 18th, lat. 0° 16·8′ N.  2. G.M.T. 1444 26th, lat. 25° 00·7′ S.
3. G.M.T. 1832 6th, lat. 51° 30·6′ S.  4. G.M.T. 0048 30th, lat. 36° 51·0′ N.
5. G.M.T. 0229 19th, lat. 50° 44·2′ S.

### Exercise 11C

1. G.M.T. 2306 4th, lat. 33° 28·5′ S.  2. G.M.T. 0721 20th, lat. 40° 41·2′ N.
3. G.M.T. 1314 19th, lat. 34° 24·7′ N.  4. G.M.T. 0505 30th, lat. 17° 45·2′ N.

## Exercise 11D
1. Sextant altitude 57° 09·0', G.M.T. 20h 10m 46s 19th, L.M.T. 18h 40m 46s 19th.
2. Sextant altitude 50°55·1', G.M.T. 17h 50m 41s 6th, L.M.T. 19h 59m 21s 6th.
3. Sextant altitude 33° 21·3', G.M.T. 08h 44m 24s 19th, L.M.T. 06h 22m 24s 19th.
4. Sextant altitude 66° 47·8', G.M.T. 14h 58m 20s 27th, L.M.T. 19h 00m 40s 27th.
5. Sextant altitude 61°14·9', G.M.T. 15h 23m 42s 20th, L.M.T. 05h 55m 26s 20th.

## Exercise 11E
1. Lat. 50°17·6' N.
2. Lat. 52° 50·5' N.
3. Lat. 55°14·1' N.
4. Lat. 49° 57·7 S.
5. Lat. 45° 28·6' S.

## Exercise 12A
1. L.H.A. 324° 17·3', C.Z.X. 63° 00·7', TZX 63° 01·4', Int. 0·7 A., Az. N. 37·0 E., P/L 127·0°/317·0° through I.T.P. 29° 30·5' S. 121° 20·4' W.
2. L.H.A. 47° 56·7', C.Z.X. 43° 24·4', TZX 43° 25·2', Int. 0·8' A., Az. N. 89·2° W., P/L 000·8°/180·8° through I.T.P. 32° 15·0' S. 48°15·1' W.
3. L.H.A. 58° 16·6', C.Z.X. 58° 17·0', T.Z.X. 58° 10·0', Int. 7·0' T., Az. N. 88·7° W., P/L 001·3°/181·3° through I.T.P. 0° 00·2' N. 161° 02·0' W.
4. L.H.A. 319° 46·4', C.Z.X. 76° 10·6'. T.Z.X. 76° 18·8', Int. 8·2' A., Az. S. 37·6° E., P/L 232·4°/052·4° through I.T.P. 43°18·5' N. 38° 31·9' W.
5. L.H.A. 300° 05·4', C.Z.X. 71° 05·2', T.Z.X. 70° 56·0', Int. 9·2' T., Az. S. 66·0° E., P/L 204·0°/024·0° through I.T.P. 44° 01·3' N. 27° 29·3' W.

## Exercise 12B
1. L.H.A. 63° 52·4', C.Z.X. 58° 56·4', T.Z.X. 58° 38·0', Int. 18·4' T., Az. N. 81·6° W., P/L 188·4°/008·4° through I.T.P. 24° 52·7' N. 144° 49·9' E.
2. L.H.A. 77° 53·9', C.Z.X. 49°10·2' T.Z.X. 49° 03·1', Int. 7·1' T., Az. N. 45·6° W., P/L 224·4°/044·4° through I.T.P. 43° 10·0' N. 177° 22·9' W.
3. L.H.A. 293° 03·1', C.Z.X. 71° 10·6', T.Z.X. 71° 11·7', Int. 1·1' A., Az. S. 74·0° E., P/L 196·0°/016·0° through I.T.P. 17° 53·9' N. 47° 31·1' W.
4. L.H.A. 311° 44·6', C.Z.X. 42° 03·7', T.Z.X. 42° 03·7', Int. Nil., Az. S. 83·9° E., P/L 186·1°/006·1° through 42° 40·0' N. 172°10·0' W.
5. L.H.A. 65° 08·4' C.Z.X. 75° 11·2', T.Z.X. 74° 44·7', Int. 26·5' T., Az. N. 69·1° W., P/L 200·9°/020·9° through I.T.P. 40° 50·0' S. 57° 29·8' W.

## Exercise 12C
1. L.H.A. 294° 40·9'. C.Z.X. 59° 46·2', T.Z.X. 59° 50·0', Int. 3·8' A., Az. N. 87·3° E., P/L 177·3°/357·3° through I.T.P. 42° 50·2' S. 41° 35·2' W.
2. L.H.A. 60°11·5', C.Z.X. 62°15·7', T.Z.X. 62°15·1', Int. 0·6' T., Az. S. 78·4° W., P/L 168·4°/348·4° through I.T.P. 25° 29·9' N. 174° 59·3' E.

## Exercise 12D
1. G.H.A. 358° 09·5°, T.Z.X. 58° 20·8', L.H.A. 30° 18·5', longitude 61° 43·4' E., Az. N. 68·9° W., P/L 201·1°/021·1°.
2. G.H.A. 182° 02·9', T.Z.X. 65° 50·3', L.H.A. 64° 48·4' longitude 117° 14·5' W., Az. S. 67·0° W., P/L 157·0°/337·0°.
3. G.H.A. 136° 06·1', T.Z.X. 46° 42·2', L.H.A. 307° 30·9', longitude 171° 24·8' E., Az. 090, P/L 000°/180°.
4. G.H.A. 302° 38·0', T.Z.X. 59° 59·8', L.H.A. 302° 34·4', longitude 0° 03·6' W., Az. S. 63·9° E., P/L 206·1°/026·1°.
5. G.H.A. 327° 46·4', T.Z.X. 70° 56·0', L.H.A. 300° 19·4', longitude 27° 27·0' W., Az. S. 65·8° E., P/L 204·2°/024·2°.

### Exercise 12E
1. G.H.A. 339° 25·2′, T.Z.X. 62° 11·2′, L.H.A. 295° 03·1′, longitude 44° 22·1′ W., Az. S. 88·9° E., P/L 181·1°/001·1°.
2. G.H.A. 204° 35·6′, T.Z.X. 75°20·2′, L.H.A. 69° 48·5′, longitude 134° 47·1′ W., Az. N. 75·1° W., P/L 194·9°/014·9°.
3. G.H.A. 244° 41·3′, T.Z.X. 57° 44·7′, L.H.A. 62° 55·1′, longitude 178° 13·8′ E., Az. N. 83·6° W., P/L 006·4°/186·4°.
4. G.H.A. 215° 44·6′, T.Z.X. 76° 46·2′, L.H.A. 294° 19·2′, longitude 78° 33·6′ E., Az. N. 66·3° E., P/L 156·3°/336·3°.
5. G.H.A. 121° 30·8′, T.Z.X. 74° 44·7′, L.H.A. 64° 30·7′ longitude 57° 00·1′ W., Az. N. 76·7° W., P/L 193·3°/013·3°.

### Exercise 12F
1. G.H.A. 116° 15·7′ T.Z.X. 57° 33·3′, L.H.A. 316° 08·3′, longitude 160° 07·4′ W., Az. N. 55·1° E., 'C' 1·01, P/L 145·1°/325·1°, obs. latitude 46° 08·0′ S., D.R. long. 161° 03·6′ W., d. long. 13·6′ E., noon position 46° 08·0′ S. 160° 50·0′ W.
2. G.H.A. 150° 35·9′ T.Z.X. 72° 34·9′ L.H.A. 319° 01·9′, longitude 168° 26·0′ E., Az. N. 39·2° E., 'C' 1·563, P/L 309·2°/129·2°, obs. latitude 37° 52·9′ S., D.R. long. 169° 01·9′ E., d. long. 9·7′ E., noon position 37° 52·9′ S. 169° 11·6′ E.

### Exercise 13A
1. L.H.A. 6° 23·8′ T.Z.X. 47° 30·0′, M.Z.X. 47° 10·9′, latitude 48° 50·5′ N., Az. S. 33·0° W., P/L 123·0°/303·0°.
2. L.H.A. 351° 54·7′, T.Z.X. 40° 54·2′, M.Z.X. 40°14·7′, latitude 41° 32·3′ N., Az. S. 12·4° E., P/L 257·6°/077·6°.
3. L.H.A. 348° 45·2′, T.Z.X. 65° 13·8′, M.Z.X. 64° 23·3′, latitude 40° 58·7′ N., Az. S. 11·4° E., P/L 258·6°/078·6°.

### Exercise 13B
1. L.H.A. 7° 09·7′, T.Z.X. 31° 04·7′, M.Z.X. 30° 29·0′, latitude 45° 09·7′ N., Az. S. 13·5° W., P/L 103·5°/283·5°.
2. L.H.A. 350° 24·9′ T.Z.X. 66° 41·5′, M.Z.X. 66° 04·8′, latitude 36° 21·4′ N., Az. S. 9·1° E., P/L 260·9°/080·9°.
3. L.H.A. 355° 23·3′, T.Z.X. 34° 38·1′, M.Z.X. 34° 20·7′, latitude 26° 07·4′ N., Az. S. 8·1° E., P/L 261·9°/081·9°.
4. L.H.A. 5° 41·9′, T.Z.X. 22° 19·0′, M.Z.X. 21° 40·3′, latitude 30°14·8′ S., Az. N. 15·0° W., P/L 255·0°/075·0°.
5. L.H.A. 357° 14·7′, T.Z.X. 64° 41·7′, M.Z.X. 64° 38·8′, latitude 18° 40·3′ S., Az. N. 2·1° E., P/L 272·1°/092·1°.

### Exercise 14A
1. L.H.A. ♈ 7° 31·5′, T. Alt. 50° 03·9′ latitude 49° 19·4′ N. P/L 090·6°/270·6°.
2. L.H.A. ♈ 282° 24·5′, T. Alt. 35° 07·6′, latitude 34° 25·7′, P/L 271°/091°.
3. L.H.A. ♈ 317° 50·4′, T. Alt. 47° 38·2′, latitude 47° 25·8′ N., P/L 271·3°/091·3°.
4. L.H.A. ♈ 15°12·5′, T. Alt. 23° 31·0′, latitude 22° 44·0′ N., P/L 090·3°/270·3°.
5. L.H.A. ♈ 85° 40·6′, T. Alt. 51° 04·6′, latitude 50° 34·5′ N., P/L 269°/089°.
6. L.H.A. ♈ 74° 35·7′, T. Alt. 32° 37·9′, latitude 32° 00·6′ N., P/L 269·3°/089·3°.
7. L.H.A. ♈ 216° 01·5′, T. Alt. 40° 28·0′, latitude 41°17·8′ N., P/L 090°/270°.

### Exercise 15A

1. Distance 1736·5 miles, initial course N. 77° 37·7′ W., final course S. 62° 13·8′ W.
2. Distance 3599 miles, initial course S. 67° 30·3′ E., position of vertex 40° 44·8′ S. 20° 17·0′ W.
3. Great circle distance 5190·4 miles, mercator distance 5594·1, saving 403·7 miles
4. Distance 4076·6 miles, initial course N. 61° 50·5′ W., position of vertex 54° 10·4′ N. 160° 22·5′ W.
   Positions along track
   | | | | | |
   |---|---|---|---|---|
   | Lat. | 52° 24·4′ N. | 54° 10·3′ N. | 52° 31·3′ N. | 46° 49·9′ N. |
   | Long. | 140° W. | 160° W. | 180° | 160° E. |
   | Course | N. 59° 13·2′ W. | N. 74° 35·2′ W. | S. 89° 18′ W. | S. 73° 13′ W. |

### Exercise 15B

1. Total distance 3613·4 miles (1296·6 + 816·8 + 1500·0), initial course S. 73° 56·6′ E.
2. Total distance 5279·6 miles (3081·8 + 1431·5 + 766·3), initial course S. 52° 00·4′ E.
3. Total distance 4803·8 miles (1144·8 + 2731·0 + 928), initial course S. 69° 23·5′ W.

### Specimen Paper 1

1. L.H.A. 311° 21·3′, C.Z.X. 49° 13·0′, T.Z.X. 49° 16·8′, intercept 3·8′ A., Az. N. 82·2° E., P/L 352·2°/172·2° through I.T.P. 5° 58·5′ S. 125° 59·2′ E.
2. G.M.T. 09h 26m 52s 19th, L.M.T. 18h 42m 32s 19th, ZX 51° 46·3′, sextant alt. 38° 20·4′.
3. Course N. 42° 51·2′ E., distance 2012·0 miles.
4. Az. S. 22·0° E, compass error 4° W., deviation 3° E.

### Specimen Paper 2

1. L.H.A. 55° 05·6′, C.Z.X. 54° 20·5′, T.Z.X. 54° 15·1′, intercept 5·4 T., Az. S. 80·8° W., P/L 350·8°/170·8° through I.T.P. 33° 04·1′ N. 131° 24·3′ W.
2. L.H.A. 358° 01·0′, T.Z.X. 48° 36·2′, M.Z.X. 48° 33·9′, latitude 26° 08·5′ N., Az. S. 2·4° E., P/L 267·6°/087·6°.
3. L.H.A. ♈ 205° 59·8′, latitude 20° 54·1′ N., Az. 359·9°.
4. Obs. pos. 39° 15·0′ S 94° 13·2′ E.

### Specimen Paper 3

1. L.H.A. 39° 54·6′, C.Z.X. 62° 06·4′, T.Z.X. 62° 10·6′, intercept 4·2′ A., Az. S. 46·4° W., P/L 136·4°/316·4° through I.T.P. 47° 32·9′ N. 45° 15·6′ W.
2. G.M.T. 1946 8th, latitude 46° 25·0′ S., P/L 090°/270°.
3. 46° 28·3′ N. 47° 00·5′ W.
4. G.M.T. 0419 2nd, amplitude W. 3·7° S., compass error 6·7° W., deviation 10·7° W.

### Specimen Paper 4

1. L.H.A. 49° 51·6′, C.Z.X. 61° 31·6′, T.Z.X. 61° 25·2′, intercept 6·4′ T., Az. S. 60·4° W., P/L 150·4°/330·4° through I.T.P. 42° 06·8′ N 50° 12·6′ W.
2. L.H.A. 358° 08·5′, T.Z.X. 45° 13·6′, M.Z.X. 45° 11·7′, latitude 38° 51·7′ S., Az. N. 2·6° E., P/L 272·6°/092·6° through 38° 51·7′ S. 138° 46·0′ E.
3. G.M.T. 1939 20th, latitude 25° 56·5′ N., P/L 090°/270°.
4. 22° 33·9′ N. 91° 15·1′ W.

# ANSWERS

### Specimen Paper 5
1. L.H.A. 317° 30·3′, C.Z.X. 61° 25·1′, T.Z.X. 61° 28·3′, intercept 3·2′ A., Az. N. 47·5° E., P/L 137·5°/317·5° through I.T.P. 29° 17·2′ S. 134° 15·3′ E.
2. G.M.T. 1451 8th, latitude 38° 58·0′ S., P/L 090°/270°.
3. 17° 35·3′ S 169° 58·6′ W.
4. 32° 46·0′ N. 31° 25·0′ W.

### Specimen Paper 6
1. G.H.A. 134° 07·8′, T.Z.X. 56° 15·0′, L.H.A. 309° 20·3′ longitude 175° 12·5′ E., Az. S. 59·1° E., P/L 210·9°/030·9°.
2. G.M.T. 1832 6th, latitude 51° 30·6′ S., P/L 090°/270°.
3. G.M.T. 13h 20m 44s 27th, L.H.A. ♈ 350° 19·3′, latitude 47° 11·7′ N., P/L 270·8°/090·8°.
4. 42° 03·5′ S. 161° 25·0′ E.

# EXTRACTS FROM THE NAUTICAL ALMANAC, 1980

# A2 ALTITUDE CORRECTION TABLES 10°–90°—SUN, STARS, PLANETS

| OCT.–MAR. SUN APR.–SEPT. || STARS AND PLANETS || DIP ||
|---|---|---|---|---|---|
| App. Alt. | Lower Upper Limb Limb | App. Alt. | Lower Upper Limb Limb | App. Alt. Corrⁿ | App. Alt. Additional Corrⁿ | Ht. of Eye Corrⁿ | Ht. of Eye Corrⁿ |

| SUN Oct.–Mar. | SUN Apr.–Sept. | STARS AND PLANETS | DIP |
|---|---|---|---|
| ° ′ ′ ′ | ° ′ ′ ′ | ° ′ ′ | 1980 | m ′ ft. | m ′ |
| 9 34 +10·8 −21·5 | 9 39 +10·6 −21·2 | 9 56 −5·3 | **VENUS** | 2·4 −2·8 8·0 | 1·0 − 1·8 |
| 9 45 +10·9 −21·4 | 9 51 +10·7 −21·1 | 10 08 −5·2 | Jan. 1–Feb. 26 | 2·6 −2·9 8·6 | 1·5 − 2·2 |
| 9 56 +11·0 −21·3 | 10 03 +10·8 −21·0 | 10 20 −5·1 | ° ′ | 2·8 −3·0 9·2 | 2·0 − 2·5 |
| 10 08 +11·1 −21·2 | 10 15 +10·9 −20·9 | 10 33 −5·0 | 42 + 0·1 | 3·0 −3·1 9·8 | 2·5 − 2·8 |
| 10 21 +11·2 −21·1 | 10 27 +11·0 −20·8 | 10 46 −4·9 | | 3·2 −3·2 10·5 | 3·0 − 3·0 |
| 10 34 +11·3 −21·0 | 10 40 +11·1 −20·7 | 11 00 −4·8 | Feb. 27–Apr. 13 | 3·4 −3·3 11·2 | See table |
| 10 47 +11·4 −20·9 | 10 54 +11·2 −20·6 | 11 14 −4·7 | ° ′ | 3·6 −3·4 11·9 | ← |
| 11 01 +11·5 −20·8 | 11 08 +11·3 −20·5 | 11 29 −4·6 | 47 + 0·2 | 3·8 −3·5 12·6 | m ′ |
| 11 15 +11·6 −20·7 | 11 23 +11·4 −20·4 | 11 45 −4·5 | Apr. 14–May 9 | 4·0 −3·6 13·3 | 20 − 7·9 |
| 11 30 +11·7 −20·6 | 11 38 +11·5 −20·3 | 12 01 −4·4 | ° ′ | 4·3 −3·7 14·1 | 22 − 8·3 |
| 11 46 +11·8 −20·5 | 11 54 +11·6 −20·2 | 12 18 −4·3 | 46 + 0·3 | 4·5 −3·8 14·9 | 24 − 8·6 |
| 12 02 +11·9 −20·4 | 12 10 +11·7 −20·1 | 12 35 −4·2 | | 4·7 −3·9 15·7 | 26 − 9·0 |
| 12 19 +12·0 −20·3 | 12 28 +11·8 −20·0 | 12 54 −4·1 | May 10–May 25 | 5·0 −4·0 16·5 | 28 − 9·3 |
| 12 37 +12·1 −20·2 | 12 46 +11·9 −19·9 | 13 13 −4·0 | ° ′ | 5·2 −4·1 17·4 | |
| 12 55 +12·2 −20·1 | 13 05 +12·0 −19·8 | 13 33 −3·9 | 11 + 0·4 | 5·5 −4·2 18·3 | 30 − 9·6 |
| 13 14 +12·3 −20·0 | 13 24 +12·1 −19·7 | 13 54 −3·8 | 41 + 0·5 | 5·8 −4·3 19·1 | 32 −10·0 |
| 13 35 +12·4 −19·9 | 13 45 +12·2 −19·6 | 14 16 −3·7 | May 26–June 3 | 6·1 −4·4 20·1 | 34 −10·3 |
| 13 56 +12·5 −19·8 | 14 07 +12·3 −19·5 | 14 40 −3·6 | ° ′ | 6·3 −4·5 21·0 | 36 −10·6 |
| 14 18 +12·6 −19·7 | 14 30 +12·4 −19·4 | 15 04 −3·5 | 6 + 0·5 | 6·6 −4·6 22·0 | 38 −10·8 |
| 14 42 +12·7 −19·6 | 14 54 +12·5 −19·3 | 15 30 −3·4 | 20 + 0·6 | 6·9 −4·7 22·9 | |
| 15 06 +12·8 −19·5 | 15 19 +12·6 −19·2 | 15 57 −3·3 | 31 + 0·7 | 7·2 −4·8 23·9 | 40 −11·1 |
| 15 32 +12·9 −19·4 | 15 46 +12·7 −19·1 | 16 26 −3·2 | June 4–June 26 | 7·5 −4·9 24·9 | 42 −11·4 |
| 15 59 +13·0 −19·3 | 16 14 +12·8 −19·0 | 16 56 −3·1 | ° ′ | 7·9 −5·0 26·0 | 44 −11·7 |
| 16 28 +13·1 −19·2 | 16 44 +12·9 −18·9 | 17 28 −3·0 | 4 + 0·6 | 8·2 −5·1 27·1 | 46 −11·9 |
| 16 59 +13·2 −19·1 | 17 15 +13·0 −18·8 | 18 02 −2·9 | 12 + 0·7 | 8·5 −5·2 28·1 | 48 −12·2 |
| 17 32 +13·3 −19·0 | 17 48 +13·1 −18·7 | 18 38 −2·8 | 22 + 0·8 | 8·8 −5·3 29·2 | ft. ′ |
| 18 06 +13·4 −18·9 | 18 24 +13·2 −18·6 | 19 17 −2·7 | June 27–July 6 | 9·2 −5·4 30·4 | 2 − 1·4 |
| 18 42 +13·5 −18·8 | 19 01 +13·3 −18·5 | 19 58 −2·6 | ° ′ | 9·5 −5·5 31·5 | 4 − 1·9 |
| 19 21 +13·6 −18·7 | 19 42 +13·4 −18·4 | 20 42 −2·5 | 6 + 0·5 | 9·9 −5·6 32·7 | 6 − 2·4 |
| 20 03 +13·7 −18·6 | 20 25 +13·5 −18·3 | 21 28 −2·4 | 20 + 0·6 | 10·3 −5·7 33·9 | 8 − 2·7 |
| 20 48 +13·8 −18·5 | 21 11 +13·6 −18·2 | 22 19 −2·3 | 31 + 0·7 | 10·6 −5·8 35·1 | 10 − 3·1 |
| 21 35 +13·9 −18·4 | 22 00 +13·7 −18·1 | 23 13 −2·2 | July 7–July 21 | 11·0 −5·9 36·3 | See table |
| 22 26 +14·0 −18·3 | 22 54 +13·8 −18·0 | 24 11 −2·1 | ° ′ | 11·4 −6·0 37·6 | ← |
| 23 22 +14·1 −18·2 | 23 51 +13·9 −17·9 | 25 14 −2·0 | 11 + 0·4 | 11·8 −6·1 38·9 | ft. ′ |
| 24 21 +14·2 −18·1 | 24 53 +14·0 −17·8 | 26 22 −1·9 | 41 + 0·5 | 12·2 −6·2 40·1 | 70 − 8·1 |
| 25 26 +14·3 −18·0 | 26 00 +14·1 −17·7 | 27 36 −1·8 | July 22–Aug. 17 | 12·6 −6·3 41·5 | 75 − 8·4 |
| 26 36 +14·4 −17·9 | 27 13 +14·2 −17·6 | 28 56 −1·7 | ° ′ | 13·0 −6·4 42·8 | 80 − 8·7 |
| 27 52 +14·5 −17·8 | 28 33 +14·3 −17·5 | 30 24 −1·6 | 46 + 0·3 | 13·4 −6·5 44·2 | 85 − 8·9 |
| 29 15 +14·6 −17·7 | 30 00 +14·4 −17·4 | 32 00 −1·5 | Aug. 18–Oct. 2 | 13·8 −6·6 45·5 | 90 − 9·2 |
| 30 46 +14·7 −17·6 | 31 35 +14·5 −17·3 | 33 45 −1·4 | ° ′ | 14·2 −6·7 46·9 | 95 − 9·5 |
| 32 26 +14·8 −17·5 | 33 20 +14·6 −17·2 | 35 40 −1·3 | 47 + 0·2 | 14·7 −6·8 48·4 | |
| 34 17 +14·9 −17·4 | 35 17 +14·7 −17·1 | 37 48 −1·2 | Oct. 3–Dec. 31 | 15·1 −6·9 49·8 | 100 − 9·7 |
| 36 20 +15·0 −17·3 | 37 26 +14·8 −17·0 | 40 08 −1·1 | ° ′ | 15·5 −7·0 51·3 | 105 − 9·9 |
| 38 36 +15·1 −17·2 | 39 50 +14·9 −16·9 | 42 44 −1·0 | 42 + 0·1 | 16·0 −7·1 52·8 | 110 −10·2 |
| 41 08 +15·2 −17·1 | 42 31 +15·0 −16·8 | 45 36 −0·9 | | 16·5 −7·2 54·3 | 115 −10·4 |
| 43 59 +15·3 −17·0 | 45 31 +15·1 −16·7 | 48 47 −0·8 | **MARS** | 16·9 −7·3 55·8 | 120 −10·6 |
| 47 10 +15·4 −16·9 | 48 55 +15·2 −16·6 | 52 18 −0·7 | Jan. 1–Apr. 28 | 17·4 −7·4 57·4 | 125 −10·8 |
| 50 46 +15·5 −16·8 | 52 44 +15·3 −16·5 | 56 11 −0·6 | ° ′ | 17·9 −7·5 58·9 | |
| 54 49 +15·6 −16·7 | 57 02 +15·4 −16·4 | 60 28 −0·5 | 41 + 0·2 | 18·4 −7·6 60·5 | 130 −11·1 |
| 59 23 +15·7 −16·6 | 61 51 +15·5 −16·3 | 65 08 −0·4 | 75 + 0·1 | 18·8 −7·7 62·1 | 135 −11·3 |
| 64 30 +15·8 −16·5 | 67 17 +15·6 −16·2 | 70 11 −0·3 | Apr. 29–Dec. 31 | 19·3 −7·8 63·8 | 140 −11·5 |
| 70 12 +15·9 −16·4 | 73 16 +15·7 −16·1 | 75 34 −0·2 | ° ′ | 19·8 −7·9 65·4 | 145 −11·7 |
| 76 26 +16·0 −16·3 | 79 43 +15·8 −16·0 | 81 13 −0·1 | 60 + 0·1 | 20·4 −8·0 67·1 | 150 −11·9 |
| 83 05 +16·1 −16·2 | 86 32 +15·9 −15·9 | 87 03 0·0 | | 20·9 −8·1 68·8 | 155 −12·1 |
| 90 00 | 90 00 | 90 00 | | 21·4 70·5 | |

App. Alt. = Apparent altitude = Sextant altitude corrected for index error and dip.
For daylight observations of Venus, see page 260.

R

## 1980 JANUARY 4, 5, 6 (FRI., SAT., SUN.)

| G.M.T. | ARIES G.H.A. | VENUS −3.4 G.H.A. Dec. | MARS +0.1 G.H.A. Dec. | JUPITER −1.9 G.H.A. Dec. | SATURN +1.1 G.H.A. Dec. | STARS Name S.H.A. Dec. |
|---|---|---|---|---|---|---|
| 4 00 | 102 46.1 | 144 39.6 S18 01.1 | 295 50.2 N 9 05.9 | 300 40.5 N 8 49.6 | 284 38.7 N 3 11.1 | Acamar 315 37.4 S40 23.4 |
| 01 | 117 48.6 | 159 39.0 18 00.2 | 310 52.3 05.8 | 315 43.0 49.6 | 299 41.2 11.1 | Achernar 335 45.6 S57 20.7 |
| 02 | 132 51.0 | 174 38.3 17 59.2 | 325 54.4 05.8 | 330 45.6 49.6 | 314 43.6 11.1 | Acrux 173 37.6 S62 59.0 |
| 03 | 147 53.5 | 189 37.7 ·· 58.3 | 340 56.5 ·· 05.7 | 345 48.1 ·· 49.7 | 329 46.1 ·· 11.1 | Adhara 255 32.1 S28 56.8 |
| 04 | 162 56.0 | 204 37.1 57.4 | 355 58.7 05.6 | 0 50.6 49.7 | 344 48.5 11.1 | Aldebaran 291 18.3 N16 28.1 |
| 05 | 177 58.4 | 219 36.5 56.4 | 11 00.8 05.6 | 15 53.1 49.8 | 359 51.0 11.1 | |
| 06 | 193 00.9 | 234 35.9 S17 55.5 | 26 02.9 N 9 05.5 | 30 55.7 N 8 49.8 | 14 53.4 N 3 11.1 | Alioth 166 43.0 N56 03.9 |
| 07 | 208 03.4 | 249 35.3 54.6 | 41 05.0 05.5 | 45 58.2 49.8 | 29 55.9 11.1 | Alkaid 153 19.1 N49 24.6 |
| 08 | 223 05.8 | 264 34.6 53.6 | 56 07.1 05.4 | 61 00.7 49.9 | 44 58.3 11.1 | Al Na'ir 28 15.9 S47 03.7 |
| F 09 | 238 08.3 | 279 34.0 ·· 52.7 | 71 09.2 ·· 05.3 | 76 03.2 ·· 49.9 | 60 00.8 ·· 11.1 | Alnilam 276 11.8 S 1 13.0 |
| R 10 | 253 10.7 | 294 33.4 51.8 | 86 11.3 05.3 | 91 05.8 49.9 | 75 03.2 11.1 | Alphard 218 20.8 S 8 34.3 |
| I 11 | 268 13.2 | 309 32.8 50.8 | 101 13.4 05.2 | 106 08.3 50.0 | 90 05.7 11.1 | |
| D 12 | 283 15.7 | 324 32.2 S17 49.9 | 116 15.5 N 9 05.2 | 121 10.8 N 8 50.0 | 105 08.1 N 3 11.1 | Alphecca 126 32.8 N26 46.9 |
| A 13 | 298 18.1 | 339 31.6 49.0 | 131 17.6 05.1 | 136 13.3 50.0 | 120 10.6 11.1 | Alpheratz 358 09.9 N28 58.9 |
| Y 14 | 313 20.6 | 354 31.0 48.0 | 146 19.7 05.0 | 151 15.9 50.1 | 135 13.0 11.1 | Altair 62 33.3 N 8 49.0 |
| 15 | 328 23.1 | 9 30.4 ·· 47.1 | 161 21.9 ·· 05.0 | 166 18.4 ·· 50.1 | 150 15.5 ·· 11.1 | Ankaa 353 40.8 S42 25.2 |
| 16 | 343 25.5 | 24 29.7 46.1 | 176 24.0 04.9 | 181 20.9 50.2 | 165 17.9 11.2 | Antares 112 57.7 S26 23.1 |
| 17 | 358 28.0 | 39 29.1 45.2 | 191 26.1 04.9 | 196 23.5 50.2 | 180 20.4 11.2 | |
| 18 | 13 30.5 | 54 28.5 S17 44.3 | 206 28.2 N 9 04.8 | 211 26.0 N 8 50.2 | 195 22.8 N 3 11.2 | Arcturus 146 19.1 N19 17.1 |
| 19 | 28 32.9 | 69 27.9 43.3 | 221 30.3 04.8 | 226 28.5 50.3 | 210 25.3 11.2 | Atria 108 22.9 S68 59.3 |
| 20 | 43 35.4 | 84 27.3 42.4 | 236 32.5 04.7 | 241 31.0 50.3 | 225 27.7 11.2 | Avior 234 27.7 S59 26.7 |
| 21 | 58 37.9 | 99 26.7 ·· 41.4 | 251 34.6 ·· 04.6 | 256 33.6 ·· 50.4 | 240 30.2 ·· 11.2 | Bellatrix 278 58.9 N 6 19.8 |
| 22 | 73 40.3 | 114 26.1 40.5 | 266 36.7 04.6 | 271 36.1 50.4 | 255 32.6 11.2 | Betelgeuse 271 28.5 N 7 24.1 |
| 23 | 88 42.8 | 129 25.5 39.5 | 281 38.8 04.5 | 286 38.6 50.4 | 270 35.1 11.2 | |
| 5 00 | 103 45.2 | 144 24.9 S17 38.6 | 296 40.9 N 9 04.5 | 301 41.2 N 8 50.5 | 285 37.5 N 3 11.2 | Canopus 264 06.9 S52 41.3 |
| 01 | 118 47.7 | 159 24.3 · 37.6 | 311 43.1 04.4 | 316 43.7 50.5 | 300 40.0 11.2 | Capella 281 11.5 N45 58.7 |
| 02 | 133 50.2 | 174 23.7 36.7 | 326 45.2 04.4 | 331 46.2 50.5 | 315 42.4 11.2 | Deneb 49 49.2 N45 12.7 |
| 03 | 148 52.6 | 189 23.1 ·· 35.7 | 341 47.3 ·· 04.3 | 346 48.8 ·· 50.6 | 330 44.9 ·· 11.2 | Denebola 182 59.5 N14 41.0 |
| 04 | 163 55.1 | 204 22.5 34.8 | 356 49.5 04.3 | 1 51.3 50.6 | 345 47.3 11.2 | Diphda 349 21.4 S18 06.0 |
| 05 | 178 57.6 | 219 21.9 33.8 | 11 51.6 04.2 | 16 53.8 50.7 | 0 49.8 11.2 | |
| 06 | 194 00.0 | 234 21.3 S17 32.9 | 26 53.7 N 9 04.2 | 31 56.4 N 8 50.7 | 15 52.2 N 3 11.3 | Dubhe 194 22.5 N61 51.3 |
| 07 | 209 02.5 | 249 20.7 31.9 | 41 55.9 04.1 | 46 58.9 50.7 | 30 54.7 11.3 | Elnath 278 44.4 N28 35.4 |
| S 08 | 224 05.0 | 264 20.1 31.0 | 56 58.0 04.1 | 62 01.4 50.8 | 45 57.1 11.3 | Eltanin 90 58.5 N51 29.5 |
| A 09 | 239 07.4 | 279 19.5 ·· 30.0 | 72 00.1 ·· 04.0 | 77 04.0 ·· 50.8 | 60 59.6 ·· 11.3 | Enif 34 12.3 N 9 47.0 |
| T 10 | 254 09.9 | 294 18.9 29.1 | 87 02.3 04.0 | 92 06.5 50.9 | 76 02.0 11.3 | Fomalhaut 15 52.2 S29 43.9 |
| U 11 | 269 12.3 | 309 18.3 28.1 | 102 04.4 03.9 | 107 09.0 50.9 | 91 04.5 11.3 | |
| R 12 | 284 14.8 | 324 17.7 S17 27.2 | 117 06.5 N 9 03.9 | 122 11.6 N 8 50.9 | 106 06.9 N 3 11.3 | Gacrux 172 29.1 S56 59.7 |
| D 13 | 299 17.3 | 339 17.1 26.2 | 132 08.7 03.8 | 137 14.1 51.0 | 121 09.4 11.3 | Gienah 176 18.4 S17 25.7 |
| A 14 | 314 19.7 | 354 16.5 25.3 | 147 10.8 03.8 | 152 16.6 51.0 | 136 11.8 11.3 | Hadar 149 24.1 S60 16.3 |
| Y 15 | 329 22.2 | 9 15.9 ·· 24.3 | 162 12.9 ·· 03.7 | 167 19.2 ·· 51.1 | 151 14.3 ·· 11.3 | Hamal 328 29.3 N23 22.1 |
| 16 | 344 24.7 | 24 15.3 23.3 | 177 15.1 03.7 | 182 21.7 51.1 | 166 16.8 11.3 | Kaus Aust. 84 17.9 S34 23.6 |
| 17 | 359 27.1 | 39 14.7 22.4 | 192 17.2 03.6 | 197 24.2 51.2 | 181 19.2 11.3 | |
| 18 | 14 29.6 | 54 14.1 S17 21.4 | 207 19.4 N 9 03.6 | 212 26.8 N 8 51.2 | 196 21.7 N 3 11.4 | Kochab 137 19.8 N74 14.1 |
| 19 | 29 32.1 | 69 13.5 20.5 | 222 21.5 03.6 | 227 29.3 51.2 | 211 24.1 11.4 | Markab 14 03.8 N15 05.9 |
| 20 | 44 34.5 | 84 12.9 19.5 | 237 23.7 03.5 | 242 31.8 51.3 | 226 26.6 11.4 | Menkar 314 41.4 N 4 00.6 |
| 21 | 59 37.0 | 99 12.3 ·· 18.5 | 252 25.8 ·· 03.5 | 257 34.4 ·· 51.3 | 241 29.0 ·· 11.4 | Menkent 148 37.7 S36 16.1 |
| 22 | 74 39.5 | 114 11.7 17.6 | 267 28.0 03.4 | 272 36.9 51.4 | 256 31.5 11.4 | Miaplacidus 221 44.1 S69 38.0 |
| 23 | 89 41.9 | 129 11.1 16.6 | 282 30.1 03.4 | 287 39.5 51.4 | 271 33.9 11.4 | |
| 6 00 | 104 44.4 | 144 10.5 S17 15.6 | 297 32.2 N 9 03.3 | 302 42.0 N 8 51.4 | 286 36.4 N 3 11.4 | Mirfak 309 16.4 N49 47.5 |
| 01 | 119 46.8 | 159 09.9 14.7 | 312 34.4 03.3 | 317 44.5 51.5 | 301 38.8 11.4 | Nunki 76 30.2 S26 19.3 |
| 02 | 134 49.3 | 174 09.4 13.7 | 327 36.5 03.2 | 332 47.1 51.5 | 316 41.3 11.4 | Peacock 53 59.8 S56 48.1 |
| 03 | 149 51.8 | 189 08.8 ·· 12.7 | 342 38.7 ·· 03.2 | 347 49.6 ·· 51.6 | 331 43.8 ·· 11.4 | Pollux 243 58.4 N28 04.4 |
| 04 | 164 54.2 | 204 08.2 11.8 | 357 40.9 03.2 | 2 52.1 51.6 | 346 46.2 11.4 | Procyon 245 26.0 N 5 16.5 |
| 05 | 179 56.7 | 219 07.6 10.8 | 12 43.0 03.1 | 17 54.7 51.7 | 1 48.7 11.5 | |
| 06 | 194 59.2 | 234 07.0 S17 09.8 | 27 45.2 N 9 03.1 | 32 57.2 N 8 51.7 | 16 51.1 N 3 11.5 | Rasalhague 96 30.4 N12 34.5 |
| 07 | 210 01.6 | 249 06.4 08.8 | 42 47.3 03.0 | 47 59.8 51.7 | 31 53.6 11.5 | Regulus 208 10.3 N12 03.8 |
| 08 | 225 04.1 | 264 05.8 07.9 | 57 49.5 03.0 | 63 02.3 51.8 | 46 56.0 11.5 | Rigel 281 36.2 S 8 13.6 |
| S 09 | 240 06.6 | 279 05.2 ·· 06.9 | 72 51.6 ·· 03.0 | 78 04.9 ·· 51.8 | 61 58.5 ·· 11.5 | Rigil Kent. 140 26.7 S60 44.8 |
| U 10 | 255 09.0 | 294 04.7 05.9 | 87 53.8 02.9 | 93 07.4 51.9 | 77 01.0 11.5 | Sabik 102 42.0 S15 41.9 |
| N 11 | 270 11.5 | 309 04.1 04.9 | 102 56.0 02.9 | 108 09.9 51.9 | 92 03.4 11.5 | |
| D 12 | 285 14.0 | 324 03.5 S17 04.0 | 117 58.1 N 9 02.9 | 123 12.5 N 8 52.0 | 107 05.9 N 3 11.5 | Schedar 350 09.5 N56 25.9 |
| A 13 | 300 16.4 | 339 02.9 03.0 | 133 00.3 02.8 | 138 15.0 52.0 | 122 08.3 11.5 | Shaula 96 56.8 S37 05.2 |
| Y 14 | 315 18.9 | 354 02.3 02.0 | 148 02.4 02.8 | 153 17.6 52.0 | 137 10.8 11.5 | Sirius 258 55.8 S16 41.5 |
| 15 | 330 21.3 | 9 01.7 ·· 01.0 | 163 04.6 ·· 02.7 | 168 20.1 ·· 52.1 | 152 13.2 ·· 11.6 | Spica 158 58.1 S11 03.3 |
| 16 | 345 23.8 | 24 01.2 17 00.1 | 178 06.8 02.7 | 183 22.6 52.1 | 167 15.7 11.6 | Suhail 223 10.7 S43 21.0 |
| 17 | 0 26.3 | 39 00.6 16 59.1 | 193 08.9 02.7 | 198 25.2 52.2 | 182 18.2 11.6 | |
| 18 | 15 28.7 | 54 00.0 S16 58.1 | 208 11.1 N 9 02.6 | 213 27.7 N 8 52.2 | 197 20.6 N 3 11.6 | Vega 80 56.6 N38 46.0 |
| 19 | 30 31.2 | 68 59.4 57.1 | 223 13.3 02.6 | 228 30.3 52.3 | 212 23.1 11.6 | Zuben'ubi 137 33.7 S15 57.4 |
| 20 | 45 33.7 | 83 58.8 56.1 | 238 15.5 02.6 | 243 32.8 52.3 | 227 25.5 11.6 | S.H.A. Mer. Pass. |
| 21 | 60 36.1 | 98 58.3 ·· 55.2 | 253 17.6 ·· 02.5 | 258 35.4 ·· 52.4 | 242 28.0 ·· 11.6 | Venus 40 39.6 14 23 |
| 22 | 75 38.6 | 113 57.7 54.2 | 268 19.8 02.5 | 273 37.9 52.4 | 257 30.5 11.6 | Mars 192 55.7 4 13 |
| 23 | 90 41.1 | 128 57.1 53.2 | 283 22.0 02.5 | 288 40.5 52.4 | 272 32.9 11.6 | Jupiter 197 55.9 3 53 |
| Mer Pass. 17 02.2 | v −0.6 d 1.0 | v 2.1 d 0.0 | v 2.5 d 0.0 | v 2.5 d 0.0 | Saturn 181 52.3 4 57 |

## 1980 JANUARY 4, 5, 6 (FRI., SAT., SUN.)

| G.M.T. | SUN G.H.A. | SUN Dec. | MOON G.H.A. | MOON v | MOON Dec. | MOON d | MOON H.P. | Lat. | Twilight Naut. | Twilight Civil | Sunrise | Moonrise 4 | Moonrise 5 | Moonrise 6 | Moonrise 7 |
|---|---|---|---|---|---|---|---|---|---|---|---|---|---|---|---|
| d h | ° ' | ° ' | ° ' | ' | ° ' | ' | ' | ° | h m | h m | h m | h m | h m | h m | h m |
| 4 00 | 178 53.3 | S22 49.6 | 339 25.3 | 11.2 | N17 23.9 | 4.5 | 55.4 | N 72 | 08 21 | 10 32 | ▓ | 15 46 | 17 38 | 19 20 | 20 58 |
| 01 | 193 53.0 | 49.3 | 353 55.5 | 11.2 | 17 19.4 | 4.5 | 55.4 | N 70 | 08 02 | 09 44 | ▓ | 16 29 | 18 04 | 19 36 | 21 07 |
| 02 | 208 52.7 | 49.1 | 8 25.7 | 11.2 | 17 14.9 | 4.6 | 55.4 | 68 | 07 48 | 09 13 | 11 36 | 16 57 | 18 23 | 19 49 | 21 14 |
| 03 | 223 52.5 | ·· 48.8 | 22 55.9 | 11.3 | 17 10.3 | 4.7 | 55.4 | 66 | 07 36 | 08 50 | 10 22 | 17 18 | 18 39 | 20 00 | 21 20 |
| 04 | 238 52.2 | 48.6 | 37 26.2 | 11.3 | 17 05.6 | 4.8 | 55.3 | 64 | 07 25 | 08 32 | 09 46 | 17 35 | 18 51 | 20 08 | 21 25 |
| 05 | 253 51.9 | 48.3 | 51 56.5 | 11.4 | 17 00.8 | 4.9 | 55.3 | 62 | 07 16 | 08 17 | 09 20 | 17 49 | 19 02 | 20 16 | 21 29 |
| 06 | 268 51.6 | S22 48.1 | 66 26.9 | 11.5 | N16 55.9 | 4.9 | 55.3 | 60 | 07 09 | 08 04 | 09 00 | 18 01 | 19 11 | 20 22 | 21 33 |
| 07 | 283 51.3 | · 47.8 | 80 57.4 | 11.5 | 16 51.0 | 5.0 | 55.3 | N 58 | 07 02 | 07 53 | 08 44 | 18 11 | 19 19 | 20 28 | 21 36 |
| 08 | 298 51.0 | 47.6 | 95 27.9 | 11.5 | 16 46.0 | 5.1 | 55.3 | 56 | 06 55 | 07 44 | 08 30 | 18 20 | 19 26 | 20 33 | 21 39 |
| F 09 | 313 50.7 | ·· 47.4 | 109 58.4 | 11.6 | 16 40.9 | 5.2 | 55.2 | 54 | 06 49 | 07 35 | 08 18 | 18 27 | 19 32 | 20 37 | 21 42 |
| R 10 | 328 50.5 | 47.1 | 124 29.0 | 11.7 | 16 35.7 | 5.2 | 55.2 | 52 | 06 44 | 07 27 | 08 08 | 18 34 | 19 38 | 20 41 | 21 45 |
| I 11 | 343 50.2 | 46.8 | 138 59.7 | 11.7 | 16 30.5 | 5.3 | 55.2 | 50 | 06 39 | 07 20 | 07 58 | 18 41 | 19 43 | 20 45 | 21 47 |
| D 12 | 358 49.9 | S22 46.6 | 153 30.4 | 11.7 | N16 25.2 | 5.4 | 55.2 | 45 | 06 28 | 07 05 | 07 38 | 18 54 | 19 54 | 20 53 | 21 52 |
| A 13 | 13 49.6 | 46.3 | 168 01.1 | 11.9 | 16 19.8 | 5.5 | 55.2 | N 40 | 06 18 | 06 52 | 07 22 | 19 06 | 20 03 | 20 59 | 21 55 |
| Y 14 | 28 49.3 | 46.1 | 182 32.0 | 11.8 | 16 14.3 | 5.5 | 55.1 | 35 | 06 09 | 06 41 | 07 09 | 19 15 | 20 10 | 21 05 | 21 59 |
| 15 | 43 49.0 | ·· 45.8 | 197 02.8 | 11.9 | 16 08.8 | 5.6 | 55.1 | 30 | 06 01 | 06 30 | 06 57 | 19 23 | 20 17 | 21 10 | 22 02 |
| 16 | 58 48.7 | 45.6 | 211 33.7 | 12.0 | 16 03.2 | 5.7 | 55.1 | 20 | 05 45 | 06 12 | 06 36 | 19 38 | 20 29 | 21 18 | 22 07 |
| 17 | 73 48.5 | 45.3 | 226 04.7 | 12.1 | 15 57.5 | 5.8 | 55.1 | N 10 | 05 29 | 05 56 | 06 18 | 19 50 | 20 39 | 21 26 | 22 12 |
| 18 | 88 48.2 | S22 45.1 | 240 35.8 | 12.0 | N15 51.7 | 5.8 | 55.1 | 0 | 05 13 | 05 39 | 06 01 | 20 02 | 20 48 | 21 33 | 22 16 |
| 19 | 103 47.9 | 44.8 | 255 06.8 | 12.2 | 15 45.9 | 5.9 | 55.1 | S 10 | 04 55 | 05 21 | 05 44 | 20 14 | 20 58 | 21 40 | 22 20 |
| 20 | 118 47.6 | 44.5 | 269 38.0 | 12.2 | 15 40.0 | 6.0 | 55.0 | 20 | 04 33 | 05 02 | 05 26 | 20 26 | 21 08 | 21 48 | 22 25 |
| 21 | 133 47.3 | ·· 44.3 | 284 09.2 | 12.2 | 15 34.0 | 6.0 | 55.0 | 30 | 04 04 | 04 38 | 05 05 | 20 40 | 21 20 | 21 56 | 22 30 |
| 22 | 148 47.0 | 44.0 | 298 40.4 | 12.3 | 15 28.0 | 6.1 | 55.0 | 35 | 03 46 | 04 23 | 04 52 | 20 49 | 21 26 | 22 01 | 22 33 |
| 23 | 163 46.8 | 43.8 | 313 11.7 | 12.3 | 15 21.9 | 6.2 | 55.0 | 40 | 03 24 | 04 05 | 04 38 | 20 58 | 21 34 | 22 06 | 22 36 |
| 5 00 | 178 46.5 | S22 43.5 | 327 43.0 | 12.4 | N15 15.7 | 6.2 | 55.0 | 45 | 02 55 | 03 44 | 04 20 | 21 09 | 21 43 | 22 13 | 22 40 |
| 01 | 193 46.2 | 43.2 | 342 14.4 | 12.5 | 15 09.5 | 6.3 | 54.9 | S 50 | 02 12 | 03 15 | 03 59 | 21 22 | 21 53 | 22 20 | 22 45 |
| 02 | 208 45.9 | 43.0 | 356 45.9 | 12.5 | 15 03.2 | 6.4 | 54.9 | 52 | 01 47 | 03 01 | 03 48 | 21 28 | 21 58 | 22 24 | 22 47 |
| 03 | 223 45.6 | ·· 42.7 | 11 17.4 | 12.6 | 14 56.8 | 6.4 | 54.9 | 54 | 01 10 | 02 44 | 03 37 | 21 35 | 22 04 | 22 28 | 22 50 |
| 04 | 238 45.3 | 42.5 | 25 49.0 | 12.6 | 14 50.4 | 6.5 | 54.9 | 56 | //// | 02 23 | 03 23 | 21 43 | 22 10 | 22 32 | 22 52 |
| 05 | 253 45.1 | 42.2 | 40 20.6 | 12.6 | 14 43.9 | 6.6 | 54.9 | 58 | //// | 01 57 | 03 07 | 21 51 | 22 16 | 22 37 | 22 55 |
| 06 | 268 44.8 | S22 41.9 | 54 52.2 | 12.8 | N14 37.3 | 6.6 | 54.9 | S 60 | //// | 01 16 | 02 48 | 22 01 | 22 24 | 22 42 | 22 58 |
| 07 | 283 44.5 | 41.7 | 69 24.0 | 12.7 | 14 30.7 | 6.7 | 54.8 | Lat. | Sunset | Twilight Civil | Twilight Naut. | Moonset 4 | Moonset 5 | Moonset 6 | Moonset 7 |
| S 08 | 298 44.2 | 41.4 | 83 55.7 | 12.8 | 14 24.0 | 6.8 | 54.8 |
| A 09 | 313 43.9 | ·· 41.1 | 98 27.5 | 12.9 | 14 17.2 | 6.8 | 54.8 |
| T 10 | 328 43.7 | 40.8 | 112 59.4 | 12.9 | 14 10.4 | 6.8 | 54.8 | ° | h m | h m | h m | h m | h m | h m | h m |
| U 11 | 343 43.4 | 40.6 | 127 31.3 | 13.0 | 14 03.6 | 7.0 | 54.8 | N 72 | ▓ | 13 38 | 15 50 | 12 03 | 11 49 | 11 40 | 11 32 |
| R 12 | 358 43.1 | S22 40.3 | 142 03.3 | 13.0 | N13 56.6 | 6.9 | 54.8 | N 70 | ▓ | 14 26 | 16 08 | 11 20 | 11 22 | 11 22 | 11 21 |
| D 13 | 13 42.8 | 40.0 | 156 35.3 | 13.1 | 13 49.7 | 7.1 | 54.7 | 68 | 12 35 | 14 57 | 16 23 | 10 51 | 11 02 | 11 08 | 11 13 |
| A 14 | 28 42.5 | 39.8 | 171 07.4 | 13.1 | 13 42.6 | 7.1 | 54.7 | 66 | 13 49 | 15 20 | 16 35 | 10 29 | 10 45 | 10 57 | 11 05 |
| Y 15 | 43 42.2 | ·· 39.5 | 185 39.5 | 13.2 | 13 35.5 | 7.1 | 54.7 | 64 | 14 25 | 15 38 | 16 45 | 10 12 | 10 32 | 10 47 | 10 59 |
| 16 | 58 42.0 | 39.2 | 200 11.7 | 13.2 | 13 28.4 | 7.2 | 54.7 | 62 | 14 50 | 15 53 | 16 54 | 09 57 | 10 21 | 10 39 | 10 54 |
| 17 | 73 41.7 | 38.9 | 214 43.9 | 13.3 | 13 21.2 | 7.3 | 54.7 | 60 | 15 10 | 16 06 | 17 02 | 09 45 | 10 11 | 10 32 | 10 49 |
| 18 | 88 41.4 | S22 38.7 | 229 16.2 | 13.3 | N13 13.9 | 7.3 | 54.7 | N 58 | 15 27 | 16 17 | 17 09 | 09 35 | 10 03 | 10 25 | 10 45 |
| 19 | 103 41.1 | 38.4 | 243 48.5 | 13.4 | 13 06.6 | 7.4 | 54.7 | 56 | 15 40 | 16 27 | 17 15 | 09 25 | 09 55 | 10 20 | 10 41 |
| 20 | 118 40.8 | 38.1 | 258 20.9 | 13.4 | 12 59.2 | 7.4 | 54.6 | 54 | 15 52 | 16 35 | 17 21 | 09 17 | 09 48 | 10 15 | 10 38 |
| 21 | 133 40.6 | ·· 37.8 | 272 53.3 | 13.5 | 12 51.8 | 7.5 | 54.6 | 52 | 16 03 | 16 43 | 17 27 | 09 10 | 09 42 | 10 10 | 10 35 |
| 22 | 148 40.3 | 37.6 | 287 25.8 | 13.5 | 12 44.3 | 7.5 | 54.6 | 50 | 16 12 | 16 50 | 17 31 | 09 03 | 09 37 | 10 06 | 10 32 |
| 23 | 163 40.0 | 37.3 | 301 58.3 | 13.5 | 12 36.8 | 7.6 | 54.6 | 45 | 16 32 | 17 06 | 17 43 | 08 49 | 09 25 | 09 57 | 10 26 |
| 6 00 | 178 39.7 | S22 37.0 | 316 30.8 | 13.7 | N12 29.2 | 7.6 | 54.6 | N 40 | 16 48 | 17 19 | 17 52 | 08 37 | 09 15 | 09 50 | 10 21 |
| 01 | 193 39.4 | 36.7 | 331 03.5 | 13.6 | 12 21.6 | 7.7 | 54.6 | 35 | 17 02 | 17 30 | 18 01 | 08 27 | 09 07 | 09 43 | 10 16 |
| 02 | 208 39.2 | 36.4 | 345 36.1 | 13.7 | 12 13.9 | 7.7 | 54.5 | 30 | 17 14 | 17 40 | 18 10 | 08 18 | 09 00 | 09 37 | 10 12 |
| 03 | 223 38.9 | ·· 36.1 | 0 08.8 | 13.8 | 12 06.2 | 7.8 | 54.5 | 20 | 17 34 | 17 58 | 18 25 | 08 03 | 08 47 | 09 27 | 10 05 |
| 04 | 238 38.6 | 35.9 | 14 41.6 | 13.7 | 11 58.4 | 7.8 | 54.5 | N 10 | 17 52 | 18 15 | 18 41 | 07 50 | 08 35 | 09 18 | 10 00 |
| 05 | 253 38.3 | 35.6 | 29 14.3 | 13.9 | 11 50.6 | 7.9 | 54.5 | 0 | 18 09 | 18 31 | 18 57 | 07 37 | 08 25 | 09 10 | 09 54 |
| 06 | 268 38.1 | S22 35.3 | 43 47.2 | 13.9 | N11 42.7 | 7.9 | 54.5 | S 10 | 18 26 | 18 49 | 19 16 | 07 24 | 08 14 | 09 02 | 09 48 |
| 07 | 283 37.8 | 35.0 | 58 20.1 | 13.9 | 11 34.8 | 8.0 | 54.5 | 20 | 18 44 | 19 08 | 19 37 | 07 11 | 08 02 | 08 53 | 09 42 |
| 08 | 298 37.5 | 34.7 | 72 53.0 | 14.0 | 11 26.8 | 8.0 | 54.5 | 30 | 19 05 | 19 33 | 20 05 | 06 55 | 07 49 | 08 43 | 09 35 |
| S 09 | 313 37.2 | ·· 34.4 | 87 26.0 | 14.0 | 11 18.8 | 8.1 | 54.5 | 35 | 19 18 | 19 47 | 20 23 | 06 46 | 07 42 | 08 37 | 09 31 |
| U 10 | 328 36.9 | 34.1 | 101 59.0 | 14.0 | 11 10.7 | 8.1 | 54.4 | 40 | 19 32 | 20 05 | 20 46 | 06 36 | 07 33 | 08 30 | 09 26 |
| N 11 | 343 36.7 | 33.9 | 116 32.0 | 14.1 | 11 02.7 | 8.2 | 54.4 | 45 | 19 49 | 20 26 | 21 15 | 06 24 | 07 23 | 08 22 | 09 21 |
| D 12 | 358 36.4 | S22 33.6 | 131 05.1 | 14.1 | N10 54.5 | 8.2 | 54.4 | S 50 | 20 11 | 20 54 | 21 57 | 06 09 | 07 10 | 08 13 | 09 15 |
| A 13 | 13 36.1 | 33.3 | 145 38.2 | 14.2 | 10 46.3 | 8.2 | 54.4 | 52 | 20 21 | 21 09 | 22 22 | 06 02 | 07 05 | 08 08 | 09 12 |
| Y 14 | 28 35.8 | 33.0 | 160 11.4 | 14.2 | 10 38.1 | 8.4 | 54.4 | 54 | 20 33 | 21 25 | 22 58 | 05 54 | 06 58 | 08 03 | 09 08 |
| 15 | 43 35.6 | ·· 32.7 | 174 44.6 | 14.3 | 10 29.9 | 8.4 | 54.4 | 56 | 20 46 | 21 46 | //// | 05 45 | 06 51 | 07 58 | 09 05 |
| 16 | 58 35.3 | 32.4 | 189 17.9 | 14.3 | 10 21.5 | 8.3 | 54.4 | 58 | 21 02 | 22 12 | //// | 05 36 | 06 43 | 07 52 | 09 01 |
| 17 | 73 35.0 | 32.1 | 203 51.2 | 14.3 | 10 13.2 | 8.4 | 54.4 | S 60 | 21 21 | 22 51 | //// | 05 24 | 06 34 | 07 45 | 08 56 |
| 18 | 88 34.7 | S22 31.8 | 218 24.5 | 14.4 | N10 04.8 | 8.4 | 54.4 | | SUN | SUN | SUN | MOON | MOON | MOON | MOON |
| 19 | 103 34.5 | 31.5 | 232 57.9 | 14.4 | 9 56.4 | 8.5 | 54.4 | Day | Eqn. of Time 00ʰ | Eqn. of Time 12ʰ | Mer. Pass. | Mer. Pass. Upper | Mer. Pass. Lower | Age | Phase |
| 20 | 118 34.2 | 31.2 | 247 31.3 | 14.5 | 9 47.9 | 8.4 | 54.4 |
| 21 | 133 33.9 | ·· 30.9 | 262 04.8 | 14.5 | 9 39.5 | 8.6 | 54.3 |
| 22 | 148 33.6 | 30.6 | 276 38.3 | 14.5 | 9 30.9 | 8.5 | 54.3 | d | m s | m s | h m | h m | h m | d | |
| 23 | 163 33.4 | 30.3 | 291 11.8 | 14.5 | 9 22.4 | 8.6 | 54.3 | 4 | 04 26 | 04 40 | 12 05 | 01 25 | 13 50 | 16 | ◐ |
| | | | | | | | | 5 | 04 54 | 05 07 | 12 05 | 02 13 | 14 37 | 17 | |
| | S.D. 16.3 | d 0.3 | S.D. 15.0 | | 14.9 | | 14.8 | 6 | 05 21 | 05 34 | 12 06 | 02 59 | 15 22 | 18 | |

## 1980 JANUARY 7, 8, 9 (MON., TUES., WED.)

| G.M.T. | ARIES G.H.A. | VENUS −3.4 G.H.A. Dec. | MARS 0.0 G.H.A. Dec. | JUPITER −1.9 G.H.A. Dec. | SATURN +1.1 G.H.A. Dec. | STARS Name | S.H.A. Dec. |
|---|---|---|---|---|---|---|---|
| d h 7 00 | ° ′ 105 43.5 | ° ′ ° ′ 143 56.5 S16 52.2 | ° ′ ° ′ 298 24.1 N 9 02.4 | ° ′ ° ′ 303 43.0 N 8 52.5 | ° ′ ° ′ 287 35.4 N 3 11.6 | Acamar | ° ′ ° ′ 315 37.4 S40 23.4 |
| 01 | 120 46.0 | 158 56.0 51.2 | 313 26.3 02.4 | 318 45.5 52.5 | 302 37.8 11.7 | Achernar | 335 45.6 S57 20.7 |
| 02 | 135 48.4 | 173 55.4 50.2 | 328 28.5 02.4 | 333 48.1 52.6 | 317 40.3 11.7 | Acrux | 173 37.5 S62 59.0 |
| 03 | 150 50.9 | 188 54.8 ·· 49.2 | 343 30.7 ·· 02.3 | 348 50.6 ·· 52.6 | 332 42.8 ·· 11.7 | Adhara | 255 32.1 S28 56.8 |
| 04 | 165 53.4 | 203 54.2 48.2 | 358 32.9 02.3 | 3 53.2 52.7 | 347 45.2 11.7 | Aldebaran | 291 18.3 N16 28.1 |
| 05 | 180 55.8 | 218 53.7 47.3 | 13 35.0 02.3 | 18 55.7 52.7 | 2 47.7 11.7 | | |
| 06 | 195 58.3 | 233 53.1 S16 46.3 | 28 37.2 N 9 02.3 | 33 58.3 N 8 52.8 | 17 50.1 N 3 11.7 | Alioth | 166 43.0 N56 03.9 |
| 07 | 211 00.8 | 248 52.5 45.3 | 43 39.4 02.2 | 49 00.8 52.8 | 32 52.6 11.7 | Alkaid | 153 19.0 N49 24.6 |
| 08 | 226 03.2 | 263 52.0 44.3 | 58 41.6 02.2 | 64 03.4 52.8 | 47 55.1 11.7 | Al Na'ir | 28 15.9 S47 03.7 |
| M 09 | 241 05.7 | 278 51.4 ·· 43.3 | 73 43.8 ·· 02.2 | 79 05.9 ·· 52.9 | 62 57.5 ·· 11.8 | Alnilam | 276 11.8 S 1 13.0 |
| O 10 | 256 08.2 | 293 50.8 42.3 | 88 46.0 02.1 | 94 08.5 52.9 | 78 00.0 11.8 | Alphard | 218 20.7 S 8 34.3 |
| N 11 | 271 10.6 | 308 50.2 41.3 | 103 48.1 02.1 | 109 11.0 53.0 | 93 02.4 11.8 | | |
| D 12 | 286 13.1 | 323 49.7 S16 40.3 | 118 50.3 N 9 02.1 | 124 13.6 N 8 53.0 | 108 04.9 N 3 11.8 | Alphecca | 126 32.8 N26 46.9 |
| A 13 | 301 15.6 | 338 49.1 39.3 | 133 52.5 02.1 | 139 16.1 53.1 | 123 07.4 11.8 | Alpheratz | 358 09.9 N28 58.9 |
| Y 14 | 316 18.0 | 353 48.5 38.3 | 148 54.7 02.0 | 154 18.7 53.1 | 138 09.8 11.8 | Altair | 62 33.3 N 8 49.0 |
| 15 | 331 20.5 | 8 48.0 ·· 37.3 | 163 56.9 ·· 02.0 | 169 21.2 ·· 53.2 | 153 12.3 ·· 11.8 | Ankaa | 353 40.8 S42 25.2 |
| 16 | 346 22.9 | 23 47.4 36.3 | 178 59.1 02.0 | 184 23.8 53.2 | 168 14.7 11.8 | Antares | 112 57.7 S26 23.1 |
| 17 | 1 25.4 | 38 46.8 35.3 | 194 01.3 02.0 | 199 26.3 53.3 | 183 17.2 11.8 | | |
| 18 | 16 27.9 | 53 46.3 S16 34.3 | 209 03.5 N 9 01.9 | 214 28.9 N 8 53.3 | 198 19.7 N 3 11.9 | Arcturus | 146 19.0 N19 17.1 |
| 19 | 31 30.3 | 68 45.7 33.3 | 224 05.7 01.9 | 229 31.4 53.4 | 213 22.1 11.9 | Atria | 108 22.8 S68 59.3 |
| 20 | 46 32.8 | 83 45.1 32.3 | 239 07.9 01.9 | 244 34.0 53.4 | 228 24.6 11.9 | Avior | 234 27.7 S59 26.7 |
| 21 | 61 35.3 | 98 44.6 ·· 31.3 | 254 10.1 ·· 01.9 | 259 36.5 ·· 53.5 | 243 27.1 ·· 11.9 | Bellatrix | 278 58.9 N 6 19.8 |
| 22 | 76 37.7 | 113 44.0 30.3 | 269 12.3 01.8 | 274 39.1 53.5 | 258 29.5 11.9 | Betelgeuse | 271 28.5 N 7 24.1 |
| 23 | 91 40.2 | 128 43.5 29.3 | 284 14.5 01.8 | 289 41.6 53.6 | 273 32.0 11.9 | | |
| 8 00 | 106 42.7 | 143 42.9 S16 28.3 | 299 16.7 N 9 01.8 | 304 44.2 N 8 53.6 | 288 34.5 N 3 11.9 | Canopus | 264 06.9 S52 41.3 |
| 01 | 121 45.1 | 158 42.3 27.3 | 314 18.9 01.8 | 319 46.7 53.6 | 303 36.9 11.9 | Capella | 281 11.6 N45 58.7 |
| 02 | 136 47.6 | 173 41.8 26.3 | 329 21.1 01.8 | 334 49.3 53.7 | 318 39.4 12.0 | Deneb | 49 49.2 N45 12.7 |
| 03 | 151 50.1 | 188 41.2 ·· 25.3 | 344 23.3 ·· 01.7 | 349 51.9 ·· 53.7 | 333 41.8 ·· 12.0 | Denebola | 182 59.4 N14 40.9 |
| 04 | 166 52.5 | 203 40.7 24.3 | 359 25.5 01.7 | 4 54.4 53.8 | 348 44.3 12.0 | Diphda | 349 21.4 S18 06.0 |
| 05 | 181 55.0 | 218 40.1 23.3 | 14 27.7 01.7 | 19 57.0 53.8 | 3 46.8 12.0 | | |
| 06 | 196 57.4 | 233 39.5 S16 22.3 | 29 29.9 N 9 01.7 | 34 59.5 N 8 53.9 | 18 49.2 N 3 12.0 | Dubhe | 194 22.5 N61 51.3 |
| 07 | 211 59.9 | 248 39.0 21.3 | 44 32.1 01.6 | 50 02.1 53.9 | 33 51.7 12.0 | Elnath | 278 44.4 N28 35.4 |
| 08 | 227 02.4 | 263 38.4 20.3 | 59 34.3 01.7 | 65 04.6 54.0 | 48 54.2 12.0 | Eltanin | 90 58.5 N51 29.5 |
| T 09 | 242 04.8 | 278 37.9 ·· 19.2 | 74 36.5 ·· 01.6 | 80 07.2 ·· 54.0 | 63 56.6 ·· 12.1 | Enif | 34 12.3 N 9 47.0 |
| U 10 | 257 07.3 | 293 37.3 18.2 | 89 38.7 01.6 | 95 09.7 54.1 | 78 59.1 12.1 | Fomalhaut | 15 52.2 S29 43.9 |
| E 11 | 272 09.8 | 308 36.8 17.2 | 104 40.9 01.6 | 110 12.3 54.1 | 94 01.6 12.1 | | |
| S 12 | 287 12.2 | 323 36.2 S16 16.2 | 119 43.2 N 9 01.6 | 125 14.8 N 8 54.2 | 109 04.0 N 3 12.1 | Gacrux | 172 29.1 S56 59.8 |
| D 13 | 302 14.7 | 338 35.7 15.2 | 134 45.4 01.6 | 140 17.4 54.2 | 124 06.5 12.1 | Gienah | 176 18.4 S17 25.7 |
| A 14 | 317 17.2 | 353 35.1 14.2 | 149 47.6 01.6 | 155 20.0 54.3 | 139 09.0 12.1 | Hadar | 149 24.1 S60 16.3 |
| Y 15 | 332 19.6 | 8 34.6 ·· 13.2 | 164 49.8 ·· 01.5 | 170 22.5 ·· 54.3 | 154 11.4 ·· 12.1 | Hamal | 328 29.3 N23 22.1 |
| 16 | 347 22.1 | 23 34.0 12.2 | 179 52.0 01.5 | 185 25.1 54.4 | 169 13.9 12.1 | Kaus Aust. | 84 17.9 S34 23.6 |
| 17 | 2 24.5 | 38 33.5 11.1 | 194 54.2 01.5 | 200 27.6 54.4 | 184 16.4 12.2 | | |
| 18 | 17 27.0 | 53 32.9 S16 10.1 | 209 56.5 N 9 01.5 | 215 30.2 N 8 54.5 | 199 18.8 N 3 12.2 | Kochab | 137 19.7 N74 14.1 |
| 19 | 32 29.5 | 68 32.4 09.1 | 224 58.7 01.5 | 230 32.8 54.5 | 214 21.3 12.2 | Markab | 14 03.8 N15 05.9 |
| 20 | 47 31.9 | 83 31.8 08.1 | 240 00.9 01.5 | 245 35.3 54.6 | 229 23.8 12.2 | Menkar | 314 41.5 N 4 00.6 |
| 21 | 62 34.4 | 98 31.3 ·· 07.1 | 255 03.1 ·· 01.5 | 260 37.9 ·· 54.6 | 244 26.2 ·· 12.2 | Menkent | 148 37.6 S36 16.1 |
| 22 | 77 36.9 | 113 30.7 06.0 | 270 05.4 01.5 | 275 40.4 54.7 | 259 28.7 12.2 | Miaplacidus | 221 44.1 S69 38.0 |
| 23 | 92 39.3 | 128 30.2 05.0 | 285 07.6 01.4 | 290 43.0 54.7 | 274 31.2 12.2 | | |
| 9 00 | 107 41.8 | 143 29.6 S16 04.0 | 300 09.8 N 9 01.4 | 305 45.6 N 8 54.8 | 289 33.6 N 3 12.3 | Mirfak | 309 16.4 N49 47.5 |
| 01 | 122 44.3 | 158 29.1 03.0 | 315 12.0 01.4 | 320 48.1 54.8 | 304 36.1 12.3 | Nunki | 76 30.1 S26 19.3 |
| 02 | 137 46.7 | 173 28.5 01.9 | 330 14.3 01.4 | 335 50.7 54.9 | 319 38.6 12.3 | Peacock | 53 59.8 S56 48.1 |
| 03 | 152 49.2 | 188 28.0 16 00.9 | 345 16.5 ·· 01.4 | 350 53.2 ·· 54.9 | 334 41.0 ·· 12.3 | Pollux | 243 58.4 N28 04.4 |
| 04 | 167 51.7 | 203 27.4 15 59.9 | 0 18.7 01.4 | 5 55.8 55.0 | 349 43.5 12.3 | Procyon | 245 25.9 N 5 16.5 |
| 05 | 182 54.1 | 218 26.9 58.9 | 15 21.0 01.4 | 20 58.4 55.0 | 4 46.0 12.3 | | |
| 06 | 197 56.6 | 233 26.3 S15 57.8 | 30 23.2 N 9 01.4 | 36 00.9 N 8 55.1 | 19 48.5 N 3 12.3 | Rasalhague | 96 30.4 N12 34.5 |
| 07 | 212 59.0 | 248 25.8 56.8 | 45 25.4 01.4 | 51 03.5 55.1 | 34 50.9 12.4 | Regulus | 208 10.3 N12 03.8 |
| W 08 | 228 01.5 | 263 25.3 55.8 | 60 27.7 01.4 | 66 06.0 55.2 | 49 53.4 12.4 | Rigel | 281 36.2 S 8 13.6 |
| E 09 | 243 04.0 | 278 24.7 ·· 54.8 | 75 29.9 ·· 01.4 | 81 08.6 ·· 55.2 | 64 55.9 ·· 12.4 | Rigil Kent. | 140 26.7 S60 44.8 |
| D 10 | 258 06.4 | 293 24.2 53.7 | 90 32.1 01.4 | 96 11.2 55.3 | 79 58.3 12.4 | Sabik | 102 42.0 S15 41.9 |
| N 11 | 273 08.9 | 308 23.6 52.7 | 105 34.4 01.4 | 111 13.7 55.4 | 95 00.8 12.4 | | |
| E 12 | 288 11.4 | 323 23.1 S15 51.7 | 120 36.6 N 9 01.4 | 126 16.3 N 8 55.4 | 110 03.3 N 3 12.5 | Schedar | 350 09.5 N56 25.9 |
| S 13 | 303 13.8 | 338 22.6 50.6 | 135 38.9 01.4 | 141 18.9 55.5 | 125 05.7 12.5 | Shaula | 96 56.8 S37 05.2 |
| D 14 | 318 16.3 | 353 22.0 49.6 | 150 41.1 01.3 | 156 21.4 55.5 | 140 08.2 12.5 | Sirius | 258 55.8 S16 41.5 |
| A 15 | 333 18.8 | 8 21.5 ·· 48.6 | 165 43.3 ·· 01.3 | 171 24.0 ·· 55.6 | 155 10.7 ·· 12.5 | Spica | 158 58.1 S11 03.3 |
| Y 16 | 348 21.2 | 23 21.0 47.5 | 180 45.6 01.3 | 186 26.6 55.6 | 170 13.2 12.5 | Suhail | 223 10.7 S43 21.1 |
| 17 | 3 23.7 | 38 20.4 46.5 | 195 47.8 01.3 | 201 29.1 55.7 | 185 15.6 12.5 | | |
| 18 | 18 26.2 | 53 19.9 S15 45.5 | 210 50.1 N 9 01.3 | 216 31.7 N 8 55.7 | 200 18.1 N 3 12.5 | Vega | 80 56.6 N38 45.9 |
| 19 | 33 28.6 | 68 19.3 44.4 | 225 52.3 01.3 | 231 34.3 55.8 | 215 20.6 12.5 | Zuben'ubi | 137 33.7 S15 57.4 |
| 20 | 48 31.1 | 83 18.8 43.4 | 240 54.6 01.3 | 246 36.8 55.8 | 230 23.0 12.6 | | S.H.A. Mer. Pass. |
| 21 | 63 33.5 | 98 18.3 ·· 42.4 | 255 56.8 ·· 01.3 | 261 39.4 ·· 55.9 | 245 25.5 ·· 12.6 | | ° ′ h m |
| 22 | 78 36.0 | 113 17.7 41.3 | 270 59.1 01.3 | 276 42.0 55.9 | 260 28.0 12.6 | Venus | 37 00.2 14 26 |
| 23 | 93 38.5 | 128 17.2 40.3 | 286 01.3 01.3 | 291 44.5 56.0 | 275 30.5 12.6 | Mars | 192 34.0 4 02 |
| | | | | | | Jupiter | 198 01.5 3 40 |
| Mer. Pass. | h m 16 50.4 | v −0.6 d 1.0 | v 2.2 d 0.0 | v 2.6 d 0.0 | v 2.5 d 0.0 | Saturn | 181 51.8 4 45 |

## 1980 JANUARY 7, 8, 9 (MON., TUES., WED.)

| G.M.T. | SUN G.H.A. | SUN Dec. | MOON G.H.A. | MOON v | MOON Dec. | MOON d | MOON H.P. | Lat. | Twilight Naut. | Twilight Civil | Sunrise | Moonrise 7 | Moonrise 8 | Moonrise 9 | Moonrise 10 |
|---|---|---|---|---|---|---|---|---|---|---|---|---|---|---|---|
| d h | ° ' | ° ' | ° ' | ' | ° ' | ' | ' | ° | h m | h m | h m | h m | h m | h m | h m |
| 7 00 | 178 33.1 | S22 30.0 | 305 45.3 14.6 | N 9 13.8 | 8.7 54.3 | | | N 72 | 08 16 | 10 22 | ▬ | 20 58 | 22 33 | 24 07 | 00 07 |
| 01 | 193 32.8 | 29.7 | 320 18.9 14.7 | 9 05.1 | 8.7 54.3 | | | N 70 | 07 59 | 09 38 | ▬ | 21 07 | 22 36 | 24 04 | 00 04 |
| 02 | 208 32.5 | 29.4 | 334 52.6 14.6 | 8 56.4 | 8.7 54.3 | | | 68 | 07 45 | 09 09 | 11 14 | 21 14 | 22 38 | 24 02 | 00 02 |
| 03 | 223 32.3 | 29.1 | 349 26.2 14.7 | 8 47.7 | 8.7 54.3 | | | 66 | 07 33 | 08 47 | 10 15 | 21 20 | 22 40 | 24 00 | 00 00 |
| 04 | 238 32.0 | 28.8 | 3 59.9 14.8 | 8 39.0 | 8.8 54.3 | | | 64 | 07 23 | 08 29 | 09 41 | 21 25 | 22 41 | 23 58 | 25 16 |
| 05 | 253 31.7 | 28.5 | 18 33.7 14.7 | 8 30.2 | 8.8 54.3 | | | 62 | 07 15 | 08 15 | 09 17 | 21 29 | 22 43 | 23 56 | 25 11 |
| 06 | 268 31.4 | S22 28.2 | 33 07.4 14.8 | N 8 21.4 | 8.8 54.3 | | | 60 | 07 07 | 08 03 | 08 58 | 21 33 | 22 44 | 23 55 | 25 07 |
| 07 | 283 31.2 | 27.9 | 47 41.2 14.8 | 8 12.6 | 8.9 54.3 | | | N 58 | 07 00 | 07 52 | 08 42 | 21 36 | 22 45 | 23 54 | 25 04 |
| 08 | 298 30.9 | 27.6 | 62 15.0 14.9 | 8 03.7 | 8.9 54.3 | | | 56 | 06 54 | 07 42 | 08 28 | 21 39 | 22 46 | 23 53 | 25 01 |
| M 09 | 313 30.6 | 27.3 | 76 48.9 14.9 | 7 54.8 | 8.9 54.3 | | | 54 | 06 49 | 07 34 | 08 17 | 21 42 | 22 47 | 23 52 | 24 58 |
| O 10 | 328 30.4 | 27.0 | 91 22.8 14.9 | 7 45.9 | 8.9 54.2 | | | 52 | 06 43 | 07 26 | 08 06 | 21 45 | 22 48 | 23 51 | 24 55 |
| N 11 | 343 30.1 | 26.7 | 105 56.7 14.9 | 7 37.0 | 9.0 54.2 | | | 50 | 06 39 | 07 20 | 07 57 | 21 47 | 22 48 | 23 50 | 24 53 |
| D 12 | 358 29.8 | S22 26.4 | 120 30.6 15.0 | N 7 28.0 | 9.0 54.2 | | | 45 | 06 28 | 07 05 | 07 38 | 21 52 | 22 50 | 23 49 | 24 48 |
| A 13 | 13 29.5 | 26.1 | 135 04.6 15.0 | 7 19.0 | 9.1 54.2 | | | N 40 | 06 18 | 06 52 | 07 22 | 21 55 | 22 51 | 23 47 | 24 44 |
| Y 14 | 28 29.3 | 25.8 | 149 38.6 15.0 | 7 09.9 | 9.0 54.2 | | | 35 | 06 09 | 06 41 | 07 09 | 21 59 | 22 53 | 23 46 | 24 41 |
| 15 | 43 29.0 | 25.5 | 164 12.6 15.0 | 7 00.9 | 9.1 54.2 | | | 30 | 06 01 | 06 31 | 06 57 | 22 02 | 22 54 | 23 45 | 24 37 |
| 16 | 58 28.7 | 25.1 | 178 46.6 15.1 | 6 51.8 | 9.1 54.2 | | | 20 | 05 46 | 06 13 | 06 37 | 22 07 | 22 55 | 23 43 | 24 32 |
| 17 | 73 28.4 | 24.8 | 193 20.7 15.1 | 6 42.7 | 9.2 54.2 | | | N 10 | 05 30 | 05 57 | 06 19 | 22 12 | 22 57 | 23 42 | 24 28 |
| 18 | 88 28.2 | S22 24.5 | 207 54.8 15.1 | N 6 33.5 | 9.1 54.2 | | | 0 | 05 14 | 05 40 | 06 03 | 22 16 | 22 58 | 23 40 | 24 23 |
| 19 | 103 27.9 | 24.2 | 222 28.9 15.2 | 6 24.4 | 9.2 54.2 | | | S 10 | 04 56 | 05 23 | 05 46 | 22 20 | 23 00 | 23 39 | 24 19 |
| 20 | 118 27.6 | 23.9 | 237 03.1 15.1 | 6 15.2 | 9.2 54.2 | | | 20 | 04 35 | 05 04 | 05 28 | 22 25 | 23 01 | 23 37 | 24 14 |
| 21 | 133 27.4 | 23.6 | 251 37.2 15.2 | 6 06.0 | 9.3 54.2 | | | 30 | 04 07 | 04 40 | 05 07 | 22 30 | 23 03 | 23 36 | 24 09 |
| 22 | 148 27.1 | 23.3 | 266 11.4 15.2 | 5 56.7 | 9.2 54.2 | | | 35 | 03 49 | 04 25 | 04 55 | 22 33 | 23 04 | 23 35 | 24 06 |
| 23 | 163 26.8 | 22.9 | 280 45.6 15.3 | 5 47.5 | 9.3 54.2 | | | 40 | 03 28 | 04 08 | 04 41 | 22 36 | 23 05 | 23 34 | 24 03 |
| | | | | | | | | 45 | 02 59 | 03 47 | 04 24 | 22 40 | 23 07 | 23 27 | 23 59 |
| 8 00 | 178 26.6 | S22 22.6 | 295 19.9 15.2 | N 5 38.2 | 9.3 54.2 | | | S 50 | 02 18 | 03 19 | 04 02 | 22 45 | 23 08 | 23 31 | 23 54 |
| 01 | 193 26.3 | 22.3 | 309 54.1 15.3 | 5 28.9 | 9.3 54.2 | | | 52 | 01 54 | 03 05 | 03 52 | 22 47 | 23 09 | 23 30 | 23 52 |
| 02 | 208 26.0 | 22.0 | 324 28.4 15.2 | 5 19.6 | 9.3 54.2 | | | 54 | 01 20 | 02 49 | 03 41 | 22 50 | 23 10 | 23 30 | 23 50 |
| 03 | 223 25.8 | 21.7 | 339 02.6 15.3 | 5 10.3 | 9.4 54.2 | | | 56 | //// | 02 29 | 03 28 | 22 52 | 23 11 | 23 29 | 23 48 |
| 04 | 238 25.5 | 21.4 | 353 36.9 15.4 | 5 00.9 | 9.4 54.2 | | | 58 | //// | 02 04 | 03 12 | 22 55 | 23 12 | 23 28 | 23 45 |
| 05 | 253 25.2 | 21.0 | 8 11.3 15.4 | 4 51.5 | 9.4 54.2 | | | S 60 | //// | 01 27 | 02 54 | 22 58 | 23 13 | 23 27 | 23 42 |
| 06 | 268 24.9 | S22 20.7 | 22 45.6 15.3 | N 4 42.1 | 9.4 54.2 | | | | | | | | | | |
| 07 | 283 24.7 | 20.4 | 37 19.9 15.4 | 4 32.7 | 9.4 54.2 | | | Lat. | Sunset | Twilight Civil | Twilight Naut. | Moonset 7 | Moonset 8 | Moonset 9 | Moonset 10 |
| 08 | 298 24.4 | 20.1 | 51 54.3 15.4 | 4 23.3 | 9.5 54.2 | | | | | | | | | | |
| T 09 | 313 24.1 | 19.7 | 66 28.7 15.4 | 4 13.8 | 9.4 54.2 | | | ° | h m | h m | h m | h m | h m | h m | h m |
| U 10 | 328 23.9 | 19.4 | 81 03.1 15.4 | 4 04.4 | 9.5 54.2 | | | N 72 | ▬ | 13 52 | 15 57 | 11 32 | 11 25 | 11 18 | 11 11 |
| E 11 | 343 23.6 | 19.1 | 95 37.5 15.4 | 3 54.9 | 9.5 54.2 | | | N 70 | ▬ | 14 35 | 16 15 | 11 21 | 11 20 | 11 18 | 11 16 |
| S 12 | 358 23.3 | S22 18.8 | 110 11.9 15.4 | N 3 45.4 | 9.5 54.2 | | | 68 | 12 59 | 15 04 | 16 29 | 11 13 | 11 16 | 11 18 | 11 21 |
| D 13 | 13 23.1 | 18.4 | 124 46.3 15.5 | 3 35.9 | 9.5 54.2 | | | 66 | 13 59 | 15 26 | 16 40 | 11 05 | 11 12 | 11 18 | 11 24 |
| A 14 | 28 22.8 | 18.1 | 139 20.8 15.4 | 3 26.4 | 9.5 54.2 | | | 64 | 14 32 | 15 44 | 16 50 | 10 59 | 11 09 | 11 18 | 11 27 |
| Y 15 | 43 22.5 | 17.8 | 153 55.2 15.5 | 3 16.9 | 9.6 54.2 | | | 62 | 14 56 | 15 58 | 16 59 | 10 54 | 11 06 | 11 18 | 11 30 |
| 16 | 58 22.3 | 17.4 | 168 29.7 15.4 | 3 07.3 | 9.6 54.2 | | | 60 | 15 16 | 16 11 | 17 06 | 10 49 | 11 04 | 11 18 | 11 33 |
| 17 | 73 22.0 | 17.1 | 183 04.1 15.5 | 2 57.7 | 9.5 54.2 | | | N 58 | 15 31 | 16 21 | 17 13 | 10 45 | 11 02 | 11 18 | 11 35 |
| 18 | 88 21.7 | S22 16.8 | 197 38.6 15.5 | N 2 48.2 | 9.6 54.2 | | | 56 | 15 45 | 16 31 | 17 19 | 10 41 | 11 00 | 11 18 | 11 37 |
| 19 | 103 21.5 | 16.5 | 212 13.1 15.4 | 2 38.6 | 9.6 54.2 | | | 54 | 15 56 | 16 39 | 17 25 | 10 38 | 10 58 | 11 18 | 11 38 |
| 20 | 118 21.2 | 16.1 | 226 47.5 15.5 | 2 29.0 | 9.6 54.2 | | | 52 | 16 07 | 16 47 | 17 30 | 10 35 | 10 57 | 11 18 | 11 40 |
| 21 | 133 20.9 | 15.8 | 241 22.0 15.5 | 2 19.4 | 9.6 54.2 | | | 50 | 16 16 | 16 54 | 17 35 | 10 32 | 10 56 | 11 18 | 11 41 |
| 22 | 148 20.7 | 15.5 | 255 56.5 15.5 | 2 09.8 | 9.7 54.2 | | | 45 | 16 35 | 17 09 | 17 45 | 10 26 | 10 53 | 11 18 | 11 44 |
| 23 | 163 20.4 | 15.1 | 270 31.0 15.5 | 2 00.1 | 9.6 54.2 | | | N 40 | 16 51 | 17 21 | 17 55 | 10 21 | 10 50 | 11 18 | 11 47 |
| 9 00 | 178 20.1 | S22 14.8 | 285 05.5 15.5 | N 1 50.5 | 9.6 54.2 | | | 35 | 17 04 | 17 32 | 18 04 | 10 16 | 10 48 | 11 18 | 11 49 |
| 01 | 193 19.9 | 14.4 | 299 40.0 15.5 | 1 40.9 | 9.7 54.2 | | | 30 | 17 16 | 17 42 | 18 12 | 10 12 | 10 46 | 11 18 | 11 51 |
| 02 | 208 19.6 | 14.1 | 314 14.5 15.5 | 1 31.2 | 9.6 54.2 | | | 20 | 17 36 | 18 00 | 18 27 | 10 06 | 10 42 | 11 19 | 11 55 |
| 03 | 223 19.4 | 13.8 | 328 49.0 15.5 | 1 21.6 | 9.7 54.2 | | | N 10 | 17 54 | 18 16 | 18 42 | 10 00 | 10 39 | 11 19 | 11 58 |
| 04 | 238 19.1 | 13.4 | 343 23.5 15.5 | 1 11.9 | 9.7 54.2 | | | 0 | 18 10 | 18 32 | 18 58 | 09 54 | 10 36 | 11 19 | 12 01 |
| 05 | 253 18.8 | 13.1 | 357 58.0 15.5 | 1 02.2 | 9.7 54.2 | | | S 10 | 18 27 | 18 50 | 19 16 | 09 48 | 10 34 | 11 18 | 12 04 |
| 06 | 268 18.6 | S22 12.8 | 12 32.5 15.4 | N 0 52.5 | 9.6 54.2 | | | 20 | 18 45 | 19 09 | 19 38 | 09 42 | 10 30 | 11 18 | 12 07 |
| 07 | 283 18.3 | 12.4 | 27 06.9 15.5 | 0 42.9 | 9.7 54.2 | | | 30 | 19 06 | 19 33 | 20 05 | 09 35 | 10 27 | 11 18 | 12 10 |
| W 08 | 298 18.0 | 12.1 | 41 41.4 15.5 | 0 33.2 | 9.7 54.3 | | | 35 | 19 18 | 19 47 | 20 23 | 09 31 | 10 25 | 11 18 | 12 12 |
| E 09 | 313 17.8 | 11.7 | 56 15.9 15.5 | 0 23.5 | 9.7 54.3 | | | 40 | 19 32 | 20 04 | 20 45 | 09 26 | 10 22 | 11 18 | 12 15 |
| D 10 | 328 17.5 | 11.4 | 70 50.4 15.4 | 0 13.8 | 9.7 54.3 | | | 45 | 19 49 | 20 25 | 21 13 | 09 21 | 10 20 | 11 18 | 12 17 |
| N 11 | 343 17.3 | 11.0 | 85 24.8 15.5 | N 0 04.1 | 9.7 54.3 | | | | | | | | | | |
| E 12 | 358 17.0 | S22 10.7 | 99 59.3 15.5 | S 0 05.6 | 9.7 54.3 | | | S 50 | 20 10 | 20 53 | 21 54 | 09 15 | 10 16 | 11 18 | 12 20 |
| S 13 | 13 16.7 | 10.4 | 114 33.8 15.4 | 0 15.3 | 9.7 54.3 | | | 52 | 20 20 | 21 07 | 22 17 | 09 12 | 10 15 | 11 18 | 12 22 |
| D 14 | 28 16.5 | 10.0 | 129 08.2 15.4 | 0 25.0 | 9.7 54.3 | | | 54 | 20 31 | 21 23 | 22 51 | 09 08 | 10 13 | 11 18 | 12 23 |
| A 15 | 43 16.2 | 09.7 | 143 42.6 15.5 | 0 34.7 | 9.7 54.3 | | | 56 | 20 44 | 21 43 | //// | 09 05 | 10 11 | 11 18 | 12 25 |
| Y 16 | 58 15.9 | 09.3 | 158 17.1 15.4 | 0 44.4 | 9.7 54.3 | | | 58 | 20 59 | 22 08 | //// | 09 01 | 10 09 | 11 18 | 12 27 |
| 17 | 73 15.7 | 09.0 | 172 51.5 15.4 | 0 54.1 | 9.7 54.3 | | | S 60 | 21 18 | 22 43 | //// | 08 56 | 10 07 | 11 18 | 12 29 |
| 18 | 88 15.4 | S22 08.6 | 187 25.9 15.4 | S 1 03.8 | 9.7 54.3 | | | | | | | | | | |
| 19 | 103 15.2 | 08.3 | 202 00.3 15.4 | 1 13.5 | 9.7 54.4 | | | | SUN | SUN | SUN | MOON | MOON | MOON | MOON |
| 20 | 118 14.9 | 07.9 | 216 34.7 15.3 | 1 23.2 | 9.7 54.4 | | | Day | Eqn. of Time 00ʰ | Eqn. of Time 12ʰ | Mer. Pass. | Mer. Pass. Upper | Mer. Pass. Lower | Age | Phase |
| 21 | 133 14.6 | 07.6 | 231 09.0 15.4 | 1 32.9 | 9.7 54.4 | | | | | | | | | | |
| 22 | 148 14.4 | 07.2 | 245 43.4 15.3 | 1 42.6 | 9.7 54.4 | | | | m s | m s | h m | h m | h m | d | |
| 23 | 163 14.1 | 06.9 | 260 17.7 15.3 | 1 52.3 | 9.7 54.4 | | | 7 | 05 47 | 06 00 | 12 06 | 03 44 | 16 05 | 19 | |
| | | | | | | | | 8 | 06 13 | 06 26 | 12 06 | 04 26 | 16 47 | 20 | ◐ |
| | S.D. 16.3 | d 0.3 | S.D. 14.8 | | 14.8 | | 14.8 | 9 | 06 39 | 06 52 | 12 07 | 05 08 | 17 29 | 21 | |

# 1980 JUNE 26, 27, 28 (THURS., FRI., SAT.)

| G.M.T. | ARIES G.H.A. | VENUS −3.6 G.H.A. / Dec. | MARS +1.1 G.H.A. / Dec. | JUPITER −1.4 G.H.A. / Dec. | SATURN +1.3 G.H.A. / Dec. | STARS Name / S.H.A. / Dec. |
|---|---|---|---|---|---|---|
| 26 00 | 274 16.2 | 196 34.3 N19 34.2 | 101 18.5 N 3 45.0 | 116 46.5 N10 35.8 | 101 29.9 N 5 30.2 | Acamar 315 37.7 S40 22.9 |
| 01 | 289 18.7 | 211 37.8   33.6 | 116 19.8   44.4 | 131 48.6   35.7 | 116 32.3   30.1 | Achernar 335 45.7 S57 20.0 |
| 02 | 304 21.2 | 226 41.4   33.1 | 131 21.1   43.9 | 146 50.7   35.5 | 131 34.6   30.0 | Acrux 173 37.3 S62 59.7 |
| 03 | 319 23.6 | 241 44.9 ·· 32.6 | 146 22.3 ·· 43.3 | 161 52.8 ·· 35.4 | 146 36.9 ·· 30.0 | Adhara 255 32.6 S28 56.8 |
| 04 | 334 26.1 | 256 48.4   32.1 | 161 23.6   42.7 | 176 54.9   35.2 | 161 39.3   29.9 | Aldebaran 291 18.5 N16 28.1 |
| 05 | 349 28.5 | 271 52.0   31.6 | 176 24.9   42.2 | 191 57.0   35.1 | 176 41.6   29.8 | |
| 06 | 4 31.0 | 286 55.5 N19 31.1 | 191 26.2 N 3 41.6 | 206 59.1 N10 34.9 | 191 43.9 N 5 29.8 | Alioth 166 42.6 N56 04.3 |
| 07 | 19 33.5 | 301 59.0   30.6 | 206 27.5   41.1 | 222 01.2   34.8 | 206 46.3   29.7 | Alkaid 153 18.5 N49 25.0 |
| T 08 | 34 35.9 | 317 02.5   30.1 | 221 28.8   40.5 | 237 03.3   34.7 | 221 48.6   29.7 | Al Na'ir 28 14.9 S47 03.1 |
| H 09 | 49 38.4 | 332 06.0 ·· 29.6 | 236 30.1 ·· 39.9 | 252 05.4 ·· 34.5 | 236 51.0 ·· 29.6 | Alnilam 276 12.2 S 1 12.9 |
| U 10 | 64 40.9 | 347 09.5   29.1 | 251 31.4   39.4 | 267 07.5   34.4 | 251 53.3   29.5 | Alphard 218 21.0 S 8 34.5 |
| R 11 | 79 43.3 | 2 13.0   28.6 | 266 32.6   38.8 | 282 09.6   34.2 | 266 55.6   29.5 | |
| S 12 | 94 45.8 | 17 16.5 N19 28.1 | 281 33.9 N 3 38.3 | 297 11.7 N10 34.1 | 281 58.0 N 5 29.4 | Alphecca 126 32.0 N26 47.0 |
| D 13 | 109 48.3 | 32 20.0   27.6 | 296 35.2   37.7 | 312 13.9   33.9 | 297 00.3   29.3 | Alpheratz 358 09.4 N28 58.7 |
| A 14 | 124 50.7 | 47 23.5   27.2 | 311 36.5   37.2 | 327 16.0   33.8 | 312 02.6   29.3 | Altair 62 32.4 N 8 49.0 |
| Y 15 | 139 53.2 | 62 27.0 ·· 26.7 | 326 37.8 ·· 36.6 | 342 18.1 ·· 33.6 | 327 05.0 ·· 29.2 | Ankaa 353 40.5 S42 24.6 |
| 16 | 154 55.7 | 77 30.5   26.2 | 341 39.1   36.0 | 357 20.2   33.5 | 342 07.3   29.1 | Antares 112 56.7 S26 23.3 |
| 17 | 169 58.1 | 92 34.0   25.7 | 356 40.3   35.5 | 12 2:.3   33.4 | 357 09.6   29.1 | |
| 18 | 185 00.6 | 107 37.4 N19 25.2 | 11 41.6 N 3 34.9 | 27 24.4 N10 33.2 | 12 12.0 N 5 29.0 | Arcturus 146 18.5 N19 17.3 |
| 19 | 200 03.0 | 122 40.9   24.7 | 26 42.9   34.4 | 42 26.5   33.1 | 27 14.3   29.0 | Atria 108 20.5 S68 59.6 |
| 20 | 215 05.5 | 137 44.4   24.2 | 41 44.2   33.8 | 57 28.6   32.9 | 42 16.6   28.9 | Avior 234 28.8 S59 27.0 |
| 21 | 230 08.0 | 152 47.8 ·· 23.7 | 56 45.5 ·· 33.2 | 72 30.7 ·· 32.8 | 57 19.0 ·· 28.8 | Bellatrix 278 59.3 N 6 19.8 |
| 22 | 245 10.4 | 167 51.3   23.2 | 71 46.8   32.7 | 87 32.8   32.6 | 72 21.3   28.8 | Betelgeuse 271 28.8 N 7 24.1 |
| 23 | 260 12.9 | 182 54.8   22.8 | 86 48.0   32.1 | 102 34.9   32.5 | 87 23.6   28.7 | |
| 27 00 | 275 15.4 | 197 58.2 N19 22.3 | 101 49.3 N 3 31.6 | 117 37.0 N10 32.3 | 102 26.0 N 5 28.6 | Canopus 264 07.8 S52 41.2 |
| 01 | 290 17.8 | 213 01.6   21.8 | 116 50.6   31.0 | 132 39.1   32.2 | 117 28.3   28.6 | Capella 281 12.0 N45 58.6 |
| 02 | 305 20.3 | 228 05.1   21.3 | 131 51.9   30.4 | 147 41.2   32.1 | 132 30.6   28.5 | Deneb 49 48.1 N45 12.5 |
| 03 | 320 22.8 | 243 08.5 ·· 20.8 | 146 53.2 ·· 29.9 | 162 43.3 ·· 31.9 | 147 33.0 ·· 28.4 | Denebola 182 59.3 N14 41.0 |
| 04 | 335 25.2 | 258 12.0   20.4 | 161 54.5   29.3 | 177 45.4   31.8 | 162 35.3   28.4 | Diphda 349 21.1 S18 05.6 |
| 05 | 350 27.7 | 273 15.4   19.9 | 176 55.7   28.8 | 192 47.5   31.6 | 177 37.6   28.3 | |
| 06 | 5 30.1 | 288 18.8 N19 19.4 | 191 57.0 N 3 28.2 | 207 49.6 N10 31.5 | 192 40.0 N 5 28.3 | Dubhe 194 22.7 N61 51.7 |
| 07 | 20 32.6 | 303 22.2   18.9 | 206 58.3   27.6 | 222 51.7   31.3 | 207 42.3   28.2 | Elnath 278 44.7 N28 35.4 |
| 08 | 35 35.1 | 318 25.7   18.5 | 221 59.6   27.1 | 237 53.8   31.2 | 222 44.6   28.1 | Eltanin 90 57.3 N51 29.6 |
| F 09 | 50 37.5 | 333 29.1 ·· 18.0 | 237 00.9 ·· 26.5 | 252 55.9 ·· 31.0 | 237 47.0 ·· 28.1 | Enif 34 11.5 N 9 47.1 |
| R 10 | 65 40.0 | 348 32.5   17.5 | 252 02.1   25.9 | 267 58.0   30.9 | 252 49.3   28.0 | Fomalhaut 15 51.5 S29 43.4 |
| I 11 | 80 42.5 | 3 35.9   17.1 | 267 03.4   25.4 | 283 00.2   30.7 | 267 51.6   27.9 | |
| D 12 | 95 44.9 | 18 39.3 N19 16.6 | 282 04.7 N 3 24.8 | 298 02.3 N10 30.6 | 282 54.0 N 5 27.9 | Gacrux 172 28.8 S57 00.4 |
| A 13 | 110 47.4 | 33 42.7   16.1 | 297 06.0   24.3 | 313 04.4   30.5 | 297 56.3   27.8 | Gienah 176 18.1 S17 26.0 |
| Y 14 | 125 49.9 | 48 46.1   15.7 | 312 07.3   23.7 | 328 06.5   30.3 | 312 58.6   27.7 | Hadar 149 23.2 S60 16.9 |
| 15 | 140 52.3 | 63 49.5 ·· 15.2 | 327 08.5 ·· 23.1 | 343 08.6 ·· 30.2 | 328 01.0 ·· 27.7 | Hamal 328 29.3 N23 22.0 |
| 16 | 155 54.8 | 78 52.9   14.7 | 342 09.8   22.6 | 358 10.7   30.0 | 343 03.3   27.6 | Kaus Aust. 84 16.7 S34 23.6 |
| 17 | 170 57.3 | 93 56.3   14.3 | 357 11.1   22.0 | 13 12.8   29.9 | 358 05.6   27.5 | |
| 18 | 185 59.7 | 108 59.6 N19 13.8 | 12 12.4 N 3 21.5 | 28 14.9 N10 29.7 | 13 08.0 N 5 27.5 | Kochab 137 18.4 N74 14.5 |
| 19 | 201 02.2 | 124 03.0   13.4 | 27 13.6   20.9 | 43 17.0   29.6 | 28 10.3   27.4 | Markab 14 03.2 N15 05.9 |
| 20 | 216 04.6 | 139 06.4   12.9 | 42 14.9   20.3 | 58 19.1   29.4 | 43 12.6   27.4 | Menkar 314 41.5 N 4 00.7 |
| 21 | 231 07.1 | 154 09.7 ·· 12.4 | 57 16.2 ·· 19.8 | 73 21.2 ·· 29.3 | 58 15.0 ·· 27.3 | Menkent 148 37.0 S36 16.5 |
| 22 | 246 09.6 | 169 13.1   12.0 | 72 17.5   19.2 | 88 23.3   29.1 | 73 17.3   27.2 | Miaplacidus 221 45.6 S69 38.4 |
| 23 | 261 12.0 | 184 16.5   11.5 | 87 18.7   18.6 | 103 25.4   29.0 | 88 19.6   27.2 | |
| 28 00 | 276 14.5 | 199 19.8 N19·.11.1 | 102 20.0 N 3 18.1 | 118 27.5 N10 28.9 | 103 22.0 N 5 27.1 | Mirfak 309 16.7 N49 47.3 |
| 01 | 291 17.0 | 214 23.2   10.6 | 117 21.3   17.5 | 133 29.6   28.7 | 118 24.3   27.0 | Nunki 76 29.0 S26 19.7 |
| 02 | 306 19.4 | 229 26.5   10.2 | 132 22.6   17.0 | 148 31.7   28.6 | 133 26.6   27.0 | Peacock 53 58.2 S56 47.7 |
| 03 | 321 21.9 | 244 29.9 ·· 09.7 | 147 23.9 ·· 16.4 | 163 33.8 ·· 28.4 | 148 28.9 ·· 26.9 | Pollux 243 58.7 N28 04.4 |
| 04 | 336 24.4 | 259 33.2   09.3 | 162 25.1   15.8 | 178 35.9   28.3 | 163 31.3   26.8 | Procyon 245 26.3 N 5 16.5 |
| 05 | 351 26.8 | 274 36.5   08.8 | 177 26.4   15.3 | 193 38.0   28.1 | 178 33.6   26.8 | |
| 06 | 6 29.3 | 289 39.9 N19 08.4 | 192 27.7 N 3 14.7 | 208 40.1 N10 28.0 | 193 35.9 N 5 26.7 | Rasalhague 96 29.4 N12 34.6 |
| 07 | 21 31.8 | 304 43.2   07.9 | 207 29.0   14.1 | 223 42.2   27.8 | 208 38.3   26.6 | Regulus 208 10.4 N12 03.9 |
| S 08 | 36 34.2 | 319 46.5   07.5 | 222 30.2   13.6 | 238 44.3   27.7 | 223 40.6   26.6 | Rigel 281 36.5 S 8 13.5 |
| A 09 | 51 36.7 | 334 49.8 ·· 07.1 | 237 31.5 ·· 13.0 | 253 46.4 ·· 27.5 | 238 42.9 ·· 26.5 | Rigil Kent. 140 25.6 S60 45.4 |
| T 10 | 66 39.1 | 349 53.1   06.6 | 252 32.8   12.4 | 268 48.5   27.4 | 253 45.3   26.4 | Sabik 102 41.0 S15 42.0 |
| U 11 | 81 41.6 | 4 56.4   06.2 | 267 34.0   11.9 | 283 50.6   27.2 | 268 47.6   26.4 | |
| R 12 | 96 44.1 | 19 59.7 N19 05.7 | 282 35.3 N 3 11.3 | 298 52.7 N10 27.1 | 283 49.9 N 5 26.3 | Schedar 350 09.2 N56 25.5 |
| D 13 | 111 46.5 | 35 03.0   05.3 | 297 36.6   10.8 | 313 54.8   26.9 | 298 52.2   26.2 | Shaula 96 55.6 S37 05.3 |
| A 14 | 126 49.0 | 50 06.3   04.9 | 312 37.9   10.2 | 328 56.9   26.8 | 313 54.6   26.2 | Sirius 258 56.2 S16 41.5 |
| Y 15 | 141 51.5 | 65 09.6 ·· 04.4 | 327 39.1 ·· 09.6 | 343 59.0 ·· 26.6 | 328 56.9 ·· 26.1 | Spica 158 57.6 S11 03.5 |
| 16 | 156 53.9 | 80 12.9   04.0 | 342 40.4   09.1 | 359 01.1   26.5 | 343 59.2   26.0 | Suhail 223 11.2 S43 21.4 |
| 17 | 171 56.4 | 95 16.2   03.6 | 357 41.7   08.5 | 14 03.2   26.4 | 359 01.6   26.0 | |
| 18 | 186 58.9 | 110 19.5 N19 03.1 | 12 43.0 N 3 07.9 | 29 05.3 N10 26.2 | 14 03.9 N 5 25.9 | Vega 80 55.5 N38 46.0 |
| 19 | 202 01.3 | 125 22.8   02.7 | 27 44.2   07.4 | 44 07.4   26.1 | 29 06.2   25.8 | Zuben'ubi 137 33.0 S15 57.6 |
| 20 | 217 03.8 | 140 26.0   02.3 | 42 45.5   06.8 | 59 09.5   25.9 | 44 08.5   25.8 | S.H.A. / Mer. Pass. |
| 21 | 232 06.2 | 155 29.3 ·· 01.9 | 57 46.8 ·· 06.2 | 74 11.6 ·· 25.8 | 59 10.9 ·· 25.7 | |
| 22 | 247 08.7 | 170 32.6   01.4 | 72 48.0   05.7 | 89 13.6   25.6 | 74 13.2   25.7 | Venus 282 42.8 10 46 |
| 23 | 262 11.2 | 185 35.8   01.0 | 87 49.3   05.1 | 104 15.7   25.5 | 89 15.5   25.6 | Mars 186 34.0 17 11 |
| Mer. Pass. | 5 38.1 | v 3.4 d 0.5 | v 1.3 d 0.6 | v 2.1 d 0.1 | v 2.3 d 0.1 | Jupiter 202 21.6 16 07 / Saturn 187 10.6 17 08 |

## 1980 JUNE 26, 27, 28 (THURS., FRI., SAT.)

| G.M.T. | SUN GHA | Dec. | MOON GHA | $v$ | Dec. | $d$ | H.P. | Lat. | Twilight Naut. | Civil | Sunrise | Moonrise 26 | 27 | 28 | 29 |
|---|---|---|---|---|---|---|---|---|---|---|---|---|---|---|---|
| d h | ° ' | ° ' | ° ' | ' | ° ' | ' | ' | ° | h m | h m | h m | h m | h m | h m | h m |
| 26 00 | 179 19.5 | N23 21.7 | 29 28.7 | 11.0 | S16 22.9 | 6.1 | 55.9 | N 72 | ☐ | ☐ | ☐ | ■ | ■ | ■ | ■ |
| 01 | 194 19.3 | 21.6 | 43 58.7 | 10.9 | 16 29.0 | 5.9 | 55.9 | 68 | ☐ | ☐ | ☐ | 21 08 | 22 54 | 23 56 | 24 01 |
| 02 | 209 19.2 | 21.6 | 58 28.6 | 10.9 | 16 34.9 | 5.9 | 56.0 | 66 | ☐ | ☐ | ☐ | 20 23 | 21 45 | 22 43 | 23 15 |
| 03 | 224 19.1 ·· | 21.5 | 72 58.5 | 10.8 | 16 40.8 | 5.9 | 56.0 | 64 | //// | //// | 01 35 | 19 54 | 21 09 | 22 07 | 22 45 |
| 04 | 239 18.9 | 21.4 | 87 28.3 | 10.8 | 16 46.7 | 5.7 | 56.0 | 62 | //// | //// | 02 12 | 19 32 | 20 43 | 21 40 | 22 22 |
| 05 | 254 18.8 | 21.3 | 101 58.1 | 10.7 | 16 52.4 | 5.7 | 56.0 | 60 | //// | 00 55 | 02 38 | 19 14 | 20 23 | 21 20 | 22 04 |
| 06 | 269 18.7 | N23 21.2 | 116 27.8 | 10.6 | S16 58.1 | 5.5 | 56.1 | N 58 | //// | 01 44 | 02 59 | 19 00 | 20 06 | 21 03 | 21 49 |
| 07 | 284 18.6 | 21.1 | 130 57.4 | 10.6 | 17 03.6 | 5.5 | 56.1 | 56 | //// | 02 13 | 03 15 | 18 47 | 19 52 | 20 49 | 21 36 |
| T 08 | 299 18.4 | 21.0 | 145 27.0 | 10.5 | 17 09.1 | 5.4 | 56.1 | 54 | 00 50 | 02 35 | 03 30 | 18 36 | 19 40 | 20 37 | 21 25 |
| H 09 | 314 18.3 ·· | 21.0 | 159 56.5 | 10.5 | 17 14.5 | 5.4 | 56.1 | 52 | 01 35 | 02 53 | 03 42 | 18 27 | 19 30 | 20 27 | 21 15 |
| U 10 | 329 18.2 | 20.9 | 174 26.0 | 10.3 | 17 19.9 | 5.2 | 56.2 | 50 | 02 03 | 03 08 | 03 53 | 18 19 | 19 21 | 20 17 | 21 07 |
| R 11 | 344 18.0 | 20.8 | 188 55.3 | 10.3 | 17 25.1 | 5.1 | 56.2 | 45 | 02 48 | 03 38 | 04 15 | 18 11 | 19 12 | 20 09 | 20 59 |
| S 12 | 359 17.9 | N23 20.7 | 203 24.6 | 10.3 | S17 30.2 | 5.1 | 56.2 | N 40 | 03 19 | 04 00 | 04 33 | 17 55 | 18 55 | 19 51 | 20 42 |
| D 13 | 14 17.8 | 20.6 | 217 53.9 | 10.2 | 17 35.3 | 5.0 | 56.2 | 35 | 03 42 | 04 18 | 04 48 | 17 42 | 18 40 | 19 36 | 20 28 |
| A 14 | 29 17.6 | 20.5 | 232 23.1 | 10.1 | 17 40.3 | 4.9 | 56.3 | 30 | 04 00 | 04 34 | 05 01 | 17 31 | 18 28 | 19 24 | 20 17 |
| Y 15 | 44 17.5 ·· | 20.4 | 246 52.2 | 10.1 | 17 45.2 | 4.8 | 56.3 | 20 | 04 29 | 04 58 | 05 23 | 17 21 | 18 18 | 19 13 | 20 06 |
| 16 | 59 17.4 | 20.3 | 261 21.3 | 10.0 | 17 50.0 | 4.7 | 56.3 | N 10 | 04 52 | 05 19 | 05 42 | 17 04 | 18 00 | 18 55 | 19 49 |
| 17 | 74 17.3 | 20.2 | 275 50.3 | 9.9 | 17 54.7 | 4.6 | 56.3 | 0 | 05 11 | 05 37 | 05 59 | 16 50 | 17 44 | 18 39 | 19 34 |
| 18 | 89 17.1 | N23 20.1 | 290 19.2 | 9.9 | S17 59.3 | 4.6 | 56.4 | S 10 | 05 28 | 05 54 | 06 17 | 16 36 | 17 29 | 18 24 | 19 19 |
| 19 | 104 17.0 | 20.0 | 304 48.1 | 9.8 | 18 03.9 | 4.4 | 56.4 | 20 | 05 44 | 06 11 | 06 35 | 16 23 | 17 14 | 18 09 | 19 05 |
| 20 | 119 16.9 | 19.9 | 319 16.9 | 9.7 | 18 08.3 | 4.3 | 56.4 | 30 | 06 00 | 06 30 | 06 56 | 16 08 | 16 59 | 17 53 | 18 50 |
| 21 | 134 16.7 ·· | 19.8 | 333 45.6 | 9.7 | 18 12.6 | 4.3 | 56.4 | 35 | 06 09 | 06 40 | 07 09 | 15 52 | 16 41 | 17 34 | 18 32 |
| 22 | 149 16.6 | 19.7 | 348 14.3 | 9.7 | 18 16.9 | 4.1 | 56.5 | 40 | 06 18 | 06 52 | 07 23 | 15 43 | 16 30 | 17 24 | 18 22 |
| 23 | 164 16.5 | 19.6 | 2 43.0 | 9.6 | 18 21.0 | 4.1 | 56.5 | 45 | 06 28 | 07 06 | 07 39 | 15 32 | 16 18 | 17 11 | 18 10 |
| 27 00 | 179 16.4 | N23 19.5 | 17 11.6 | 9.5 | S18 25.1 | 4.0 | 56.5 | S 50 | 06 40 | 07 22 | 08 00 | 15 19 | 16 04 | 16 57 | 17 57 |
| 01 | 194 16.2 | 19.4 | 31 40.1 | 9.4 | 18 29.1 | 3.8 | 56.6 | 52 | 06 45 | 07 29 | 08 10 | 15 04 | 15 47 | 16 39 | 17 40 |
| 02 | 209 16.1 | 19.3 | 46 08.5 | 9.4 | 18 32.9 | 3.8 | 56.6 | 54 | 06 51 | 07 37 | 08 21 | 14 57 | 15 39 | 16 31 | 17 32 |
| 03 | 224 16.0 ·· | 19.2 | 60 36.9 | 9.4 | 18 36.7 | 3.7 | 56.6 | 56 | 06 57 | 07 46 | 08 33 | 14 49 | 15 30 | 16 22 | 17 23 |
| 04 | 239 15.8 | 19.1 | 75 05.3 | 9.3 | 18 40.4 | 3.6 | 56.6 | 58 | 07 04 | 07 56 | 08 48 | 14 40 | 15 20 | 16 11 | 17 13 |
| 05 | 254 15.7 | 19.0 | 89 33.6 | 9.2 | 18 44.0 | 3.4 | 56.7 | S 60 | 07 11 | 08 08 | 09 05 | 14 30 | 15 09 | 15 59 | 17 02 |
| 06 | 269 15.6 | N23 18.9 | 104 01.8 | 9.2 | S18 47.4 | 3.4 | 56.7 | | | | | 14 18 | 14 56 | 15 45 | 16 49 |
| 07 | 284 15.5 | 18.8 | 118 30.0 | 9.1 | 18 50.8 | 3.3 | 56.7 | Lat. | Sunset | Twilight Civil | Naut. | Moonset 26 | 27 | 28 | 29 |
| 08 | 299 15.3 | 18.7 | 132 58.1 | 9.0 | 18 54.1 | 3.1 | 56.7 | | | | | | | | |
| F 09 | 314 15.2 ·· | 18.6 | 147 26.1 | 9.0 | 18 57.2 | 3.1 | 56.8 | ° | h m | h m | h m | h m | h m | h m | h m |
| R 10 | 329 15.1 | 18.5 | 161 54.1 | 9.0 | 19 00.3 | 3.0 | 56.8 | N 72 | ☐ | ☐ | ☐ | ■ | ■ | ■ | ■ |
| I 11 | 344 14.9 | 18.4 | 176 22.1 | 8.9 | 19 03.3 | 2.8 | 56.8 | 70 | ☐ | ☐ | ☐ | ■ | ■ | ■ | ■ |
| D 12 | 359 14.8 | N23 18.3 | 190 50.0 | 8.8 | S19 06.1 | 2.8 | 56.8 | N 70 | ☐ | ☐ | ☐ | 00 24 | 00 23 | 00 27 | 01 21 |
| A 13 | 14 14.7 | 18.2 | 205 17.8 | 8.8 | 19 08.9 | 2.6 | 56.9 | 68 | ☐ | ☐ | ☐ | 00 54 | 01 08 | 01 36 | 02 33 |
| Y 14 | 29 14.6 | 18.1 | 219 45.6 | 8.8 | 19 11.5 | 2.6 | 56.9 | 66 | ☐ | ☐ | ☐ | 01 16 | 01 37 | 02 13 | 03 09 |
| 15 | 44 14.4 ·· | 18.0 | 234 13.4 | 8.7 | 19 14.1 | 2.4 | 56.9 | 64 | 22 30 | //// | //// | 01 34 | 02 00 | 02 39 | 03 36 |
| 16 | 59 14.3 | 17.9 | 248 41.1 | 8.6 | 19 16.5 | 2.4 | 56.9 | 62 | 21 53 | //// | //// | 01 48 | 02 18 | 02 59 | 03 56 |
| 17 | 74 14.2 | 17.8 | 263 08.7 | 8.6 | 19 18.9 | 2.2 | 57.0 | 60 | 21 27 | 23 10 | //// | 02 01 | 02 33 | 03 16 | 04 12 |
| 18 | 89 14.0 | N23 17.7 | 277 36.3 | 8.6 | S19 21.1 | 2.1 | 57.0 | N 58 | 21 07 | 22 22 | //// | 02 11 | 02 46 | 03 30 | 04 26 |
| 19 | 104 13.9 | 17.5 | 292 03.9 | 8.5 | 19 23.2 | 2.0 | 57.0 | 56 | 20 50 | 21 52 | //// | 02 21 | 02 57 | 03 42 | 04 38 |
| 20 | 119 13.8 | 17.4 | 306 31.4 | 8.5 | 19 25.2 | 1.9 | 57.0 | 54 | 20 36 | 21 30 | 23 14 | 02 29 | 03 06 | 03 52 | 04 49 |
| 21 | 134 13.7 ·· | 17.3 | 320 58.9 | 8.4 | 19 27.1 | 1.8 | 57.1 | 52 | 20 24 | 21 13 | 22 30 | 02 37 | 03 15 | 04 02 | 04 58 |
| 22 | 149 13.5 | 17.2 | 335 26.3 | 8.3 | 19 28.9 | 1.6 | 57.1 | 50 | 20 13 | 20 58 | 22 03 | 02 43 | 03 23 | 04 10 | 05 06 |
| 23 | 164 13.4 | 17.1 | 349 53.6 | 8.4 | 19 30.5 | 1.6 | 57.1 | 45 | 19 51 | 20 28 | 21 18 | 02 58 | 03 39 | 04 28 | 05 24 |
| 28 00 | 179 13.3 | N23 17.0 | 4 21.0 | 8.2 | S19 32.1 | 1.5 | 57.1 | N 40 | 19 33 | 20 06 | 20 47 | 03 09 | 03 53 | 04 42 | 05 39 |
| 01 | 194 13.2 | 16.9 | 18 48.2 | 8.3 | 19 33.6 | 1.3 | 57.2 | 35 | 19 18 | 19 48 | 20 24 | 03 20 | 04 04 | 04 55 | 05 51 |
| 02 | 209 13.0 | 16.7 | 33 15.5 | 8.2 | 19 34.9 | 1.2 | 57.2 | 30 | 19 05 | 19 32 | 20 06 | 03 29 | 04 15 | 05 06 | 06 01 |
| 03 | 224 12.9 ·· | 16.6 | 47 42.7 | 8.1 | 19 36.1 | 1.1 | 57.2 | 20 | 18 43 | 19 08 | 19 37 | 03 44 | 04 32 | 05 24 | 06 20 |
| 04 | 239 12.8 | 16.5 | 62 09.8 | 8.2 | 19 37.2 | 1.0 | 57.3 | N 10 | 18 24 | 18 47 | 19 14 | 03 57 | 04 47 | 05 40 | 06 35 |
| 05 | 254 12.7 | 16.4 | 76 37.0 | 8.1 | 19 38.2 | 0.9 | 57.3 | 0 | 18 07 | 18 29 | 18 55 | 04 10 | 05 01 | 05 55 | 06 50 |
| 06 | 269 12.5 | N23 16.3 | 91 04.1 | 8.0 | S19 39.1 | 0.8 | 57.3 | S 10 | 17 49 | 18 12 | 18 38 | 04 23 | 05 16 | 06 10 | 07 05 |
| 07 | 284 12.4 | 16.1 | 105 31.1 | 8.0 | 19 39.9 | 0.7 | 57.3 | 20 | 17 31 | 17 55 | 18 22 | 04 36 | 05 31 | 06 26 | 07 21 |
| S 08 | 299 12.3 | 16.0 | 119 58.1 | 8.0 | 19 40.6 | 0.5 | 57.4 | 30 | 17 10 | 17 36 | 18 06 | 04 52 | 05 48 | 06 44 | 07 39 |
| A 09 | 314 12.1 ·· | 15.9 | 134 25.1 | 7.9 | 19 41.1 | 0.4 | 57.4 | 35 | 16 57 | 17 26 | 17 57 | 05 01 | 05 58 | 06 55 | 07 49 |
| T 10 | 329 12.0 | 15.8 | 148 52.0 | 7.9 | 19 41.5 | 0.3 | 57.4 | 40 | 16 43 | 17 14 | 17 48 | 05 11 | 06 10 | 07 07 | 08 01 |
| U 11 | 344 11.9 | 15.7 | 163 18.9 | 7.9 | 19 41.8 | 0.2 | 57.4 | 45 | 16 27 | 17 00 | 17 38 | 05 23 | 06 24 | 07 22 | 08 15 |
| R 12 | 359 11.8 | N23 15.5 | 177 45.8 | 7.9 | S19 42.0 | 0.1 | 57.4 | S 50 | 16 06 | 16 44 | 17 26 | 05 38 | 06 40 | 07 39 | 08 33 |
| D 13 | 14 11.6 | 15.4 | 192 12.7 | 7.8 | 19 42.1 | 0.0 | 57.5 | 52 | 15 56 | 16 37 | 17 21 | 05 45 | 06 48 | 07 48 | 08 41 |
| A 14 | 29 11.5 | 15.3 | 206 39.5 | 7.8 | 19 42.1 | 0.2 | 57.5 | 54 | 15 45 | 16 29 | 17 15 | 05 52 | 06 57 | 07 57 | 08 50 |
| Y 15 | 44 11.4 ·· | 15.2 | 221 06.3 | 7.7 | 19 41.9 | 0.3 | 57.5 | 56 | 15 33 | 16 20 | 17 09 | 06 01 | 07 07 | 08 07 | 09 00 |
| 16 | 59 11.3 | 15.0 | 235 33.0 | 7.7 | 19 41.6 | 0.3 | 57.5 | 58 | 15 18 | 16 10 | 17 02 | 06 11 | 07 18 | 08 19 | 09 11 |
| 17 | 74 11.1 | 14.9 | 249 59.7 | 7.7 | 19 41.3 | 0.6 | 57.5 | S 60 | 15 01 | 15 58 | 16 55 | 06 22 | 07 31 | 08 33 | 09 24 |
| 18 | 89 11.0 | N23 14.8 | 264 26.4 | 7.7 | S19 40.7 | 0.7 | 57.6 | | | SUN | | | MOON | | |
| 19 | 104 10.9 | 14.6 | 278 53.1 | 7.7 | 19 40.1 | 0.7 | 57.6 | Day | Eqn. of Time 00ʰ | 12ʰ | Mer. Pass. | Mer. Pass. Upper | Lower | Age | Phase |
| 20 | 119 10.8 | 14.5 | 293 19.8 | 7.6 | 19 39.4 | 0.9 | 57.6 | | | | | | | | |
| 21 | 134 10.6 ·· | 14.4 | 307 46.4 | 7.6 | 19 38.5 | 1.0 | 57.7 | | m s | m s | h m | h m | h m | d | |
| 22 | 149 10.5 | 14.3 | 322 13.0 | 7.6 | 19 37.5 | 1.1 | 57.7 | 26 | 02 42 | 02 48 | 12 03 | 22 49 | 10 23 | 14 | |
| 23 | 164 10.4 | 14+1 | 336 39.6 | 7.6 | 19 36.4 | 1.2 | 57.7 | 27 | 02 54 | 03 00 | 12 03 | 23 42 | 11 15 | 15 | ○ |
| | | | | | | | | 28 | 03 07 | 03 13 | 12 03 | 24 37 | 12 09 | 16 | |
| | S.D. 15.8 | $d$ 0.1 | S.D. | 15.3 | 15.5 | | 15.7 | | | | | | | | |

# 1980 SEPTEMBER 18, 19, 20 (THURS., FRI., SAT.)

| G.M.T. | ARIES G.H.A. | VENUS −3.8 G.H.A. / Dec. | MARS +1.5 G.H.A. / Dec. | JUPITER −1.2 G.H.A. / Dec. | SATURN +1.2 G.H.A. / Dec. | STARS Name / S.H.A. / Dec. |
|---|---|---|---|---|---|---|
| d h | ° ′ | ° ′ ° ′ | ° ′ ° ′ | ° ′ ° ′ | ° ′ ° ′ | ° ′ ° ′ |
| 18 00 | 357 03.9 | 223 55.7 N16 29.9 | 136 34.2 S16 19.0 | 184 13.8 N 4 13.8 | 176 38.2 N 2 03.8 | Acamar   315 37.0 S40 22.8 |
| 01 | 12 06.3 | 238 55.4       29.3 | 151 35.0       19.5 | 199 15.8       13.6 | 191 40.4       03.7 | Achernar 335 44.8 S57 20.0 |
| 02 | 27 08.8 | 253 55.0       28.8 | 166 35.8       20.1 | 214 17.8       13.3 | 206 42.5       03.5 | Acrux    173 37.9 S62 59.4 |
| 03 | 42 11.3 | 268 54.7  ··   28.2 | 181 36.5  ··   20.6 | 229 19.7  ··   13.1 | 221 44.7  ··   03.4 | Adhara   255 32.2 S28 56.5 |
| 04 | 57 13.7 | 283 54.4       27.7 | 196 37.3       21.1 | 244 21.7       12.9 | 236 46.9       03.3 | Aldebaran 291 17.9 N16 28.2 |
| 05 | 72 16.2 | 298 54.0       27.1 | 211 38.1       21.7 | 259 23.7       12.7 | 251 49.1       03.2 |  |
| 06 | 87 18.6 | 313 53.7 N16 26.6 | 226 38.9 S16 22.2 | 274 25.6 N 4 12.5 | 266 51.2 N 2 03.1 | Alioth   166 43.1 N56 04.1 |
| 07 | 102 21.1 | 328 53.4      26.0 | 241 39.7       22.7 | 289 27.6       12.3 | 281 53.4       02.9 | Alkaid   153 19.0 N49 24.9 |
| T 08 | 117 23.6 | 343 53.0      25.5 | 256 40.5       23.3 | 304 29.6       12.1 | 296 55.6       02.8 | Al Na'ir  28 14.6 S47 03.3 |
| H 09 | 132 26.0 | 358 52.7  ··  24.9 | 271 41.3  ··   23.8 | 319 31.5  ··   11.9 | 311 57.8  ··   02.7 | Alnilam  276 11.7 S 1 12.8 |
| U 10 | 147 28.5 | 13 52.4       24.4 | 286 42.0       24.4 | 334 33.5       11.6 | 327 00.0       02.6 | Alphard  218 20.8 S 8 34.3 |
| R 11 | 162 31.0 | 28 52.0       23.8 | 301 42.8       24.9 | 349 35.5       11.4 | 342 02.1       02.4 |  |
| S 12 | 177 33.4 | 43 51.7 N16 23.2 | 316 43.6 S16 25.4 | 4 37.4 N 4 11.2 | 357 04.3 N 2 02.3 | Alphecca  126 32.3 N26 47.1 |
| D 13 | 192 35.9 | 58 51.4       22.7 | 331 44.4       26.0 | 19 39.4       11.0 | 12 06.5       02.2 | Alpheratz 358 09.0 N28 59.1 |
| A 14 | 207 38.4 | 73 51.0       22.1 | 346 45.2       26.5 | 34 41.4       10.8 | 27 08.7       02.1 | Altair    62 32.4 N 8 49.2 |
| Y 15 | 222 40.8 | 88 50.7  ··   21.6 | 1 46.0  ··    27.0 | 49 43.3  ··   10.6 | 42 10.9  ··   02.0 | Ankaa    353 39.9 S42 24.6 |
| 16 | 237 43.3 | 103 50.4      21.0 | 16 46.7       27.6 | 64 45.3       10.4 | 57 13.0       01.8 | Antares  112 57.0 S26 23.3 |
| 17 | 252 45.8 | 118 50.0      20.4 | 31 47.5       28.1 | 79 47.2       10.2 | 72 15.2       01.7 |  |
| 18 | 267 48.2 | 133 49.7 N16 19.9 | 46 48.3 S16 28.6 | 94 49.2 N 4 10.0 | 87 17.4 N 2 01.6 | Arcturus 146 18.8 N19 17.3 |
| 19 | 282 50.7 | 148 49.4      19.3 | 61 49.1       29.2 | 109 51.2       09.7 | 102 19.6       01.5 | Atria    108 21.3 S68 59.8 |
| 20 | 297 53.1 | 163 49.0      18.7 | 76 49.9       29.7 | 124 53.1       09.5 | 117 21.8       01.3 | Avior    234 28.6 S59 26.6 |
| 21 | 312 55.6 | 178 48.7  ··  18.2 | 91 50.6  ··   30.2 | 139 55.1  ··   09.3 | 132 23.9  ··   01.2 | Bellatrix 278 58.7 N 6 20.0 |
| 22 | 327 58.1 | 193 48.4      17.6 | 106 51.4       30.8 | 154 57.1       09.1 | 147 26.1       01.1 | Betelgeuse 271 28.3 N 7 24.3 |
| 23 | 343 00.5 | 208 48.0      17.0 | 121 52.2       31.3 | 169 59.0       08.9 | 162 28.3       01.0 |  |
| 19 00 | 358 03.0 | 223 47.7 N16 16.5 | 136 53.0 S16 31.9 | 185 01.0 N 4 08.7 | 177 30.5 N 2 00.9 | Canopus  264 07.3 S52 40.9 |
| 01 | 13 05.5 | 238 47.4      15.9 | 151 53.8       32.4 | 200 03.0       08.5 | 192 32.7       00.7 | Capella  281 11.2 N45 58.5 |
| 02 | 28 07.9 | 253 47.0      15.3 | 166 54.5       32.9 | 215 04.9       08.3 | 207 34.8       00.6 | Deneb     49 48.1 N45 12.9 |
| 03 | 43 10.4 | 268 46.7  ··  14.8 | 181 55.3  ··   33.5 | 230 06.9  ··   08.0 | 222 37.0  ··   00.5 | Denebola 182 59.4 N14 41.0 |
| 04 | 58 12.9 | 283 46.3      14.2 | 196 56.1       34.0 | 245 08.9       07.8 | 237 39.2       00.4 | Diphda   349 20.6 S18 05.5 |
| 05 | 73 15.3 | 298 46.0      13.6 | 211 56.9       34.5 | 260 10.8       07.6 | 252 41.4       00.2 |  |
| 06 | 88 17.8 | 313 45.7 N16 13.0 | 226 57.7 S16 35.1 | 275 12.8 N 4 07.4 | 267 43.6 N 2 00.1 | Dubhe    194 22.8 N61 51.4 |
| 07 | 103 20.3 | 328 45.3      12.5 | 241 58.4       35.6 | 290 14.8       07.2 | 282 45.7       2 00.0 | Elnath   278 44.1 N28 35.4 |
| 08 | 118 22.7 | 343 45.0      11.9 | 256 59.2       36.1 | 305 16.7       07.0 | 297 47.9      1 59.9 | Eltanin   90 57.7 N51 29.9 |
| F 09 | 133 25.2 | 358 44.7  ··  11.3 | 272 00.0  ··   36.7 | 320 18.7  ··   06.8 | 312 50.1  ··   59.8 | Enif      34 11.3 N 9 47.3 |
| R 10 | 148 27.6 | 13 44.3       10.7 | 287 00.8       37.2 | 335 20.7       06.6 | 327 52.3       59.6 | Fomalhaut 15 51.1 S29 43.4 |
| I 11 | 163 30.1 | 28 44.0       10.2 | 302 01.5       37.7 | 350 22.6       06.3 | 342 54.5       59.5 |  |
| D 12 | 178 32.6 | 43 43.7 N16 09.6 | 317 02.3 S16 38.2 | 5 24.6 N 4 06.1 | 357 56.6 N 1 59.4 | Gacrux   172 29.3 S57 00.2 |
| A 13 | 193 35.0 | 58 43.3       09.0 | 332 03.1       38.8 | 20 26.6       05.9 | 12 58.8       59.3 | Gienah   176 18.3 S17 25.9 |
| Y 14 | 208 37.5 | 73 43.0       08.4 | 347 03.9       39.3 | 35 28.5       05.7 | 28 01.0       59.1 | Hadar    149 23.8 S60 16.8 |
| 15 | 223 40.0 | 88 42.7  ··   07.8 | 2 04.6  ··    39.8 | 50 30.5  ··   05.5 | 43 03.2  ··   59.0 | Hamal    328 28.6 N23 22.2 |
| 16 | 238 42.4 | 103 42.3      07.3 | 17 05.4       40.4 | 65 32.5       05.3 | 58 05.3       58.9 | Kaus Aust. 84 16.8 S34 23.7 |
| 17 | 253 44.9 | 118 42.0      06.7 | 32 06.2       40.9 | 80 34.4       05.1 | 73 07.5       58.8 |  |
| 18 | 268 47.4 | 133 41.6 N16 06.1 | 47 07.0 S16 41.4 | 95 36.4 N 4 04.9 | 88 09.7 N 1 58.7 | Kochab   137 19.9 N74 14.4 |
| 19 | 283 49.8 | 148 41.3      05.5 | 62 07.7       42.0 | 110 38.4       04.7 | 103 11.9       58.5 | Markab    14 02.9 N15 06.2 |
| 20 | 298 52.3 | 163 41.0      04.9 | 77 08.5       42.5 | 125 40.3       04.4 | 118 14.1       58.4 | Menkar   314 40.9 N 4 00.9 |
| 21 | 313 54.7 | 178 40.6  ··  04.3 | 92 09.3  ··   43.0 | 140 42.3  ··   04.2 | 133 16.2  ··   58.3 | Menkent  148 37.3 S36 16.4 |
| 22 | 328 57.2 | 193 40.3      03.8 | 107 10.1       43.6 | 155 44.2       04.0 | 148 18.4       58.2 | Miaplacidus 221 45.6 S69 38.0 |
| 23 | 343 59.7 | 208 40.0      03.2 | 122 10.8       44.1 | 170 46.2       03.8 | 163 20.6       58.0 |  |
| 20 00 | 359 02.1 | 223 39.6 N16 02.6 | 137 11.6 S16 44.6 | 185 48.2 N 4 03.6 | 178 22.8 N 1 57.9 | Mirfak   309 15.8 N49 47.4 |
| 01 | 14 04.6 | 238 39.3      02.0 | 152 12.4       45.1 | 200 50.1       03.4 | 193 25.0       57.8 | Nunki     76 29.1 S26 19.2 |
| 02 | 29 07.1 | 253 39.0      01.4 | 167 13.1       45.7 | 215 52.1       03.2 | 208 27.1       57.7 | Peacock   53 58.1 S56 48.0 |
| 03 | 44 09.5 | 268 38.6  ··  00.8 | 182 13.9  ··   46.2 | 230 54.1  ··   03.0 | 223 29.3  ··   57.6 | Pollux   243 58.4 N28 04.3 |
| 04 | 59 12.0 | 283 38.3   16 00.2 | 197 14.7       46.7 | 245 56.0       02.7 | 238 31.5       57.4 | Procyon  245 26.0 N 5 16.6 |
| 05 | 74 14.5 | 298 37.9   15 59.6 | 212 15.5       47.3 | 260 58.0       02.5 | 253 33.7       57.3 |  |
| 06 | 89 16.9 | 313 37.6 N15 59.0 | 227 16.2 S16 47.8 | 276 00.0 N 4 02.3 | 268 35.9 N 1 57.2 | Rasalhague 96 29.6 N12 34.7 |
| 07 | 104 19.4 | 328 37.3      58.4 | 242 17.0       48.3 | 291 01.9       02.1 | 283 38.1       57.1 | Regulus  208 10.3 N12 03.8 |
| S 08 | 119 21.9 | 343 36.9      57.8 | 257 17.8       48.8 | 306 03.9       01.9 | 298 40.2       56.9 | Rigel    281 36.0 S 8 13.3 |
| A 09 | 134 24.3 | 358 36.6  ··  57.2 | 272 18.5  ··   49.4 | 321 05.9  ··   01.7 | 313 42.4  ··   56.8 | Rigil Kent. 140 26.3 S60 45.3 |
| T 10 | 149 26.8 | 13 36.3       56.6 | 287 19.3       49.9 | 336 07.8       01.5 | 328 44.6       56.7 | Sabik    102 41.2 S15 42.0 |
| U 11 | 164 29.2 | 28 35.9       56.0 | 302 20.1       50.4 | 351 09.8       01.3 | 343 46.8       56.6 |  |
| R 12 | 179 31.7 | 43 35.6 N15 55.5 | 317 20.8 S16 51.0 | 6 11.8 N 4 01.1 | 358 48.9 N 1 56.5 | Schedar  350 08.4 N56 25.8 |
| D 13 | 194 34.2 | 58 35.2       54.9 | 332 21.6       51.5 | 21 13.7       00.8 | 13 51.1       56.3 | Shaula    96 55.8 S37 05.4 |
| A 14 | 209 36.6 | 73 34.9       54.3 | 347 22.4       52.0 | 36 15.7       00.6 | 28 53.3       56.2 | Sirius   258 55.8 S16 41.2 |
| Y 15 | 224 39.1 | 88 34.6  ··   53.7 | 2 23.2  ··    52.5 | 51 17.7  ··   00.4 | 43 55.5  ··   56.1 | Spica    158 57.8 S11 03.4 |
| 16 | 239 41.6 | 103 34.2      53.1 | 17 23.9       53.1 | 66 19.6       00.2 | 58 57.6       56.0 | Suhail   223 11.1 S43 21.0 |
| 17 | 254 44.0 | 118 33.9      52.4 | 32 24.7       53.6 | 81 21.6    4 00.0 | 73 59.8       55.8 |  |
| 18 | 269 46.5 | 133 33.5 N15 51.8 | 47 25.5 S16 54.1 | 96 23.6 N 3 59.8 | 89 02.0 N 1 55.7 | Vega      80 55.8 N38 46.3 |
| 19 | 284 49.0 | 148 33.2      51.2 | 62 26.2       54.6 | 111 25.5       59.6 | 104 04.2       55.6 | Zuben'ubi 137 33.3 S15 57.5 |
| 20 | 299 51.4 | 163 32.9      50.6 | 77 27.0       55.2 | 126 27.5       59.4 | 119 06.4       55.5 |  |
| 21 | 314 53.9 | 178 32.5  ··  50.0 | 92 27.8  ··   55.7 | 141 29.5  ··   59.1 | 134 08.5  ··   55.3 |           S.H.A.   Mer. Pass. |
| 22 | 329 56.4 | 193 32.2      49.4 | 107 28.5       56.2 | 156 31.4       58.9 | 149 10.7       55.2 | Venus   225 44.7    9 05 |
| 23 | 344 58.8 | 208 31.9      48.8 | 122 29.3       56.7 | 171 33.4       58.7 | 164 12.9       55.1 | Mars    138 50.0   14 52 |
|  |  |  |  |  |  | Jupiter 186 58.0   11 38 |
| Mer. Pass | h m 0 07.8 | v −0.3  d 0.6 | v 0.8  d 0.5 | v 2.0  d 0.2 | v 2.2  d 0.1 | Saturn  179 27.5   12 08 |

## 1980 SEPTEMBER 18, 19, 20 (THURS., FRI., SAT.)

| G.M.T. | SUN G.H.A. | SUN Dec. | MOON G.H.A. | MOON $v$ | MOON Dec. | MOON $d$ | MOON H.P. | Lat. | Twilight Naut. | Twilight Civil | Sunrise | Moonrise 18 | Moonrise 19 | Moonrise 20 | Moonrise 21 |
|---|---|---|---|---|---|---|---|---|---|---|---|---|---|---|---|
| d h | ° ' | ° ' | ° ' | ' | ° ' | ' | ' | ° | h m | h m | h m | h m | h m | h m | h m |
| | | | | | | | | N 72 | 02 39 | 04 16 | 05 25 | ■ | ■ | 19 42 | 19 07 |
| 18 00 | 181 26.7 N | 1 53.8 | 86 55.5 | 9.8 | S19 29.0 | 1.8 | 56.2 | N 70 | 03 03 | 04 26 | 05 28 | 18 24 | 18 40 | 18 39 | 18 36 |
| 01 | 196 26.9 | 52.8 | 101 24.3 | 9.8 | 19 30.8 | 1.7 | 56.2 | 68 | 03 21 | 04 34 | 05 31 | 17 02 | 17 43 | 18 03 | 18 14 |
| 02 | 211 27.1 | 51.8 | 115 53.1 | 9.7 | 19 32.5 | 1.5 | 56.3 | 66 | 03 35 | 04 41 | 05 33 | 16 24 | 17 09 | 17 38 | 17 56 |
| 03 | 226 27.4 ·· | 50.9 | 130 21.8 | 9.6 | 19 34.0 | 1.5 | 56.3 | 64 | 03 47 | 04 46 | 05 35 | 15 57 | 16 44 | 17 18 | 17 41 |
| 04 | 241 27.6 | 49.9 | 144 50.4 | 9.7 | 19 35.5 | 1.4 | 56.3 | 62 | 03 56 | 04 51 | 05 36 | 15 37 | 16 25 | 17 01 | 17 29 |
| 05 | 256 27.8 | 48.9 | 159 19.1 | 9.5 | 19 36.9 | 1.3 | 56.4 | 60 | 04 04 | 04 55 | 05 37 | 15 20 | 16 09 | 16 48 | 17 18 |
| 06 | 271 28.0 N | 1 48.0 | 173 47.6 | 9.5 | S19 38.2 | 1.1 | 56.4 | N 58 | 04 11 | 04 59 | 05 38 | 15 05 | 15 55 | 16 36 | 17 09 |
| 07 | 286 28.2 | 47.0 | 188 16.1 | 9.5 | 19 39.3 | 1.1 | 56.4 | 56 | 04 17 | 05 02 | 05 39 | 14 53 | 15 43 | 16 26 | 17 01 |
| T 08 | 301 28.5 | 46.0 | 202 44.6 | 9.4 | 19 40.4 | 1.0 | 56.5 | 54 | 04 22 | 05 05 | 05 40 | 14 43 | 15 33 | 16 17 | 16 54 |
| H 09 | 316 28.7 ·· | 45.1 | 217 13.0 | 9.4 | 19 41.4 | 0.8 | 56.5 | 52 | 04 27 | 05 07 | 05 41 | 14 33 | 15 24 | 16 09 | 16 47 |
| U 10 | 331 28.9 | 44.1 | 231 41.4 | 9.3 | 19 42.2 | 0.8 | 56.5 | 50 | 04 31 | 05 09 | 05 42 | 14 25 | 15 16 | 16 01 | 16 41 |
| R 11 | 346 29.1 | 43.1 | 246 09.7 | 9.3 | 19 43.0 | 0.6 | 56.6 | 45 | 04 39 | 05 14 | 05 43 | 14 07 | 14 59 | 15 46 | 16 29 |
| S 12 | 1 29.4 N | 1 42.2 | 260 38.0 | 9.2 | S19 43.6 | 0.5 | 56.6 | N 40 | 04 46 | 05 18 | 05 45 | 13 52 | 14 44 | 15 33 | 16 18 |
| D 13 | 16 29.6 | 41.2 | 275 06.2 | 9.1 | 19 44.1 | 0.5 | 56.6 | 35 | 04 51 | 05 20 | 05 46 | 13 40 | 14 32 | 15 22 | 16 09 |
| A 14 | 31 29.8 | 40.2 | 289 34.3 | 9.2 | 19 44.6 | 0.3 | 56.7 | 30 | 04 55 | 05 23 | 05 47 | 13 29 | 14 22 | 15 13 | 16 01 |
| Y 15 | 46 30.0 ·· | 39.3 | 304 02.5 | 9.0 | 19 44.9 | 0.2 | 56.7 | 20 | 05 01 | 05 26 | 05 48 | 13 11 | 14 04 | 14 56 | 15 47 |
| 16 | 61 30.3 | 38.3 | 318 30.5 | 9.0 | 19 45.1 | 0.1 | 56.7 | N 10 | 05 04 | 05 28 | 05 49 | 12 54 | 13 48 | 14 42 | 15 35 |
| 17 | 76 30.5 | 37.3 | 332 58.5 | 9.0 | 19 45.2 | 0.0 | 56.8 | 0 | 05 06 | 05 30 | 05 50 | 12 39 | 13 33 | 14 28 | 15 24 |
| 18 | 91 30.7 N | 1 36.3 | 347 26.5 | 9.0 | S19 45.2 | 0.2 | 56.8 | S 10 | 05 06 | 05 30 | 05 51 | 12 24 | 13 18 | 14 15 | 15 13 |
| 19 | 106 30.9 | 35.4 | 1 54.5 | 8.8 | 19 45.0 | 0.2 | 56.9 | 20 | 05 05 | 05 30 | 05 52 | 12 08 | 13 03 | 14 00 | 15 01 |
| 20 | 121 31.1 | 34.4 | 16 22.3 | 8.9 | 19 44.8 | 0.3 | 56.9 | 30 | 05 02 | 05 29 | 05 53 | 11 50 | 12 44 | 13 44 | 14 47 |
| 21 | 136 31.4 ·· | 33.4 | 30 50.2 | 8.8 | 19 44.5 | 0.5 | 56.9 | 35 | 04 59 | 05 28 | 05 54 | 11 39 | 12 34 | 13 34 | 14 39 |
| 22 | 151 31.6 | 32.5 | 45 18.0 | 8.7 | 19 44.0 | 0.6 | 57.0 | 40 | 04 56 | 05 27 | 05 54 | 11 27 | 12 22 | 13 23 | 14 30 |
| 23 | 166 31.8 | 31.5 | 59 45.7 | 8.7 | 19 43.4 | 0.7 | 57.0 | 45 | 04 51 | 05 26 | 05 55 | 11 13 | 12 08 | 13 10 | 14 19 |
| 19 00 | 181 32.0 N | 1 30.5 | 74 13.4 | 8.7 | S19 42.7 | 0.7 | 57.0 | S 50 | 04 45 | 05 23 | 05 55 | 10 55 | 11 50 | 12 54 | 14 06 |
| 01 | 196 32.3 | 29.6 | 88 41.1 | 8.6 | 19 42.0 | 1.0 | 57.1 | 52 | 04 43 | 05 22 | 05 56 | 10 47 | 11 42 | 12 47 | 14 00 |
| 02 | 211 32.5 | 28.6 | 103 08.7 | 8.6 | 19 41.0 | 1.0 | 57.1 | 54 | 04 39 | 05 21 | 05 56 | 10 37 | 11 33 | 12 39 | 13 53 |
| 03 | 226 32.7 ·· | 27.6 | 117 36.3 | 8.6 | 19 40.0 | 1.1 | 57.2 | 56 | 04 36 | 05 19 | 05 56 | 10 27 | 11 23 | 12 29 | 13 46 |
| 04 | 241 32.9 | 26.6 | 132 03.9 | 8.5 | 19 38.9 | 1.3 | 57.2 | 58 | 04 31 | 05 18 | 05 57 | 10 15 | 11 11 | 12 19 | 13 37 |
| 05 | 256 33.1 | 25.7 | 146 31.4 | 8.4 | 19 37.6 | 1.3 | 57.2 | S 60 | 04 26 | 05 16 | 05 57 | 10 01 | 10 57 | 12 07 | 13 28 |
| 06 | 271 33.4 N | 1 24.7 | 160 58.8 | 8.4 | S19 36.3 | 1.5 | 57.3 | Lat. | Sunset | Twilight Civil | Twilight Naut. | Moonset 18 | Moonset 19 | Moonset 20 | Moonset 21 |
| 07 | 286 33.6 | 23.7 | 175 26.2 | 8.4 | 19 34.8 | 1.6 | 57.3 | | | | | | | | |
| 08 | 301 33.8 | 22.8 | 189 53.6 | 8.4 | 19 33.2 | 1.7 | 57.4 | | | | | | | | |
| F 09 | 316 34.0 ·· | 21.8 | 204 21.0 | 8.3 | 19 31.5 | 1.8 | 57.4 | ° | h m | h m | h m | h m | h m | h m | h m |
| R 10 | 331 34.3 | 20.8 | 218 48.3 | 8.3 | 19 29.7 | 2.0 | 57.4 | | | | | | | | |
| I 11 | 346 34.5 | 19.9 | 233 15.6 | 8.2 | 19 27.7 | 2.0 | 57.5 | N 72 | 18 19 | 19 29 | 21 03 | ■ | ■ | 21 52 | 24 25 |
| D 12 | 1 34.7 N | 1 18.9 | 247 42.8 | 8.2 | S19 25.7 | 2.2 | 57.5 | N 70 | 18 17 | 19 19 | 20 40 | 19 20 | 20 58 | 22 55 | 24 54 |
| A 13 | 16 34.9 | 17.9 | 262 10.0 | 8.2 | 19 23.5 | 2.3 | 57.6 | 68 | 18 15 | 19 11 | 20 23 | 20 42 | 21 55 | 23 30 | 25 16 |
| Y 14 | 31 35.1 | 16.9 | 276 37.2 | 8.1 | 19 21.2 | 2.4 | 57.6 | 66 | 18 13 | 19 04 | 20 09 | 21 20 | 22 28 | 23 55 | 25 33 |
| 15 | 46 35.4 ·· | 16.0 | 291 04.3 | 8.1 | 19 18.8 | 2.5 | 57.6 | 64 | 18 11 | 18 59 | 19 58 | 21 47 | 22 52 | 24 14 | 00 14 |
| 16 | 61 35.6 | 15.0 | 305 31.4 | 8.1 | 19 16.3 | 2.7 | 57.7 | 62 | 18 10 | 18 54 | 19 49 | 22 08 | 23 12 | 24 30 | 00 30 |
| 17 | 76 35.8 | 14.0 | 319 58.5 | 8.0 | 19 13.6 | 2.7 | 57.7 | 60 | 18 09 | 18 50 | 19 41 | 22 25 | 23 27 | 24 43 | 00 43 |
| 18 | 91 36.0 N | 1 13.1 | 334 25.5 | 8.0 | S19 10.9 | 2.9 | 57.8 | N 58 | 18 08 | 18 47 | 19 34 | 22 39 | 23 41 | 24 54 | 00 54 |
| 19 | 106 36.3 | 12.1 | 348 52.5 | 8.0 | 19 08.0 | 3.0 | 57.8 | 56 | 18 07 | 18 44 | 19 29 | 22 51 | 23 52 | 25 04 | 01 04 |
| 20 | 121 36.5 | 11.1 | 3 19.5 | 7.9 | 19 05.0 | 3.1 | 57.8 | 54 | 18 06 | 18 41 | 19 24 | 23 02 | 24 02 | 00 02 | 01 13 |
| 21 | 136 36.7 ·· | 10.2 | 17 46.4 | 8.0 | 19 01.9 | 3.3 | 57.9 | 52 | 18 05 | 18 39 | 19 19 | 23 11 | 24 11 | 00 11 | 01 20 |
| 22 | 151 36.9 | 09.2 | 32 13.4 | 7.8 | 18 58.6 | 3.3 | 57.9 | 50 | 18 05 | 18 37 | 19 15 | 23 20 | 24 19 | 00 19 | 01 27 |
| 23 | 166 37.1 | 08.2 | 46 40.2 | 7.9 | 18 55.3 | 3.5 | 58.0 | 45 | 18 03 | 18 33 | 19 07 | 23 37 | 24 36 | 00 36 | 01 42 |
| 20 00 | 181 37.4 N | 1 07.2 | 61 07.1 | 7.8 | S18 51.8 | 3.6 | 58.0 | N 40 | 18 02 | 18 29 | 19 01 | 23 52 | 24 50 | 00 50 | 01 54 |
| 01 | 196 37.6 | 06.3 | 75 33.9 | 7.8 | 18 48.2 | 3.7 | 58.0 | 35 | 18 01 | 18 26 | 18 56 | 24 04 | 00 04 | 01 02 | 02 04 |
| 02 | 211 37.8 | 05.3 | 90 00.7 | 7.8 | 18 44.5 | 3.8 | 58.1 | 30 | 18 00 | 18 24 | 18 52 | 24 15 | 00 15 | 01 12 | 02 13 |
| 03 | 226 38.0 ·· | 04.3 | 104 27.5 | 7.8 | 18 40.7 | 3.9 | 58.1 | 20 | 17 59 | 18 21 | 18 47 | 24 34 | 00 34 | 01 30 | 02 29 |
| 04 | 241 38.3 | 03.4 | 118 54.3 | 7.7 | 18 36.8 | 4.1 | 58.2 | N 10 | 17 58 | 18 19 | 18 43 | 24 50 | 00 50 | 01 45 | 02 42 |
| 05 | 256 38.5 | 02.4 | 133 21.0 | 7.7 | 18 32.7 | 4.2 | 58.2 | 0 | 17 57 | 18 18 | 18 42 | 00 12 | 01 05 | 01 59 | 02 55 |
| 06 | 271 38.7 N | 1 01.4 | 147 47.7 | 7.7 | S18 28.5 | 4.3 | 58.2 | S 10 | 17 56 | 18 17 | 18 41 | 00 27 | 01 20 | 02 14 | 03 07 |
| 07 | 286 38.9 | 1 00.4 | 162 14.4 | 7.7 | 18 24.2 | 4.4 | 58.3 | 20 | 17 55 | 18 17 | 18 43 | 00 43 | 01 36 | 02 29 | 03 21 |
| S 08 | 301 39.1 | 0 59.5 | 176 41.1 | 7.6 | 18 19.8 | 4.5 | 58.3 | 30 | 17 55 | 18 19 | 18 46 | 01 01 | 01 54 | 02 46 | 03 36 |
| A 09 | 316 39.4 ·· | 58.5 | 191 07.7 | 7.6 | 18 15.3 | 4.6 | 58.4 | 35 | 17 54 | 18 20 | 18 49 | 01 12 | 02 05 | 02 56 | 03 44 |
| T 10 | 331 39.6 | 57.5 | 205 34.3 | 7.6 | 18 10.7 | 4.8 | 58.4 | 40 | 17 54 | 18 21 | 18 52 | 01 24 | 02 17 | 03 08 | 03 54 |
| U 11 | 346 39.8 | 56.6 | 220 00.9 | 7.6 | 18 05.9 | 4.9 | 58.4 | 45 | 17 53 | 18 23 | 18 57 | 01 38 | 02 32 | 03 21 | 04 06 |
| R 12 | 1 40.0 N | 0 55.6 | 234 27.5 | 7.6 | S18 01.0 | 5.0 | 58.5 | S 50 | 17 53 | 18 25 | 19 03 | 01 56 | 02 49 | 03 37 | 04 20 |
| D 13 | 16 40.3 | 54.6 | 248 54.1 | 7.5 | 17 56.0 | 5.1 | 58.5 | 52 | 17 53 | 18 26 | 19 06 | 02 04 | 02 57 | 03 45 | 04 26 |
| A 14 | 31 40.5 | 53.6 | 263 20.6 | 7.5 | 17 50.9 | 5.3 | 58.6 | 54 | 17 53 | 18 28 | 19 09 | 02 13 | 03 07 | 03 54 | 04 34 |
| Y 15 | 46 40.7 ·· | 52.7 | 277 47.1 | 7.5 | 17 45.6 | 5.3 | 58.6 | 56 | 17 52 | 18 29 | 19 13 | 02 23 | 03 17 | 04 03 | 04 42 |
| 16 | 61 40.9 | 51.7 | 292 13.6 | 7.5 | 17 40.3 | 5.5 | 58.6 | 58 | 17 52 | 18 31 | 19 18 | 02 35 | 03 29 | 04 14 | 04 51 |
| 17 | 76 41.1 | 50.7 | 306 40.1 | 7.5 | 17 34.8 | 5.6 | 58.7 | S 60 | 17 52 | 18 33 | 19 23 | 02 49 | 03 43 | 04 26 | 05 01 |
| 18 | 91 41.4 N | 0 49.8 | 321 06.6 | 7.5 | S17 29.2 | 5.7 | 58.7 | | SUN | | | MOON | | | |
| 19 | 106 41.6 | 48.8 | 335 33.1 | 7.4 | 17 23.5 | 5.8 | 58.8 | Day | Eqn. of Time 00ʰ | Eqn. of Time 12ʰ | Mer. Pass. | Mer. Pass. Upper | Mer. Pass. Lower | Age | Phase |
| 20 | 121 41.8 | 47.8 | 349 59.5 | 7.5 | 17 17.7 | 5.9 | 58.8 | | | | | | | | |
| 21 | 136 42.0 ·· | 46.8 | 4 26.0 | 7.4 | 17 11.8 | 6.1 | 58.8 | | | | | | | | |
| 22 | 151 42.2 | 45.9 | 18 52.4 | 7.4 | 17 05.7 | 6.1 | 58.9 | | m s | m s | h m | h m | h m | d | |
| 23 | 166 42.5 | 44.9 | 33 18.8 | 7.4 | 16 59.6 | 6.3 | 58.9 | 18 | 05 46 | 05 57 | 11 54 | 18 52 | 06 26 | 09 | ◐ |
| | | | | | | | | 19 | 06 08 | 06 18 | 11 54 | 19 46 | 07 19 | 10 | |
| | S.D. 16.0 | $d$ 1.0 | S.D. 15.4 | | 15.7 | | 15.9 | 20 | 06 29 | 06 40 | 11 53 | 20 42 | 08 14 | 11 | |

## 1980 SEPT. 30, OCT. 1, 2 (TUES., WED., THURS.)

| G.M.T. | ARIES G.H.A. | VENUS −3.7 G.H.A. / Dec. | MARS +1.5 G.H.A. / Dec. | JUPITER −1.2 G.H.A. / Dec. | SATURN +1.2 G.H.A. / Dec. | STARS Name / S.H.A. / Dec. |
|---|---|---|---|---|---|---|
| 30 00 | 8 53.5 | 222 17.4 N13 18.6 | 140 08.3 S18 45.8 | 193 40.6 N 3 13.0 | 187 05.8 N 1 28.7 | Acamar 315 36.9 S40 22.8 |
| 01 | 23 56.0 | 237 17.0 · 17.8 | 155 09.0 · 46.3 | 208 42.5 · 12.8 | 202 07.9 · 28.6 | Achernar 335 44.7 S57 20.0 |
| 02 | 38 58.5 | 252 16.7 · 17.1 | 170 09.7 · 46.8 | 223 44.5 · 12.5 | 217 10.1 · 28.4 | Acrux 173 37.9 S62 59.4 |
| 03 | 54 00.9 | 267 16.3 · · 16.3 | 185 10.4 · · 47.3 | 238 46.5 · · 12.3 | 232 12.3 · · 28.3 | Adhara 255 32.1 S28 56.5 |
| 04 | 69 03.4 | 282 16.0 · 15.5 | 200 11.1 · 47.7 | 253 48.4 · 12.1 | 247 14.5 · 28.2 | Aldebaran 291 17.8 N16 28.2 |
| 05 | 84 05.8 | 297 15.7 · 14.7 | 215 11.8 · 48.2 | 268 50.4 · 11.9 | 262 16.7 · 28.1 | |
| 06 | 99 08.3 | 312 15.3 N13 14.0 | 230 12.5 S18 48.7 | 283 52.4 N 3 11.7 | 277 18.8 N 1 28.0 | Alioth 166 43.1 N56 04.0 |
| 07 | 114 10.8 | 327 15.0 · 13.2 | 245 13.2 · 49.2 | 298 54.4 · 11.5 | 292 21.0 · 27.8 | Alkaid 153 19.0 N49 24.8 |
| T 08 | 129 13.2 | 342 14.6 · 12.4 | 260 13.9 · 49.7 | 313 56.3 · 11.3 | 307 23.2 · 27.7 | Al Na'ir 28 14.6 S47 03.4 |
| U 09 | 144 15.7 | 357 14.3 · · 11.6 | 275 14.6 · · 50.1 | 328 58.3 · · 11.1 | 322 25.4 · · 27.6 | Alnilam 276 11.6 S 1 12.8 |
| E 10 | 159 18.2 | 12 13.9 · 10.8 | 290 15.3 · 50.6 | 344 00.3 · 10.9 | 337 27.6 · 27.5 | Alphard 218 20.8 S 8 34.3 |
| S 11 | 174 20.6 | 27 13.6 · 10.1 | 305 16.0 · 51.1 | 359 02.2 · 10.7 | 352 29.7 · 27.4 | |
| D 12 | 189 23.1 | 42 13.3 N13 09.3 | 320 16.7 S18 51.6 | 14 04.2 N 3 10.5 | 7 31.9 N 1 27.2 | Alphecca 126 32.4 N26 47.1 |
| A 13 | 204 25.6 | 57 12.9 · 08.5 | 335 17.4 · 52.0 | 29 06.2 · 10.2 | 22 34.1 · 27.1 | Alpheratz 358 08.9 N28 59.1 |
| Y 14 | 219 28.0 | 72 12.6 · 07.7 | 350 18.1 · 52.5 | 44 08.2 · 10.0 | 37 36.3 · 27.0 | Altair 62 32.4 N 8 49.2 |
| 15 | 234 30.5 | 87 12.2 · · 06.9 | 5 18.8 · · 53.0 | 59 10.1 · · 09.8 | 52 38.5 · · 26.9 | Ankaa 353 39.8 S42 24.6 |
| 16 | 249 33.0 | 102 11.9 · 06.1 | 20 19.5 · 53.5 | 74 12.1 · 09.6 | 67 40.6 · 26.7 | Antares 112 57.0 S26 23.3 |
| 17 | 264 35.4 | 117 11.5 · 05.3 | 35 20.2 · 53.9 | 89 14.1 · 09.4 | 82 42.8 · 26.6 | |
| 18 | 279 37.9 | 132 11.2 N13+04.6 | 50 20.9 S18 54.4 | 104 16.0 N 3 09.2 | 97 45.0 N 1 26.5 | Arcturus 146 18.8 N19 17.2 |
| 19 | 294 40.3 | 147 10.9 · 03.8 | 65 21.6 · 54.9 | 119 18.0 · 09.0 | 112 47.2 · 26.4 | Atria 108 21.5 S68 59.7 |
| 20 | 309 42.8 | 162 10.5 · 03.0 | 80 22.2 · 55.4 | 134 20.0 · 08.8 | 127 49.4 · 26.3 | Avior 234 28.4 S59 26.6 |
| 21 | 324 45.3 | 177 10.2 · · 02.2 | 95 22.9 · · 55.8 | 149 22.0 · · 08.6 | 142 51.5 · · 26.1 | Bellatrix 278 58.6 N 6 20.0 |
| 22 | 339 47.7 | 192 09.8 · 01.4 | 110 23.6 · 56.3 | 164 23.9 · 08.4 | 157 53.7 · 26.0 | Betelgeuse 271 28.2 N 7 24.3 |
| 23 | 354 50.2 | 207 09.5 13 00.6 | 125 24.3 · 56.8 | 179 25.9 · 08.2 | 172 55.9 · 25.9 | |
| 1 00 | 9 52.7 | 222 09.1 N12 59.8 | 140 25.0 S18 57.3 | 194 27.9 N 3 08.0 | 187 58.1 N 1 25.8 | Canopus 264 07.2 S52 40.8 |
| 01 | 24 55.1 | 237 08.8 · 59.0 | 155 25.7 · 57.7 | 209 29.8 · 07.7 | 203 00.3 · 25.7 | Capella 281 11.1 N45 58.6 |
| 02 | 39 57.6 | 252 08.5 · 58.2 | 170 26.4 · 58.2 | 224 31.8 · 07.5 | 218 02.4 · 25.5 | Deneb 49 48.2 N45 13.0 |
| 03 | 55 00.1 | 267 08.1 · · 57.4 | 185 27.1 · · 58.7 | 239 33.8 · · 07.3 | 233 04.6 · · 25.4 | Denebola 182 59.4 N14 41.0 |
| 04 | 70 02.5 | 282 07.8 · 56.7 | 200 27.8 · 59.1 | 254 35.8 · 07.1 | 248 06.8 · 25.3 | Diphda 349 20.5 S18 05.5 |
| 05 | 85 05.0 | 297 07.4 · 55.9 | 215 28.5 18 59.6 | 269 37.7 · 06.9 | 263 09.0 · 25.2 | |
| 06 | 100 07.5 | 312 07.1 N12 55.1 | 230 29.2 S19 00.1 | 284 39.7 N 3 06.7 | 278 11.1 N 1 25.1 | Dubhe 194 22.8 N61 51.3 |
| 07 | 115 09.9 | 327 06.7 · 54.3 | 245 29.9 · 00.6 | 299 41.7 · 06.5 | 293 13.3 · 24.9 | Elnath 278 44.0 N28 35.4 |
| W 08 | 130 12.4 | 342 06.4 · 53.5 | 260 30.5 · 01.0 | 314 43.6 · 06.3 | 308 15.5 · 24.8 | Eltanin 90 57.8 N51 29.9 |
| E 09 | 145 14.8 | 357 06.1 · · 52.7 | 275 31.2 · · 01.5 | 329 45.6 · · 06.1 | 323 17.7 · · 24.7 | Enif 34 11.3 N 9 47.3 |
| D 10 | 160 17.3 | 12 05.7 · 51.9 | 290 31.9 · 02.0 | 344 47.6 · 05.9 | 338 19.9 · 24.6 | Fomalhaut 15 51.1 S29 43.5 |
| N 11 | 175 19.8 | 27 05.4 · 51.1 | 305 32.6 · 02.4 | 359 49.6 · 05.7 | 353 22.0 · 24.5 | |
| E 12 | 190 22.2 | 42 05.0 N12 50.3 | 320 33.3 S19 02.9 | 14 51.5 N 3 05.4 | 8 24.2 N 1 24.3 | Gacrux 172 29.3 S57 00.2 |
| S 13 | 205 24.7 | 57 04.7 · 49.5 | 335 34.0 · 03.4 | 29 53.5 · 05.2 | 23 26.4 · 24.2 | Gienah 176 18.3 S17 25.9 |
| D 14 | 220 27.2 | 72 04.3 · 48.7 | 350 34.7 · 03.9 | 44 55.5 · 05.0 | 38 28.6 · 24.1 | Hadar 149 23.9 S60 16.8 |
| A 15 | 235 29.6 | 87 04.0 · · 47.9 | 5 35.4 · · 04.3 | 59 57.5 · · 04.8 | 53 30.8 · · 24.0 | Hamal 328 28.6 N23 22.3 |
| Y 16 | 250 32.1 | 102 03.7 · 47.1 | 20 36.0 · 04.8 | 74 59.4 · 04.6 | 68 32.9 · 23.9 | Kaus Aust. 84 16.9 S34 23.7 |
| 17 | 265 34.6 | 117 03.3 · 46.3 | 35 36.7 · 05.3 | 90 01.4 · 04.4 | 83 35.1 · 23.7 | |
| 18 | 280 37.0 | 132 03.0 N12 45.5 | 50 37.4 S19 05.7 | 105 03.4 N 3 04.2 | 98 37.3 N 1 23.6 | Kochab 137 20.0 N74 14.4 |
| 19 | 295 39.5 | 147 02.6 · 44.7 | 65 38.1 · 06.2 | 120 05.3 · 04.0 | 113 39.5 · 23.5 | Markab 14 02.9 N15 06.2 |
| 20 | 310 41.9 | 162 02.3 · 43.9 | 80 38.8 · 06.7 | 135 07.3 · 03.8 | 128 41.7 · 23.4 | Menkar 314 40.9 N 4 00.9 |
| 21 | 325 44.4 | 177 01.9 · · 43.1 | 95 39.5 · · 07.1 | 150 09.3 · · 03.6 | 143 43.8 · · 23.3 | Menkent 148 37.4 S36 16.4 |
| 22 | 340 46.9 | 192 01.6 · 42.2 | 110 40.2 · 07.6 | 165 11.3 · 03.4 | 158 46.0 · 23.1 | Miaplacidus 221 45.5 S69 38.0 |
| 23 | 355 49.3 | 207 01.3 · 41.4 | 125 40.8 · 08.1 | 180 13.2 · 03.2 | 173 48.2 · 23.0 | |
| 2 00 | 10 51.8 | 222 00.9 N12 40.6 | 140 41.5 S19 08.5 | 195 15.2 N 3 02.9 | 188 50.4 N 1 22.9 | Mirfak 309 15.7 N49 47.4 |
| 01 | 25 54.3 | 237 00.6 · 39.8 | 155 42.2 · 09.0 | 210 17.2 · 02.7 | 203 52.6 · 22.8 | Nunki 76 29.1 S26 19.2 |
| 02 | 40 56.7 | 252 00.2 · 39.0 | 170 42.9 · 09.5 | 225 19.2 · 02.5 | 218 54.7 · 22.7 | Peacock 53 58.1 S56 48.0 |
| 03 | 55 59.2 | 266 59.9 · · 38.2 | 185 43.6 · · 09.9 | 240 21.1 · · 02.3 | 233 56.9 · · 22.5 | Pollux 243 58.3 N28 04.3 |
| 04 | 71 01.7 | 281 59.5 · 37.4 | 200 44.3 · 10.4 | 255 23.1 · 02.1 | 248 59.1 · 22.4 | Procyon 245 27.9 N 5 16.6 |
| 05 | 86 04.1 | 296 59.2 · 36.6 | 215 44.9 · 10.9 | 270 25.1 · 01.9 | 264 01.3 · 22.3 | |
| 06 | 101 06.6 | 311 58.9 N12 35.8 | 230 45.6 S19 11.3 | 285 27.0 N 3 01.7 | 279 03.5 N 1 22.2 | Rasalhague 96 29.7 N12 34.7 |
| 07 | 116 09.1 | 326 58.5 · 35.0 | 245 46.3 · 11.8 | 300 29.0 · 01.5 | 294 05.6 · 22.1 | Regulus 208 10.3 N12 03.8 |
| T 08 | 131 11.5 | 341 58.2 · 34.1 | 260 47.0 · 12.3 | 315 31.0 · 01.3 | 309 07.8 · 21.9 | Rigel 281 35.9 S 8 13.3 |
| H 09 | 146 14.0 | 356 57.8 · · 33.3 | 275 47.7 · · 12.7 | 330 33.0 · · 01.1 | 324 10.0 · · 21.8 | Rigil Kent. 140 26.4 S60 45.3 |
| U 10 | 161 16.4 | 11 57.5 · 32.5 | 290 48.4 · 13.2 | 345 34.9 · 00.9 | 339 12.2 · 21.7 | Sabik 102 41.3 S15 42.0 |
| R 11 | 176 18.9 | 26 57.1 · 31.7 | 305 49.0 · 13.7 | 0 36.9 · 00.6 | 354 14.4 · 21.6 | |
| S 12 | 191 21.4 | 41 56.8 N12 30.9 | 320 49.7 S19 14.1 | 15 38.9 N 3 00.4 | 9 16.5 N 1 21.5 | Schedar 350 08.4 N56 25.9 |
| D 13 | 206 23.8 | 56 56.5 · 30.1 | 335 50.4 · 14.6 | 30 40.9 · 00.2 | 24 18.7 · 21.3 | Shaula 96 55.9 S37 05.4 |
| A 14 | 221 26.3 | 71 56.1 · 29.3 | 350 51.1 · 15.1 | 45 42.8 3 00.0 | 39 20.9 · 21.2 | Sirius 258 55.7 S16 41.2 |
| Y 15 | 236 28.8 | 86 55.8 · · 28.4 | 5 51.8 · · 15.5 | 60 44.8 2 59.8 | 54 23.1 · · 21.1 | Spica 158 57.8 S11 03.4 |
| 16 | 251 31.2 | 101 55.4 · 27.6 | 20 52.4 · 16.0 | 75 46.8 · 59.6 | 69 25.3 · 21.0 | Suhail 223 11.0 S43 21.0 |
| 17 | 266 33.7 | 116 55.1 · 26.8 | 35 53.1 · 16.5 | 90 48.8 · 59.4 | 84 27.4 · 20.9 | |
| 18 | 281 36.2 | 131 54.8 N12 26.0 | 50 53.8 S19 16.9 | 105 50.7 N 2 59.2 | 99 29.6 N 1 20.7 | Vega 80 55.8 N38 46.3 |
| 19 | 296 38.6 | 146 54.4 · 25.2 | 65 54.5 · 17.4 | 120 52.7 · 59.0 | 114 31.8 · 20.6 | Zuben'ubi 137 33.3 S15 57.5 |
| 20 | 311 41.1 | 161 54.1 · 24.3 | 80 55.2 · 17.8 | 135 54.7 · 58.8 | 129 34.0 · 20.5 | |
| 21 | 326 43.6 | 176 53.7 · · 23.5 | 95 55.8 · · 18.3 | 150 56.6 · · 58.6 | 144 36.2 · · 20.4 | S.H.A. / Mer. Pass. |
| 22 | 341 46.0 | 191 53.4 · 22.7 | 110 56.5 · 18.8 | 165 58.6 · 58.4 | 159 38.3 · 20.3 | Venus 212 16.5 9 12 |
| 23 | 356 48.5 | 206 53.0 · 21.9 | 125 57.2 · 19.2 | 181 00.6 · 58.1 | 174 40.5 · 20.1 | Mars 130 32.3 14 38 |
| Mer. Pass. 23 16.7 | | v −0.3 d 0.8 | v 0.7 d 0.5 | v 2.0 d 0.2 | v 2.2 d 0.1 | Jupiter 184 35.2 11 01 / Saturn 178 05.4 11 26 |

## 1980 SEPT. 30, OCT. 1, 2 (TUES., WED., THURS.)

| G.M.T. | SUN G.H.A. | Dec. | MOON G.H.A. | v | Dec. | d | H.P. | Lat. | Twilight Naut. | Civil | Sunrise | Moonrise 30 | 1 | 2 | 3 |
|---|---|---|---|---|---|---|---|---|---|---|---|---|---|---|---|
| | ° ' | ° ' | ° ' | ' | ° ' | ' | ' | ° | h m | h m | h m | h m | h m | h m | h m |
| 30 00 | 182 29.1 S | 2 46.3 | 286 32.7 | 7.4 N18 57.7 | | 3.5 | 58.4 | N 72 | 03 50 | 05 12 | 06 19 | ▭ | ▭ | ▭ | 22 00 |
| 01 | 197 29.3 | 47.2 | 300 59.1 | 7.3 | 19 01.2 | 3.4 | 58.4 | N 70 | 04 02 | 05 15 | 06 16 | 18 24 | 19 21 | 21 01 | 22 44 |
| 02 | 212 29.5 | 48.2 | 315 25.4 | 7.5 | 19 04.6 | 3.2 | 58.3 | 68 | 04 11 | 05 18 | 06 13 | 19 30 | 20 27 | 21 45 | 23 13 |
| 03 | 227 29.7 ·· | 49.2 | 329 51.9 | 7.4 | 19 07.8 | 3.1 | 58.3 | 66 | 04 19 | 05 20 | 06 11 | 20 06 | 21 02 | 22 14 | 23 35 |
| 04 | 242 29.9 | 50.2 | 344 18.3 | 7.5 | 19 10.9 | 3.0 | 58.3 | 64 | 04 25 | 05 22 | 06 09 | 20 32 | 21 28 | 22 36 | 23 52 |
| 05 | 257 30.1 | 51.1 | 358 44.8 | 7.5 | 19 13.9 | 2.8 | 58.2 | 62 | 04 31 | 05 23 | 06 07 | 20 52 | 21 48 | 22 54 | 24 07 |
| 06 | 272 30.3 S | 2 52.1 | 13 11.3 | 7.5 N19 16.7 | | 2.8 | 58.2 | 60 | 04 35 | 05 24 | 06 06 | 21 08 | 22 04 | 23 09 | 24 19 |
| 07 | 287 30.5 | 53.1 | 27 37.8 | 7.5 | 19 19.5 | 2.6 | 58.1 | N 58 | 04 39 | 05 25 | 06 04 | 21 22 | 22 18 | 23 21 | 24 29 |
| T 08 | 302 30.7 | 54.0 | 42 04.3 | 7.6 | 19 22.1 | 2.5 | 58.1 | 56 | 04 43 | 05 26 | 06 03 | 21 34 | 22 30 | 23 32 | 24 38 |
| U 09 | 317 30.9 ·· | 55.0 | 56 30.9 | 7.6 | 19 24.6 | 2.4 | 58.1 | 54 | 04 45 | 05 27 | 06 02 | 21 45 | 22 40 | 23 41 | 24 46 |
| E 10 | 332 31.1 | 56.0 | 70 57.5 | 7.6 | 19 27.0 | 2.2 | 58.0 | 52 | 04 48 | 05 27 | 06 01 | 21 54 | 22 49 | 23 49 | 24 53 |
| S 11 | 347 31.3 | 56.9 | 85 24.1 | 7.7 | 19 29.2 | 2.1 | 58.0 | 50 | 04 50 | 05 28 | 06 00 | 22 02 | 22 57 | 23 57 | 24 59 |
| D 12 | 2 31.5 S | 2 57.9 | 99 50.8 | 7.7 N19 31.3 | | 2.1 | 58.0 | 45 | 04 55 | 05 29 | 05 58 | 22 20 | 23 15 | 24 13 | 00 13 |
| A 13 | 17 31.7 | 58.9 | 114 17.5 | 7.7 | 19 33.4 | 1.9 | 57.9 | N 40 | 04 58 | 05 29 | 05 56 | 22 35 | 23 29 | 24 26 | 00 26 |
| Y 14 | 32 31.9 | 2 59.9 | 128 44.2 | 7.7 | 19 35.3 | 1.7 | 57.9 | 35 | 05 00 | 05 29 | 05 55 | 22 47 | 23 41 | 24 37 | 00 37 |
| 15 | 47 32.1 | 3 00.8 | 143 10.9 | 7.8 | 19 37.0 | 1.7 | 57.8 | 30 | 05 02 | 05 29 | 05 53 | 22 58 | 23 52 | 24 47 | 00 47 |
| 16 | 62 32.3 | 01.8 | 157 37.7 | 7.8 | 19 38.7 | 1.5 | 57.8 | 20 | 05 03 | 05 29 | 05 51 | 23 16 | 24 10 | 00 10 | 01 04 |
| 17 | 77 32.5 | 02.8 | 172 04.5 | 7.9 | 19 40.2 | 1.5 | 57.8 | N 10 | 05 03 | 05 28 | 05 49 | 23 32 | 24 26 | 00 26 | 01 19 |
| 18 | 92 32.7 S | 3 03.7 | 186 31.4 | 7.9 N19 41.7 | | 1.3 | 57.7 | 0 | 05 02 | 05 26 | 05 46 | 23 47 | 24 41 | 00 41 | 01 32 |
| 19 | 107 32.9 | 04.7 | 200 58.3 | 7.9 | 19 43.0 | 1.1 | 57.7 | S 10 | 04 59 | 05 23 | 05 44 | 24 02 | 00 02 | 00 56 | 01 46 |
| 20 | 122 33.1 | 05.7 | 215 25.2 | 8.0 | 19 44.1 | 1.1 | 57.7 | 20 | 04 54 | 05 19 | 05 41 | 24 19 | 00 19 | 01 12 | 02 01 |
| 21 | 137 33.3 ·· | 06.7 | 229 52.2 | 8.0 | 19 45.2 | 1.0 | 57.6 | 30 | 04 46 | 05 14 | 05 38 | 24 37 | 00 37 | 01 30 | 02 17 |
| 22 | 152 33.5 | 07.6 | 244 19.2 | 8.0 | 19 46.2 | 0.8 | 57.6 | 35 | 04 41 | 05 11 | 05 36 | 24 48 | 00 48 | 01 40 | 02 27 |
| 23 | 167 33.7 | 08.6 | 258 46.2 | 8.1 | 19 47.0 | 0.7 | 57.6 | 40 | 04 35 | 05 07 | 05 34 | 00 02 | 01 00 | 01 53 | 02 38 |
| | | | | | | | | 45 | 04 28 | 05 02 | 05 32 | 00 16 | 01 15 | 02 07 | 02 51 |
| 1 00 | 182 33.9 S | 3 09.6 | 273 13.3 | 8.1 N19 47.7 | | 0.6 | 57.5 | S 50 | 04 18 | 04 56 | 05 29 | 00 33 | 01 33 | 02 24 | 03 07 |
| 01 | 197 34.1 | 10.5 | 287 40.4 | 8.1 | 19 48.3 | 0.5 | 57.5 | 52 | 04 13 | 04 54 | 05 28 | 00 41 | 01 42 | 02 33 | 03 14 |
| 02 | 212 34.3 | 11.5 | 302 07.5 | 8.2 | 19 48.8 | 0.3 | 57.5 | 54 | 04 07 | 04 50 | 05 26 | 00 50 | 01 51 | 02 42 | 03 23 |
| 03 | 227 34.5 ·· | 12.5 | 316 34.7 | 8.3 | 19 49.1 | 0.3 | 57.4 | 56 | 04 01 | 04 47 | 05 24 | 01 00 | 02 02 | 02 52 | 03 32 |
| 04 | 242 34.7 | 13.4 | 331 02.0 | 8.2 | 19 49.4 | 0.1 | 57.4 | 58 | 03 54 | 04 43 | 05 23 | 01 12 | 02 14 | 03 04 | 03 42 |
| 05 | 257 34.9 | 14.4 | 345 29.2 | 8.3 | 19 49.5 | 0.0 | 57.3 | S 60 | 03 46 | 04 38 | 05 20 | 01 25 | 02 28 | 03 17 | 03 54 |
| 06 | 272 35.1 S | 3 15.4 | 359 56.5 | 8.4 N19 49.5 | | 0.1 | 57.3 | | | | | | | | |
| W 07 | 287 35.3 | 16.3 | 14 23.9 | 8.4 | 19 49.4 | 0.2 | 57.3 | Lat. | Sunset | Twilight Civil | Naut. | Moonset 30 | 1 | 2 | 3 |
| E 08 | 302 35.5 | 17.3 | 28 51.3 | 8.5 | 19 49.2 | 0.3 | 57.2 | | | | | | | | |
| D 09 | 317 35.7 ·· | 18.3 | 43 18.8 | 8.4 | 19 48.9 | 0.4 | 57.2 | | h m | h m | h m | h m | h m | h m | h m |
| N 10 | 332 35.9 | 19.3 | 57 46.2 | 8.6 | 19 48.5 | 0.6 | 57.2 | | | | | | | | |
| E 11 | 347 36.1 | 20.2 | 72 13.8 | 8.6 | 19 47.9 | 0.6 | 57.1 | N 72 | 17 18 | 18 25 | 19 46 | ▭ | ▭ | ▭ | 18 29 |
| S 12 | 2 36.3 S | 3 21.2 | 86 41.4 | 8.6 N19 47.3 | | 0.8 | 57.1 | N 70 | 17 21 | 18 22 | 19 34 | 16 39 | 17 36 | 17 44 | 17 44 |
| D 13 | 17 36.5 | 22.2 | 101 09.0 | 8.7 | 19 46.5 | 0.9 | 57.1 | 68 | 17 24 | 18 19 | 19 25 | 15 33 | 16 30 | 17 00 | 17 14 |
| A 14 | 32 36.7 | 23.1 | 115 36.7 | 8.7 | 19 45.6 | 1.0 | 57.0 | 66 | 17 27 | 18 18 | 19 18 | 14 58 | 15 54 | 16 30 | 16 52 |
| Y 15 | 47 36.9 ·· | 24.1 | 130 04.4 | 8.8 | 19 44.6 | 1.1 | 57.0 | 64 | 17 29 | 18 16 | 19 12 | 14 32 | 15 28 | 16 07 | 16 34 |
| 16 | 62 37.1 | 25.1 | 144 32.2 | 8.8 | 19 43.5 | 1.2 | 57.0 | 62 | 17 31 | 18 15 | 19 06 | 14 12 | 15 08 | 15 50 | 16 19 |
| 17 | 77 37.3 | 26.0 | 159 00.0 | 8.9 | 19 42.3 | 1.3 | 56.9 | 60 | 17 32 | 18 14 | 19 02 | 13 56 | 14 52 | 15 35 | 16 07 |
| 18 | 92 37.5 S | 3 27.0 | 173 27.9 | 8.9 N19 41.0 | | 1.5 | 56.9 | N 58 | 17 34 | 18 13 | 18 58 | 13 42 | 14 38 | 15 22 | 15 56 |
| 19 | 107 37.7 | 28.0 | 187 55.8 | 8.9 | 19 39.5 | 1.5 | 56.9 | 56 | 17 35 | 18 12 | 18 55 | 13 30 | 14 26 | 15 11 | 15 46 |
| 20 | 122 37.9 | 28.9 | 202 23.7 | 9.1 | 19 38.0 | 1.7 | 56.8 | 54 | 17 36 | 18 11 | 18 52 | 13 20 | 14 16 | 15 01 | 15 38 |
| 21 | 137 38.1 ·· | 29.9 | 216 51.8 | 9.0 | 19 36.3 | 1.7 | 56.8 | 52 | 17 37 | 18 11 | 18 50 | 13 10 | 14 06 | 14 53 | 15 31 |
| 22 | 152 38.3 | 30.9 | 231 19.8 | 9.2 | 19 34.6 | 1.9 | 56.8 | 50 | 17 38 | 18 10 | 18 48 | 13 02 | 13 58 | 14 45 | 15 24 |
| 23 | 167 38.5 | 31.9 | 245 48.0 | 9.1 | 19 32.7 | 2.0 | 56.7 | 45 | 17 40 | 18 10 | 18 44 | 12 45 | 13 40 | 14 28 | 15 09 |
| 2 00 | 182 38.7 S | 3 32.8 | 260 16.1 | 9.3 N19 30.7 | | 2.0 | 56.7 | N 40 | 17 42 | 18 09 | 18 41 | 12 30 | 13 26 | 14 15 | 14 57 |
| 01 | 197 38.9 | 33.8 | 274 44.4 | 9.2 | 19 28.7 | 2.2 | 56.7 | 35 | 17 44 | 18 09 | 18 38 | 12 18 | 13 14 | 14 03 | 14 47 |
| 02 | 212 39.1 | 34.8 | 289 12.6 | 9.4 | 19 26.5 | 2.3 | 56.6 | 30 | 17 45 | 18 09 | 18 37 | 12 07 | 13 03 | 13 53 | 14 38 |
| 03 | 227 39.3 ·· | 35.7 | 303 41.0 | 9.3 | 19 24.2 | 2.4 | 56.6 | 20 | 17 48 | 18 10 | 18 36 | 11 49 | 12 44 | 13 36 | 14 23 |
| 04 | 242 39.5 | 36.7 | 318 09.3 | 9.5 | 19 21.8 | 2.5 | 56.6 | N 10 | 17 50 | 18 11 | 18 36 | 11 33 | 12 28 | 13 20 | 14 09 |
| 05 | 257 39.7 | 37.7 | 332 37.8 | 9.5 | 19 19.3 | 2.6 | 56.6 | 0 | 17 53 | 18 14 | 18 38 | 11 18 | 12 13 | 13 06 | 13 56 |
| 06 | 272 39.9 S | 3 38.6 | 347 06.3 | 9.5 N19 16.7 | | 2.7 | 56.5 | S 10 | 17 55 | 18 16 | 18 41 | 11 04 | 11 58 | 12 52 | 13 43 |
| 07 | 287 40.1 | 39.6 | 1 34.8 | 9.6 | 19 14.0 | 2.8 | 56.5 | 20 | 17 58 | 18 20 | 18 46 | 10 48 | 11 42 | 12 36 | 13 30 |
| T 08 | 302 40.3 | 40.6 | 16 03.4 | 9.7 | 19 11.2 | 2.9 | 56.5 | 30 | 18 01 | 18 25 | 18 53 | 10 29 | 11 23 | 12 18 | 13 14 |
| H 09 | 317 40.5 ·· | 41.5 | 30 32.1 | 9.7 | 19 08.3 | 2.9 | 56.4 | 35 | 18 03 | 18 29 | 18 59 | 10 19 | 11 12 | 12 08 | 13 05 |
| U 10 | 332 40.7 | 42.5 | 45 00.8 | 9.7 | 19 05.4 | 3.1 | 56.4 | 40 | 18 05 | 18 33 | 19 05 | 10 07 | 11 00 | 11 56 | 12 54 |
| R 11 | 347 40.9 | 43.5 | 59 29.5 | 9.9 | 19 02.3 | 3.2 | 56.4 | 45 | 18 08 | 18 38 | 19 13 | 09 52 | 10 45 | 11 42 | 12 41 |
| S 12 | 2 41.1 S | 3 44.4 | 73 58.4 | 9.9 N18 59.1 | | 3.3 | 56.3 | S 50 | 18 11 | 18 44 | 19 23 | 09 35 | 10 27 | 11 25 | 12 26 |
| D 13 | 17 41.3 | 45.4 | 88 27.3 | 9.9 | 18 55.8 | 3.4 | 56.3 | 52 | 18 13 | 18 47 | 19 28 | 09 27 | 10 19 | 11 17 | 12 19 |
| A 14 | 32 41.5 | 46.4 | 102 56.2 | 10.0 | 18 52.4 | 3.5 | 56.3 | 54 | 18 14 | 18 50 | 19 33 | 19 17 | 10 09 | 11 08 | 12 11 |
| Y 15 | 47 41.7 ·· | 47.3 | 117 25.2 | 10.0 | 18 48.9 | 3.6 | 56.2 | 56 | 18 16 | 18 54 | 19 40 | 09 07 | 09 59 | 10 58 | 12 02 |
| 16 | 62 41.9 | 48.3 | 131 54.2 | 10.1 | 18 45.3 | 3.7 | 56.2 | 58 | 18 18 | 18 58 | 19 47 | 08 55 | 09 47 | 10 46 | 11 52 |
| 17 | 77 42.1 | 49.3 | 146 23.3 | 10.2 | 18 41.6 | 3.7 | 56.2 | S 60 | 18 20 | 19 03 | 19 55 | 08 41 | 09 32 | 10 33 | 11 41 |
| 18 | 92 42.3 S | 3 50.2 | 160 52.5 | 10.2 N18 37.9 | | 3.9 | 56.2 | | | SUN | | | MOON | | |
| 19 | 107 42.5 | 51.2 | 175 21.7 | 10.3 | 18 34.0 | 3.9 | 56.1 | Day | Eqn. of Time | | Mer. | Mer. Pass. | | Age | Phase |
| 20 | 122 42.7 | 52.2 | 189 51.0 | 10.3 | 18 30.1 | 4.1 | 56.1 | | 00ʰ | 12ʰ | Pass. | Upper | Lower | | |
| 21 | 137 42.8 ·· | 53.1 | 204 20.3 | 10.4 | 18 26.0 | 4.1 | 56.1 | | | | | | | | |
| 22 | 152 43.0 | 54.1 | 218 49.7 | 10.5 | 18 21.9 | 4.3 | 56.0 | | m s | m s | h m | h m | h m | d | |
| 23 | 167 43.2 | 55.1 | 233 19.2 | 10.5 | 18 17.6 | 4.3 | 56.0 | 30 | 09 56 | 10 06 | 11 50 | 05 05 | 17 33 | 21 | ◐ |
| | | | | | | | | 1 | 10 15 | 10 25 | 11 50 | 06 00 | 18 27 | 22 | |
| | S.D. 16.0 | d 1.0 | S.D. 15.8 | | 15.6 | | 15.4 | 2 | 10 34 | 10 44 | 11 49 | 06 53 | 19 19 | 23 | |

1980 DECEMBER 17, 18, 19 (WED., THURS., FRI.)

| G.M.T. | ARIES G.H.A. | VENUS −3.4 G.H.A. Dec. | MARS +1.4 G.H.A. Dec. | JUPITER −1.5 G.H.A. Dec. | SATURN +1.1 G.H.A. Dec. | STARS Name S.H.A. Dec. |
|---|---|---|---|---|---|---|
| 17 00 | 85 46.4 | 209 15.7 S18 26.4 | 154 53.8 S23 11.2 | 257 49.8 S 2 04.4 | 256 45.3 S 1 24.1 | Acamar 315 36.8 S40 23.1 |
| 01 | 100 48.8 | 224 15.0 27.2 | 169 54.2 10.9 | 272 52.0 04.5 | 271 47.6 24.2 | Achernar 335 44.9 S57 20.4 |
| 02 | 115 51.3 | 239 14.3 27.9 | 184 54.6 10.7 | 287 54.2 04.6 | 286 49.9 24.2 | Acrux 173 37.1 S62 59.2 |
| 03 | 130 53.7 | 254 13.6 ·· 28.7 | 199 54.9 ·· 10.4 | 302 56.4 ·· 04.7 | 301 52.3 ·· 24.3 | Adhara 255 31.6 S28 56.8 |
| 04 | 145 56.2 | 269 12.8 29.4 | 214 55.3 10.1 | 317 58.6 04.8 | 316 54.6 24.3 | Aldebaran 291 17.5 N16 28.2 |
| 05 | 160 58.7 | 284 12.1 30.2 | 229 55.7 09.8 | 333 00.8 04.9 | 331 56.9 24.4 | |
| 06 | 176 01.1 | 299 11.4 S18 30.9 | 244 56.1 S23 09.6 | 348 03.0 S 2 05.0 | 346 59.3 S 1 24.4 | Alioth 166 42.6 N56 03.6 |
| W 07 | 191 03.6 | 314 10.7 31.7 | 259 56.4 09.3 | 3 05.2 05.1 | 2 01.6 24.4 | Alkaid 153 18.7 N49 24.4 |
| E 08 | 206 06.1 | 329 10.0 32.4 | 274 56.8 09.0 | 18 07.4 05.2 | 17 03.9 24.5 | Al Na'ir 28 15.0 S47 03.5 |
| D 09 | 221 08.5 | 344 09.2 ·· 33.2 | 289 57.2 ·· 08.7 | 33 09.6 ·· 05.3 | 32 06.3 ·· 24.5 | Alnilam 276 11.2 S 1 12.9 |
| N 10 | 236 11.0 | 359 08.5 33.9 | 304 57.6 08.5 | 48 11.9 05.4 | 47 08.6 24.6 | Alphard 218 20.2 S 8 34.5 |
| E 11 | 251 13.5 | 14 07.8 34.7 | 319 57.9 08.2 | 63 14.1 05.5 | 62 10.9 24.6 | |
| S 12 | 266 15.9 | 29 07.1 S18 35.4 | 334 58.3 S23 07.9 | 78 16.3 S 2 05.6 | 77 13.3 S 1 24.7 | Alphecca 126 32.3 N26 46.8 |
| D 13 | 281 18.4 | 44 06.3 36.2 | 349 58.7 07.6 | 93 18.5 05.7 | 92 15.6 24.7 | Alpheratz 358 09.1 N28 59.2 |
| A 14 | 296 20.8 | 59 05.6 36.9 | 4 59.0 07.4 | 108 20.7 05.8 | 107 17.9 24.8 | Altair 62 32.7 N 8 49.1 |
| Y 15 | 311 23.3 | 74 04.9 ·· 37.7 | 19 59.4 ·· 07.1 | 123 22.9 ·· 05.9 | 122 20.3 ·· 24.8 | Ankaa 353 40.0 S42 24.9 |
| 16 | 326 25.8 | 89 04.2 38.4 | 34 59.8 06.8 | 138 25.1 06.0 | 137 22.6 24.8 | Antares 112 57.0 S26 23.3 |
| 17 | 341 28.2 | 104 03.4 39.2 | 50 00.2 06.5 | 153 27.3 06.1 | 152 24.9 24.9 | |
| 18 | 356 30.7 | 119 02.7 S18 39.9 | 65 00.5 S23 06.2 | 168 29.5 S 2 06.2 | 167 27.3 S 1 24.9 | Arcturus 146 18.6 N19 16.9 |
| 19 | 11 33.2 | 134 02.0 40.6 | 80 00.9 06.0 | 183 31.7 06.3 | 182 29.6 25.0 | Atria 108 21.7 S68 59.5 |
| 20 | 26 35.6 | 149 01.3 41.4 | 95 01.3 05.7 | 198 33.9 06.4 | 197 31.9 25.0 | Avior 234 27.6 S59 26.7 |
| 21 | 41 38.1 | 164 00.5 ·· 42.1 | 110 01.7 ·· 05.4 | 213 36.1 ·· 06.5 | 212 34.2 ·· 25.1 | Bellatrix 278 58.2 N 6 19.9 |
| 22 | 56 40.6 | 178 59.8 42.9 | 125 02.0 05.1 | 228 38.3 06.6 | 227 36.6 25.1 | Betelgeuse 271 27.8 N 7 24.2 |
| 23 | 71 43.0 | 193 59.1 43.6 | 140 02.4 04.8 | 243 40.5 06.7 | 242 38.9 25.2 | |
| 18 00 | 86 45.5 | 208 58.3 S18 44.3 | 155 02.8 S23 04.5 | 258 42.7 S 2 06.8 | 257 41.2 S 1 25.2 | Canopus 264 06.6 S52 41.1 |
| 01 | 101 48.0 | 223 57.6 45.1 | 170 03.2 04.3 | 273 45.0 06.9 | 272 43.6 25.2 | Capella 281 10.6 N45 58.7 |
| 02 | 116 50.4 | 238 56.9 45.8 | 185 03.5 04.0 | 288 47.2 07.0 | 287 45.9 25.3 | Deneb 49 48.7 N45 12.9 |
| 03 | 131 52.9 | 253 56.1 ·· 46.6 | 200 03.9 ·· 03.7 | 303 49.4 ·· 07.1 | 302 48.2 ·· 25.3 | Denebola 182 58.9 N14 40.7 |
| 04 | 146 55.3 | 268 55.4 47.3 | 215 04.3 03.4 | 318 51.6 07.2 | 317 50.6 25.4 | Diphda 349 20.6 S18 05.7 |
| 05 | 161 57.8 | 283 54.7 48.0 | 230 04.7 03.1 | 333 53.8 07.3 | 332 52.9 25.4 | |
| 06 | 177 00.3 | 298 53.9 S18 48.8 | 245 05.1 S23 02.8 | 348 56.0 S 2 07.4 | 347 55.3 S 1 25.5 | Dubhe 194 21.9 N61 51.0 |
| 07 | 192 02.7 | 313 53.2 49.5 | 260 05.4 02.6 | 3 58.2 07.5 | 2 57.6 25.5 | Elnath 278 43.6 N28 35.4 |
| T 08 | 207 05.2 | 328 52.5 50.2 | 275 05.8 02.3 | 19 00.4 07.6 | 17 59.9 25.6 | Eltanin 90 58.2 N51 29.6 |
| H 09 | 222 07.7 | 343 51.7 ·· 51.0 | 290 06.2 ·· 02.0 | 34 02.6 ·· 07.7 | 33 02.3 ·· 25.6 | Enif 34 11.6 N 9 47.3 |
| U 10 | 237 10.1 | 358 51.0 51.7 | 305 06.6 01.7 | 49 04.8 07.8 | 48 04.6 25.6 | Fomalhaut 15 51.3 S29 43.6 |
| R 11 | 252 12.6 | 13 50.3 52.4 | 320 06.9 01.4 | 64 07.0 07.9 | 63 06.9 25.7 | |
| S 12 | 267 15.1 | 28 49.5 S18 53.1 | 335 07.3 S23 01.1 | 79 09.3 S 2 08.0 | 78 09.3 S 1 25.7 | Gacrux 172 28.6 S57 00.0 |
| D 13 | 282 17.5 | 43 48.8 53.9 | 350 07.7 00.8 | 94 11.5 08.1 | 93 11.6 25.8 | Gienah 176 17.8 S17 26.0 |
| A 14 | 297 20.0 | 58 48.1 54.6 | 5 08.1 00.5 | 109 13.7 08.2 | 108 13.9 25.8 | Hadar 149 23.5 S60 16.5 |
| Y 15 | 312 22.5 | 73 47.3 ·· 55.3 | 20 08.4 ·· 00.3 | 124 15.9 ·· 08.3 | 123 16.3 ·· 25.9 | Hamal 328 28.5 N23 22.4 |
| 16 | 327 24.9 | 88 46.6 56.0 | 35 08.8 23 00.0 | 139 18.1 08.4 | 138 18.6 25.9 | Kaus Aust. 84 17.1 S34 23.6 |
| 17 | 342 27.4 | 103 45.8 56.8 | 50 09.2 22 59.7 | 154 20.3 08.5 | 153 20.9 25.9 | |
| 18 | 357 29.8 | 118 45.1 S18 57.5 | 65 09.6 S22 59.4 | 169 22.5 S 2 08.6 | 168 23.3 S 1 26.0 | Kochab 137 20.0 N74 13.9 |
| 19 | 12 32.3 | 133 44.4 58.2 | 80 09.9 59.1 | 184 24.7 08.7 | 183 25.6 26.0 | Markab 14 03.1 N15 06.2 |
| 20 | 27 34.8 | 148 43.6 58.9 | 95 10.3 58.8 | 199 27.0 08.8 | 198 27.9 26.1 | Menkar 314 40.7 N 4 00.8 |
| 21 | 42 37.2 | 163 42.9 18 59.6 | 110 10.7 ·· 58.5 | 214 29.2 ·· 08.9 | 213 30.3 ·· 26.1 | Menkent 148 37.1 S36 16.3 |
| 22 | 57 39.7 | 178 42.1 19 00.4 | 125 11.1 58.2 | 229 31.4 09.0 | 228 32.6 26.2 | Miaplacidus 221 44.2 S69 38.1 |
| 23 | 72 42.2 | 193 41.4 01.1 | 140 11.5 57.9 | 244 33.6 09.1 | 243 35.0 26.2 | |
| 19 00 | 87 44.6 | 208 40.7 S19 01.8 | 155 11.8 S22 57.6 | 259 35.8 S 2 09.2 | 258 37.3 S 1 26.2 | Mirfak 309 15.4 N49 47.7 |
| 01 | 102 47.1 | 223 39.9 02.5 | 170 12.2 57.3 | 274 38.0 09.3 | 273 39.6 26.3 | Nunki 76 29.4 S26 19.2 |
| 02 | 117 49.6 | 238 39.2 03.2 | 185 12.6 57.0 | 289 40.2 09.4 | 288 42.0 26.3 | Peacock 53 58.7 S56 48.0 |
| 03 | 132 52.0 | 253 38.4 ·· 03.9 | 200 13.0 ·· 56.7 | 304 42.4 ·· 09.5 | 303 44.3 ·· 26.4 | Pollux 243 57.7 N28 04.3 |
| 04 | 147 54.5 | 268 37.7 04.7 | 215 13.3 56.5 | 319 44.7 09.6 | 318 46.6 26.4 | Procyon 245 25.3 N 5 16.4 |
| 05 | 162 56.9 | 283 36.9 05.4 | 230 13.7 56.2 | 334 46.9 09.7 | 333 49.0 26.4 | |
| 06 | 177 59.4 | 298 36.2 S19 06.1 | 245 14.1 S22 55.9 | 349 49.1 S 2 09.8 | 348 51.3 S 1 26.5 | Rasalhague 96 29.8 N12 34.5 |
| 07 | 193 01.9 | 313 35.4 06.8 | 260 14.5 55.6 | 4 51.3 09.9 | 3 53.7 26.5 | Regulus 208 09.7 N12 03.6 |
| 08 | 208 04.3 | 328 34.7 07.5 | 275 14.9 55.3 | 19 53.5 10.0 | 18 56.0 26.6 | Rigel 281 35.5 S 8 13.5 |
| F 09 | 223 06.8 | 343 34.0 ·· 08.2 | 290 15.2 ·· 55.0 | 34 55.7 ·· 10.1 | 33 58.3 ·· 26.6 | Rigil Kent. 140 26.1 S60 45.0 |
| R 10 | 238 09.3 | 358 33.2 08.9 | 305 15.6 54.7 | 49 58.0 10.1 | 49 00.7 26.7 | Sabik 102 41.3 S15 42.0 |
| I 11 | 253 11.7 | 13 32.5 09.6 | 320 16.0 54.4 | 65 00.2 10.2 | 64 03.0 26.7 | |
| D 12 | 268 14.2 | 28 31.7 S19 10.3 | 335 16.4 S22 54.1 | 80 02.4 S 2 10.3 | 79 05.3 S 1 26.7 | Schedar 350 08.6 N56 26.2 |
| A 13 | 283 16.7 | 43 31.0 11.0 | 350 16.7 53.8 | 95 04.6 10.4 | 94 07.7 26.8 | Shaula 96 56.0 S37 05.3 |
| Y 14 | 298 19.1 | 58 30.2 11.7 | 5 17.1 53.5 | 110 06.8 10.5 | 109 10.0 26.8 | Sirius 258 55.2 S16 41.5 |
| 15 | 313 21.6 | 73 29.5 ·· 12.4 | 20 17.5 ·· 53.2 | 125 09.0 ·· 10.6 | 124 12.4 ·· 26.9 | Spica 158 57.5 S11 03.5 |
| 16 | 328 24.1 | 88 28.7 13.1 | 35 17.9 52.9 | 140 11.3 10.7 | 139 14.7 26.9 | Suhail 223 10.3 S43 21.1 |
| 17 | 343 26.5 | 103 28.0 13.8 | 50 18.3 52.6 | 155 13.5 10.8 | 154 17.0 26.9 | |
| 18 | 358 29.0 | 118 27.2 S19 14.6 | 65 18.6 S22 52.3 | 170 15.7 S 2 10.9 | 169 19.4 S 1 27.0 | Vega 80 56.2 N38 46.1 |
| 19 | 13 31.4 | 133 26.5 15.2 | 80 19.0 52.0 | 185 17.9 11.0 | 184 21.7 27.0 | Zuben'ubi 137 33.1 S15 57.6 |
| 20 | 28 33.9 | 148 25.7 15.9 | 95 19.4 51.7 | 200 20.1 11.1 | 199 24.1 27.1 | S.H.A. Mer. Pass. |
| 21 | 43 36.4 | 163 25.0 ·· 16.6 | 110 19.8 ·· 51.4 | 215 22.3 ·· 11.2 | 214 26.4 ·· 27.1 | |
| 22 | 58 38.8 | 178 24.2 17.3 | 125 20.2 51.1 | 230 24.6 11.3 | 229 28.7 27.2 | Venus 122 12.8 10 05 |
| 23 | 73 41.3 | 193 23.4 18.0 | 140 20.5 50.8 | 245 26.8 11.4 | 244 31.1 27.2 | Mars 68 17.3 13 39 |
| | h m | | | | | Jupiter 171 57.3 6 44 |
| Mer. Pass. 18 10.0 | | v −0.7 d 0.7 | v 0.4 d 0.3 | v 2.2 d 0.1 | v 2.3 d 0.0 | Saturn 170 55.8 6 48 |

## 1980 DECEMBER 17, 18, 19 (WED., THURS., FRI.)

| G.M.T. | SUN G.H.A. | Dec. | MOON G.H.A. | $v$ | Dec. | $d$ | H.P. | Lat. | Twilight Naut. | Civil | Sunrise | Moonrise 17 | 18 | 19 | 20 |
|---|---|---|---|---|---|---|---|---|---|---|---|---|---|---|---|
| d h | ° ' | ° ' | ° ' | ' | ° ' | ' | ' | ° | h m | h m | h m | h m | h m | h m | h m |
| 17 00 | 180 59.3 | S23 21.2 | 64 58.1 | 9.6 | N 3 35.9 | 12.3 | 59.7 | N 72 | 08 24 | 10 54 | ▬ | 12 51 | 12 38 | 12 19 | 11 36 |
| 01 | 195 59.0 | 21.3 | 79 26.7 | 9.5 | 3 48.2 | 12.3 | 59.7 | N 70 | 08 04 | 09 52 | ▬ | 13 00 | 12 55 | 12 49 | 12 43 |
| 02 | 210 58.7 | 21.4 | 93 55.2 | 9.5 | 4 00.5 | 12.3 | 59.8 | 68 | 07 48 | 09 17 | ▬ | 13 07 | 13 08 | 13 12 | 13 20 |
| 03 | 225 58.4 | ·· 21.5 | 108 23.7 | 9.5 | 4 12.8 | 12.3 | 59.8 | 66 | 07 35 | 08 52 | 10 32 | 13 12 | 13 19 | 13 30 | 13 46 |
| 04 | 240 58.1 | 21.5 | 122 52.2 | 9.5 | 4 25.1 | 12.2 | 59.8 | 64 | 07 24 | 08 32 | 09 50 | 13 17 | 13 29 | 13 44 | 14 07 |
| 05 | 255 57.8 | 21.6 | 137 20.7 | 9.4 | 4 37.3 | 12.3 | 59.8 | 62 | 07 14 | 08 16 | 09 22 | 13 22 | 13 37 | 13 56 | 14 23 |
| 06 | 270 57.5 | S23 21.7 | 151 49.1 | 9.3 | N 4 49.6 | 12.2 | 59.8 | 60 | 07 06 | 08 02 | 09 00 | 13 25 | 13 44 | 14 07 | 14 37 |
| W 07 | 285 57.2 | 21.8 | 166 17.4 | 9.3 | 5 01.8 | 12.1 | 59.8 | N 58 | 06 58 | 07 51 | 08 43 | 13 29 | 13 50 | 14 16 | 14 49 |
| E 08 | 300 56.9 | 21.9 | 180 45.7 | 9.3 | 5 13.9 | 12.2 | 59.8 | 56 | 06 51 | 07 41 | 08 28 | 13 32 | 13 56 | 14 24 | 14 59 |
| D 09 | 315 56.5 | ·· 22.0 | 195 14.0 | 9.3 | 5 26.1 | 12.1 | 59.9 | 54 | 06 45 | 07 31 | 08 15 | 13 35 | 14 01 | 14 31 | 15 08 |
| N 10 | 330 56.2 | 22.1 | 209 42.3 | 9.2 | 5 38.2 | 12.2 | 59.9 | 52 | 06 39 | 07 23 | 08 04 | 13 37 | 14 05 | 14 38 | 15 17 |
| E 11 | 345 55.9 | 22.2 | 224 10.5 | 9.1 | 5 50.4 | 12.0 | 59.9 | 50 | 06 34 | 07 16 | 07 54 | 13 39 | 14 09 | 14 43 | 15 24 |
| S 12 | 0 55.6 | S23 22.2 | 238 38.6 | 9.1 | N 6 02.4 | 12.1 | 59.9 | 45 | 06 22 | 07 00 | 07 34 | 13 44 | 14 18 | 14 56 | 15 40 |
| D 13 | 15 55.3 | 22.3 | 253 06.7 | 9.1 | 6 14.5 | 12.0 | 59.9 | N 40 | 06 12 | 06 46 | 07 17 | 13 49 | 14 26 | 15 07 | 15 53 |
| A 14 | 30 55.0 | 22.4 | 267 34.8 | 9.0 | 6 26.5 | 12.0 | 59.9 | 35 | 06 03 | 06 34 | 07 03 | 13 52 | 14 32 | 15 16 | 16 04 |
| Y 15 | 45 54.7 | ·· 22.5 | 282 02.8 | 9.0 | 6 38.5 | 11.9 | 59.9 | 30 | 05 54 | 06 24 | 06 50 | 13 55 | 14 38 | 15 24 | 16 14 |
| 16 | 60 54.4 | 22.6 | 296 30.8 | 8.9 | 6 50.4 | 12.0 | 59.9 | 20 | 05 37 | 06 05 | 06 29 | 14 01 | 14 48 | 15 37 | 16 31 |
| 17 | 75 54.1 | 22.6 | 310 58.7 | 8.9 | 7 02.4 | 11.8 | 60.0 | N 10 | 05 21 | 05 48 | 06 10 | 14 06 | 14 56 | 15 50 | 16 45 |
| 18 | 90 53.8 | S23 22.7 | 325 26.6 | 8.9 | N 7 14.2 | 11.9 | 60.0 | 0 | 05 04 | 05 30 | 05 53 | 14 11 | 15 05 | 16 01 | 16 59 |
| 19 | 105 53.5 | 22.8 | 339 54.5 | 8.8 | 7 26.1 | 11.8 | 60.0 | S 10 | 04 45 | 05 12 | 05 35 | 14 15 | 15 13 | 16 13 | 17 13 |
| 20 | 120 53.2 | 22.9 | 354 22.3 | 8.7 | 7 37.9 | 11.8 | 60.0 | 20 | 04 23 | 04 52 | 05 16 | 14 21 | 15 22 | 16 25 | 17 28 |
| 21 | 135 52.9 | ·· 22.9 | 8 50.0 | 8.7 | 7 49.7 | 11.7 | 60.0 | 30 | 03 54 | 04 27 | 04 54 | 14 27 | 15 32 | 16 39 | 17 46 |
| 22 | 150 52.6 | 23.0 | 23 17.7 | 8.7 | 8 01.4 | 11.7 | 60.0 | 35 | 03 35 | 04 12 | 04 41 | 14 30 | 15 38 | 16 47 | 17 56 |
| 23 | 165 52.3 | 23.1 | 37 45.4 | 8.6 | 8 13.1 | 11.6 | 60.0 | 40 | 03 12 | 03 54 | 04 26 | 14 34 | 15 45 | 16 57 | 18 07 |
| | | | | | | | | 45 | 02 41 | 03 31 | 04 08 | 14 38 | 15 53 | 17 08 | 18 21 |
| 18 00 | 180 52.0 | S23 23.2 | 52 13.0 | 8.6 | N 8 24.7 | 11.6 | 60.0 | S 50 | 01 56 | 03 01 | 03 46 | 14 44 | 16 03 | 17 22 | 18 37 |
| 01 | 195 51.6 | 23.2 | 66 40.6 | 8.5 | 8 36.3 | 11.5 | 60.0 | 52 | 01 28 | 02 46 | 03 35 | 14 46 | 16 07 | 17 28 | 18 45 |
| 02 | 210 51.3 | 23.3 | 81 08.1 | 8.4 | 8 47.8 | 11.5 | 60.1 | 54 | 00 42 | 02 28 | 03 23 | 14 49 | 16 12 | 17 35 | 18 54 |
| 03 | 225 51.0 | ·· 23.4 | 95 35.5 | 8.4 | 8 59.3 | 11.4 | 60.1 | 56 | //// | 02 06 | 03 08 | 14 52 | 16 18 | 17 43 | 19 04 |
| 04 | 240 50.7 | 23.4 | 110 02.9 | 8.4 | 9 10.8 | 11.3 | 60.1 | 58 | //// | 01 36 | 02 52 | 14 56 | 16 24 | 17 52 | 19 15 |
| 05 | 255 50.4 | 23.5 | 124 30.3 | 8.3 | 9 22.1 | 11.4 | 60.1 | S 60 | //// | 00 46 | 02 31 | 15 00 | 16 31 | 18 02 | 19 28 |
| 06 | 270 50.1 | S23 23.6 | 138 57.6 | 8.2 | N 9 33.5 | 11.3 | 60.1 | Lat. | Sunset | Twilight Civil | Naut. | Moonset 17 | 18 | 19 | 20 |
| 07 | 285 49.8 | 23.7 | 153 24.8 | 8.3 | 9 44.8 | 11.2 | 60.1 |
| T 08 | 300 49.5 | 23.7 | 167 52.1 | 8.1 | 9 56.0 | 11.2 | 60.1 |
| H 09 | 315 49.2 | ·· 23.8 | 182 19.2 | 8.1 | 10 07.2 | 11.1 | 60.1 |
| U 10 | 330 48.9 | 23.9 | 196 46.3 | 8.1 | 10 18.3 | 11.0 | 60.1 | ° | h m | h m | h m | h m | h m | h m | h m |
| R 11 | 345 48.6 | 23.9 | 211 13.4 | 8.0 | 10 29.3 | 11.0 | 60.1 | N 72 | ▬ | 12 59 | 15 30 | 02 35 | 04 40 | 06 55 | 09 39 |
| S 12 | 0 48.3 | S23 24.0 | 225 40.4 | 7.9 | N10 40.3 | 10.9 | 60.1 | N 70 | ▬ | 14 01 | 15 50 | 02 29 | 04 25 | 06 26 | 08 33 |
| D 13 | 15 48.0 | 24.0 | 240 07.3 | 7.9 | 10 51.2 | 10.9 | 60.1 | 68 | ▬ | 14 36 | 16 06 | 02 24 | 04 13 | 06 05 | 07 57 |
| A 14 | 30 47.6 | 24.1 | 254 34.2 | 7.8 | 11 02.1 | 10.8 | 60.1 | 66 | 13 21 | 15 02 | 16 19 | 02 20 | 04 03 | 05 48 | 07 32 |
| Y 15 | 45 47.3 | ·· 24.2 | 269 01.0 | 7.8 | 11 12.9 | 10.7 | 60.1 | 64 | 14 03 | 15 21 | 16 30 | 02 17 | 03 55 | 05 35 | 07 12 |
| 16 | 60 47.0 | 24.2 | 283 27.8 | 7.8 | 11 23.6 | 10.6 | 60.2 | 62 | 14 32 | 15 37 | 16 39 | 02 14 | 03 48 | 05 23 | 06 56 |
| 17 | 75 46.7 | 24.3 | 297 54.6 | 7.6 | 11 34.2 | 10.6 | 60.2 | 60 | 14 53 | 15 51 | 16 48 | 02 12 | 03 42 | 05 14 | 06 43 |
| 18 | 90 46.4 | S23 24.3 | 312 21.2 | 7.7 | N11 44.8 | 10.5 | 60.2 | N 58 | 15 11 | 16 03 | 16 55 | 02 09 | 03 37 | 05 05 | 06 31 |
| 19 | 105 46.1 | 24.4 | 326 47.9 | 7.5 | 11 55.3 | 10.5 | 60.2 | 56 | 15 26 | 16 13 | 17 02 | 02 07 | 03 32 | 04 58 | 06 21 |
| 20 | 120 45.8 | 24.5 | 341 14.4 | 7.6 | 12 05.8 | 10.3 | 60.2 | 54 | 15 38 | 16 22 | 17 08 | 02 06 | 03 28 | 04 51 | 06 13 |
| 21 | 135 45.5 | ·· 24.5 | 355 41.0 | 7.4 | 12 16.1 | 10.3 | 60.2 | 52 | 15 49 | 16 30 | 17 14 | 02 04 | 03 25 | 04 46 | 06 05 |
| 22 | 150 45.2 | 24.6 | 10 07.4 | 7.4 | 12 26.4 | 10.2 | 60.2 | 50 | 15 59 | 16 38 | 17 19 | 02 03 | 03 21 | 04 40 | 05 58 |
| 23 | 165 44.9 | 24.6 | 24 33.8 | 7.4 | 12 36.6 | 10.2 | 60.2 | 45 | 16 20 | 16 54 | 17 31 | 01 59 | 03 14 | 04 29 | 05 43 |
| 19 00 | 180 44.6 | S23 24.7 | 39 00.2 | 7.3 | N12 46.8 | 10.0 | 60.3 | N 40 | 16 37 | 17 07 | 17 41 | 01 57 | 03 08 | 04 19 | 05 30 |
| 01 | 195 44.3 | 24.7 | 53 26.5 | 7.3 | 12 56.8 | 10.0 | 60.2 | 35 | 16 51 | 17 19 | 17 51 | 01 55 | 03 02 | 04 11 | 05 20 |
| 02 | 210 44.0 | 24.8 | 67 52.8 | 7.3 | 13 06.8 | 9.9 | 60.2 | 30 | 17 03 | 17 30 | 18 00 | 01 53 | 02 58 | 04 04 | 05 11 |
| 03 | 225 43.6 | ·· 24.8 | 82 19.0 | 7.1 | 13 16.7 | 9.8 | 60.2 | 20 | 17 25 | 17 49 | 18 16 | 01 49 | 02 50 | 03 52 | 04 55 |
| 04 | 240 43.3 | 24.9 | 96 45.1 | 7.1 | 13 26.5 | 9.7 | 60.2 | N 10 | 17 43 | 18 06 | 18 32 | 01 46 | 02 43 | 03 41 | 04 41 |
| 05 | 255 43.0 | 24.9 | 111 11.2 | 7.1 | 13 36.2 | 9.6 | 60.2 | 0 | 18 01 | 18 23 | 18 49 | 01 43 | 02 36 | 03 31 | 04 29 |
| 06 | 270 42.7 | S23 25.0 | 125 37.3 | 6.9 | N13 45.8 | 9.6 | 60.2 | S 10 | 18 18 | 18 41 | 19 08 | 01 40 | 02 30 | 03 21 | 04 16 |
| 07 | 285 42.4 | 25.0 | 140 03.2 | 7.0 | 13 55.4 | 9.4 | 60.2 | 20 | 18 37 | 19 02 | 19 31 | 01 37 | 02 23 | 03 11 | 04 02 |
| 08 | 300 42.1 | 25.1 | 154 29.2 | 6.9 | 14 04.8 | 9.4 | 60.2 | 30 | 18 59 | 19 27 | 20 00 | 01 34 | 02 15 | 02 59 | 03 47 |
| F 09 | 315 41.8 | ·· 25.1 | 168 55.1 | 6.8 | 14 14.2 | 9.2 | 60.3 | 35 | 19 12 | 19 42 | 20 19 | 01 32 | 02 10 | 02 52 | 03 37 |
| R 10 | 330 41.5 | 25.2 | 183 20.9 | 6.8 | 14 23.4 | 9.2 | 60.2 | 40 | 19 27 | 20 00 | 20 42 | 01 29 | 02 05 | 02 44 | 03 27 |
| I 11 | 345 41.2 | 25.2 | 197 46.7 | 6.7 | 14 32.6 | 9.1 | 60.2 | 45 | 19 45 | 20 23 | 21 12 | 01 27 | 01 59 | 02 35 | 03 15 |
| D 12 | 0 40.9 | S23 25.3 | 212 12.4 | 6.7 | N14 41.7 | 9.0 | 60.2 | S 50 | 20 08 | 20 52 | 21 58 | 01 23 | 01 52 | 02 23 | 03 01 |
| A 13 | 15 40.6 | 25.3 | 226 38.1 | 6.7 | 14 50.7 | 8.8 | 60.2 | 52 | 20 19 | 21 08 | 22 26 | 01 22 | 01 48 | 02 18 | 02 54 |
| Y 14 | 30 40.2 | 25.3 | 241 03.8 | 6.6 | 14 59.5 | 8.8 | 60.2 | 54 | 20 31 | 21 26 | 23 13 | 01 20 | 01 45 | 02 13 | 02 46 |
| 15 | 45 39.9 | ·· 25.4 | 255 29.4 | 6.5 | 15 08.3 | 8.7 | 60.2 | 56 | 20 45 | 21 48 | //// | 01 19 | 01 41 | 02 07 | 02 38 |
| 16 | 60 39.6 | 25.4 | 269 54.9 | 6.5 | 15 17.0 | 8.6 | 60.2 | 58 | 21 02 | 22 18 | //// | 01 17 | 01 36 | 02 00 | 02 28 |
| 17 | 75 39.3 | 25.5 | 284 20.4 | 6.4 | 15 25.6 | 8.5 | 60.2 | S 60 | 21 23 | 23 09 | //// | 01 15 | 01 32 | 01 52 | 02 18 |
| 18 | 90 39.0 | S23 25.5 | 298 45.8 | 6.4 | N15 34.1 | 8.4 | 60.2 | | SUN | | | MOON | | | |
| 19 | 105 38.7 | 25.5 | 313 11.2 | 6.4 | 15 42.5 | 8.2 | 60.1 | Day | Eqn. of Time 00ʰ | 12ʰ | Mer. Pass. | Mer. Pass. Upper | Lower | Age | Phase |
| 20 | 120 38.4 | 25.6 | 327 36.6 | 6.3 | 15 50.7 | 8.2 | 60.1 |
| 21 | 135 38.1 | ·· 25.6 | 342 01.9 | 6.3 | 15 58.9 | 8.1 | 60.1 |
| 22 | 150 37.8 | 25.6 | 356 27.2 | 6.2 | 16 07.0 | 7.9 | 60.1 | | m s | m s | h m | h m | h m | d | |
| 23 | 165 37.5 | 25.7 | 10 52.4 | 6.2 | 16 14.9 | 7.9 | 60.1 | 17 | 03 58 | 03 43 | 11 56 | 20 23 | 07 57 | 10 | ◐ |
| | | | | | | | | 18 | 03 28 | 03 14 | 11 57 | 21 18 | 08 50 | 11 | |
| | S.D. 16.3 | $d$ 0.1 | S.D. 16.3 | | 16.4 | | 16.4 | 19 | 02 59 | 02 44 | 11 57 | 22 15 | 09 46 | 12 | |

# INCREMENTS AND CORRECTIONS

## 0ᵐ

| m | SUN PLANETS | ARIES | MOON | v or Corrⁿ d | v or Corrⁿ d | v or Corrⁿ d |
|---|---|---|---|---|---|---|
|  | ° ′ | ° ′ | ° ′ | ′ ′ | ′ ′ | ′ ′ |
| 00 | 0 00·0 | 0 00·0 | 0 00·0 | 0·0 0·0 | 6·0 0·1 | 12·0 0·1 |
| 01 | 0 00·3 | 0 00·3 | 0 00·2 | 0·1 0·0 | 6·1 0·1 | 12·1 0·1 |
| 02 | 0 00·5 | 0 00·5 | 0 00·5 | 0·2 0·0 | 6·2 0·1 | 12·2 0·1 |
| 03 | 0 00·8 | 0 00·8 | 0 00·7 | 0·3 0·0 | 6·3 0·1 | 12·3 0·1 |
| 04 | 0 01·0 | 0 01·0 | 0 01·0 | 0·4 0·0 | 6·4 0·1 | 12·4 0·1 |
| 05 | 0 01·3 | 0 01·3 | 0 01·2 | 0·5 0·0 | 6·5 0·1 | 12·5 0·1 |
| 06 | 0 01·5 | 0 01·5 | 0 01·4 | 0·6 0·0 | 6·6 0·1 | 12·6 0·1 |
| 07 | 0 01·8 | 0 01·8 | 0 01·7 | 0·7 0·0 | 6·7 0·1 | 12·7 0·1 |
| 08 | 0 02·0 | 0 02·0 | 0 01·9 | 0·8 0·0 | 6·8 0·1 | 12·8 0·1 |
| 09 | 0 02·3 | 0 02·3 | 0 02·1 | 0·9 0·0 | 6·9 0·1 | 12·9 0·1 |
| 10 | 0 02·5 | 0 02·5 | 0 02·4 | 1·0 0·0 | 7·0 0·1 | 13·0 0·1 |
| 11 | 0 02·8 | 0 02·8 | 0 02·6 | 1·1 0·0 | 7·1 0·1 | 13·1 0·1 |
| 12 | 0 03·0 | 0 03·0 | 0 02·9 | 1·2 0·0 | 7·2 0·1 | 13·2 0·1 |
| 13 | 0 03·3 | 0 03·3 | 0 03·1 | 1·3 0·0 | 7·3 0·1 | 13·3 0·1 |
| 14 | 0 03·5 | 0 03·5 | 0 03·3 | 1·4 0·0 | 7·4 0·1 | 13·4 0·1 |
| 15 | 0 03·8 | 0 03·8 | 0 03·6 | 1·5 0·0 | 7·5 0·1 | 13·5 0·1 |
| 16 | 0 04·0 | 0 04·0 | 0 03·8 | 1·6 0·0 | 7·6 0·1 | 13·6 0·1 |
| 17 | 0 04·3 | 0 04·3 | 0 04·1 | 1·7 0·0 | 7·7 0·1 | 13·7 0·1 |
| 18 | 0 04·5 | 0 04·5 | 0 04·3 | 1·8 0·0 | 7·8 0·1 | 13·8 0·1 |
| 19 | 0 04·8 | 0 04·8 | 0 04·5 | 1·9 0·0 | 7·9 0·1 | 13·9 0·1 |
| 20 | 0 05·0 | 0 05·0 | 0 04·8 | 2·0 0·0 | 8·0 0·1 | 14·0 0·1 |
| 21 | 0 05·3 | 0 05·3 | 0 05·0 | 2·1 0·0 | 8·1 0·1 | 14·1 0·1 |
| 22 | 0 05·5 | 0 05·5 | 0 05·2 | 2·2 0·0 | 8·2 0·1 | 14·2 0·1 |
| 23 | 0 05·8 | 0 05·8 | 0 05·5 | 2·3 0·0 | 8·3 0·1 | 14·3 0·1 |
| 24 | 0 06·0 | 0 06·0 | 0 05·7 | 2·4 0·0 | 8·4 0·1 | 14·4 0·1 |
| 25 | 0 06·3 | 0 06·3 | 0 06·0 | 2·5 0·0 | 8·5 0·1 | 14·5 0·1 |
| 26 | 0 06·5 | 0 06·5 | 0 06·2 | 2·6 0·0 | 8·6 0·1 | 14·6 0·1 |
| 27 | 0 06·8 | 0 06·8 | 0 06·4 | 2·7 0·0 | 8·7 0·1 | 14·7 0·1 |
| 28 | 0 07·0 | 0 07·0 | 0 06·7 | 2·8 0·0 | 8·8 0·1 | 14·8 0·1 |
| 29 | 0 07·3 | 0 07·3 | 0 06·9 | 2·9 0·0 | 8·9 0·1 | 14·9 0·1 |
| 30 | 0 07·5 | 0 07·5 | 0 07·2 | 3·0 0·0 | 9·0 0·1 | 15·0 0·1 |
| 31 | 0 07·8 | 0 07·8 | 0 07·4 | 3·1 0·0 | 9·1 0·1 | 15·1 0·1 |
| 32 | 0 08·0 | 0 08·0 | 0 07·6 | 3·2 0·0 | 9·2 0·1 | 15·2 0·1 |
| 33 | 0 08·3 | 0 08·3 | 0 07·9 | 3·3 0·0 | 9·3 0·1 | 15·3 0·1 |
| 34 | 0 08·5 | 0 08·5 | 0 08·1 | 3·4 0·0 | 9·4 0·1 | 15·4 0·1 |
| 35 | 0 08·8 | 0 08·8 | 0 08·4 | 3·5 0·0 | 9·5 0·1 | 15·5 0·1 |
| 36 | 0 09·0 | 0 09·0 | 0 08·6 | 3·6 0·0 | 9·6 0·1 | 15·6 0·1 |
| 37 | 0 09·3 | 0 09·3 | 0 08·8 | 3·7 0·0 | 9·7 0·1 | 15·7 0·1 |
| 38 | 0 09·5 | 0 09·5 | 0 09·1 | 3·8 0·0 | 9·8 0·1 | 15·8 0·1 |
| 39 | 0 09·8 | 0 09·8 | 0 09·3 | 3·9 0·0 | 9·9 0·1 | 15·9 0·1 |
| 40 | 0 10·0 | 0 10·0 | 0 09·5 | 4·0 0·0 | 10·0 0·1 | 16·0 0·1 |
| 41 | 0 10·3 | 0 10·3 | 0 09·8 | 4·1 0·0 | 10·1 0·1 | 16·1 0·1 |
| 42 | 0 10·5 | 0 10·5 | 0 10·0 | 4·2 0·0 | 10·2 0·1 | 16·2 0·1 |
| 43 | 0 10·8 | 0 10·8 | 0 10·3 | 4·3 0·0 | 10·3 0·1 | 16·3 0·1 |
| 44 | 0 11·0 | 0 11·0 | 0 10·5 | 4·4 0·0 | 10·4 0·1 | 16·4 0·1 |
| 45 | 0 11·3 | 0 11·3 | 0 10·7 | 4·5 0·0 | 10·5 0·1 | 16·5 0·1 |
| 46 | 0 11·5 | 0 11·5 | 0 11·0 | 4·6 0·0 | 10·6 0·1 | 16·6 0·1 |
| 47 | 0 11·8 | 0 11·8 | 0 11·2 | 4·7 0·0 | 10·7 0·1 | 16·7 0·1 |
| 48 | 0 12·0 | 0 12·0 | 0 11·5 | 4·8 0·0 | 10·8 0·1 | 16·8 0·1 |
| 49 | 0 12·3 | 0 12·3 | 0 11·7 | 4·9 0·0 | 10·9 0·1 | 16·9 0·1 |
| 50 | 0 12·5 | 0 12·5 | 0 11·9 | 5·0 0·0 | 11·0 0·1 | 17·0 0·1 |
| 51 | 0 12·8 | 0 12·8 | 0 12·2 | 5·1 0·0 | 11·1 0·1 | 17·1 0·1 |
| 52 | 0 13·0 | 0 13·0 | 0 12·4 | 5·2 0·0 | 11·2 0·1 | 17·2 0·1 |
| 53 | 0 13·3 | 0 13·3 | 0 12·6 | 5·3 0·0 | 11·3 0·1 | 17·3 0·1 |
| 54 | 0 13·5 | 0 13·5 | 0 12·9 | 5·4 0·0 | 11·4 0·1 | 17·4 0·1 |
| 55 | 0 13·8 | 0 13·8 | 0 13·1 | 5·5 0·0 | 11·5 0·1 | 17·5 0·1 |
| 56 | 0 14·0 | 0 14·0 | 0 13·4 | 5·6 0·0 | 11·6 0·1 | 17·6 0·1 |
| 57 | 0 14·3 | 0 14·3 | 0 13·6 | 5·7 0·0 | 11·7 0·1 | 17·7 0·1 |
| 58 | 0 14·5 | 0 14·5 | 0 13·8 | 5·8 0·0 | 11·8 0·1 | 17·8 0·1 |
| 59 | 0 14·8 | 0 14·8 | 0 14·1 | 5·9 0·0 | 11·9 0·1 | 17·9 0·1 |
| 60 | 0 15·0 | 0 15·0 | 0 14·3 | 6·0 0·1 | 12·0 0·1 | 18·0 0·2 |

## 1ᵐ

| m | SUN PLANETS | ARIES | MOON | v or Corrⁿ d | v or Corrⁿ d | v or Corrⁿ d |
|---|---|---|---|---|---|---|
|  | ° ′ | ° ′ | ° ′ | ′ ′ | ′ ′ | ′ ′ |
| 00 | 0 15·0 | 0 15·0 | 0 14·3 | 0·0 0·0 | 6·0 0·2 | 12·0 0·3 |
| 01 | 0 15·3 | 0 15·3 | 0 14·6 | 0·1 0·0 | 6·1 0·2 | 12·1 0·3 |
| 02 | 0 15·5 | 0 15·5 | 0 14·8 | 0·2 0·0 | 6·2 0·2 | 12·2 0·3 |
| 03 | 0 15·8 | 0 15·8 | 0 15·0 | 0·3 0·0 | 6·3 0·2 | 12·3 0·3 |
| 04 | 0 16·0 | 0 16·0 | 0 15·3 | 0·4 0·0 | 6·4 0·2 | 12·4 0·3 |
| 05 | 0 16·3 | 0 16·3 | 0 15·5 | 0·5 0·0 | 6·5 0·2 | 12·5 0·3 |
| 06 | 0 16·5 | 0 16·5 | 0 15·7 | 0·6 0·0 | 6·6 0·2 | 12·6 0·3 |
| 07 | 0 16·8 | 0 16·8 | 0 16·0 | 0·7 0·0 | 6·7 0·2 | 12·7 0·3 |
| 08 | 0 17·0 | 0 17·0 | 0 16·2 | 0·8 0·0 | 6·8 0·2 | 12·8 0·3 |
| 09 | 0 17·3 | 0 17·3 | 0 16·5 | 0·9 0·0 | 6·9 0·2 | 12·9 0·3 |
| 10 | 0 17·5 | 0 17·5 | 0 16·7 | 1·0 0·0 | 7·0 0·2 | 13·0 0·3 |
| 11 | 0 17·8 | 0 17·8 | 0 16·9 | 1·1 0·0 | 7·1 0·2 | 13·1 0·3 |
| 12 | 0 18·0 | 0 18·0 | 0 17·2 | 1·2 0·0 | 7·2 0·2 | 13·2 0·3 |
| 13 | 0 18·3 | 0 18·3 | 0 17·4 | 1·3 0·0 | 7·3 0·2 | 13·3 0·3 |
| 14 | 0 18·5 | 0 18·6 | 0 17·7 | 1·4 0·0 | 7·4 0·2 | 13·4 0·3 |
| 15 | 0 18·8 | 0 18·8 | 0 17·9 | 1·5 0·0 | 7·5 0·2 | 13·5 0·3 |
| 16 | 0 19·0 | 0 19·1 | 0 18·1 | 1·6 0·0 | 7·6 0·2 | 13·6 0·3 |
| 17 | 0 19·3 | 0 19·3 | 0 18·4 | 1·7 0·0 | 7·7 0·2 | 13·7 0·3 |
| 18 | 0 19·5 | 0 19·6 | 0 18·6 | 1·8 0·0 | 7·8 0·2 | 13·8 0·3 |
| 19 | 0 19·8 | 0 19·8 | 0 18·9 | 1·9 0·0 | 7·9 0·2 | 13·9 0·3 |
| 20 | 0 20·0 | 0 20·1 | 0 19·1 | 2·0 0·1 | 8·0 0·2 | 14·0 0·4 |
| 21 | 0 20·3 | 0 20·3 | 0 19·3 | 2·1 0·1 | 8·1 0·2 | 14·1 0·4 |
| 22 | 0 20·5 | 0 20·6 | 0 19·6 | 2·2 0·1 | 8·2 0·2 | 14·2 0·4 |
| 23 | 0 20·8 | 0 20·8 | 0 19·8 | 2·3 0·1 | 8·3 0·2 | 14·3 0·4 |
| 24 | 0 21·0 | 0 21·1 | 0 20·0 | 2·4 0·1 | 8·4 0·2 | 14·4 0·4 |
| 25 | 0 21·3 | 0 21·3 | 0 20·3 | 2·5 0·1 | 8·5 0·2 | 14·5 0·4 |
| 26 | 0 21·5 | 0 21·6 | 0 20·5 | 2·6 0·1 | 8·6 0·2 | 14·6 0·4 |
| 27 | 0 21·8 | 0 21·8 | 0 20·8 | 2·7 0·1 | 8·7 0·2 | 14·7 0·4 |
| 28 | 0 22·0 | 0 22·1 | 0 21·0 | 2·8 0·1 | 8·8 0·2 | 14·8 0·4 |
| 29 | 0 22·3 | 0 22·3 | 0 21·2 | 2·9 0·1 | 8·9 0·2 | 14·9 0·4 |
| 30 | 0 22·5 | 0 22·6 | 0 21·5 | 3·0 0·1 | 9·0 0·2 | 15·0 0·4 |
| 31 | 0 22·8 | 0 22·8 | 0 21·7 | 3·1 0·1 | 9·1 0·2 | 15·1 0·4 |
| 32 | 0 23·0 | 0 23·1 | 0 22·0 | 3·2 0·1 | 9·2 0·2 | 15·2 0·4 |
| 33 | 0 23·3 | 0 23·3 | 0 22·2 | 3·3 0·1 | 9·3 0·2 | 15·3 0·4 |
| 34 | 0 23·5 | 0 23·6 | 0 22·4 | 3·4 0·1 | 9·4 0·2 | 15·4 0·4 |
| 35 | 0 23·8 | 0 23·8 | 0 22·7 | 3·5 0·1 | 9·5 0·2 | 15·5 0·4 |
| 36 | 0 24·0 | 0 24·1 | 0 22·9 | 3·6 0·1 | 9·6 0·2 | 15·6 0·4 |
| 37 | 0 24·3 | 0 24·3 | 0 23·1 | 3·7 0·1 | 9·7 0·2 | 15·7 0·4 |
| 38 | 0 24·5 | 0 24·6 | 0 23·4 | 3·8 0·1 | 9·8 0·2 | 15·8 0·4 |
| 39 | 0 24·8 | 0 24·8 | 0 23·6 | 3·9 0·1 | 9·9 0·2 | 15·9 0·4 |
| 40 | 0 25·0 | 0 25·1 | 0 23·9 | 4·0 0·1 | 10·0 0·3 | 16·0 0·4 |
| 41 | 0 25·3 | 0 25·3 | 0 24·1 | 4·1 0·1 | 10·1 0·3 | 16·1 0·4 |
| 42 | 0 25·5 | 0 25·6 | 0 24·3 | 4·2 0·1 | 10·2 0·3 | 16·2 0·4 |
| 43 | 0 25·8 | 0 25·8 | 0 24·6 | 4·3 0·1 | 10·3 0·3 | 16·3 0·4 |
| 44 | 0 26·0 | 0 26·1 | 0 24·8 | 4·4 0·1 | 10·4 0·3 | 16·4 0·4 |
| 45 | 0 26·3 | 0 26·3 | 0 25·1 | 4·5 0·1 | 10·5 0·3 | 16·5 0·4 |
| 46 | 0 26·5 | 0 26·6 | 0 25·3 | 4·6 0·1 | 10·6 0·3 | 16·6 0·4 |
| 47 | 0 26·8 | 0 26·8 | 0 25·5 | 4·7 0·1 | 10·7 0·3 | 16·7 0·4 |
| 48 | 0 27·0 | 0 27·1 | 0 25·8 | 4·8 0·1 | 10·8 0·3 | 16·8 0·4 |
| 49 | 0 27·3 | 0 27·3 | 0 26·0 | 4·9 0·1 | 10·9 0·3 | 16·9 0·4 |
| 50 | 0 27·5 | 0 27·6 | 0 26·2 | 5·0 0·1 | 11·0 0·3 | 17·0 0·4 |
| 51 | 0 27·8 | 0 27·8 | 0 26·5 | 5·1 0·1 | 11·1 0·3 | 17·1 0·4 |
| 52 | 0 28·0 | 0 28·1 | 0 26·7 | 5·2 0·1 | 11·2 0·3 | 17·2 0·4 |
| 53 | 0 28·3 | 0 28·3 | 0 27·0 | 5·3 0·1 | 11·3 0·3 | 17·3 0·4 |
| 54 | 0 28·5 | 0 28·6 | 0 27·2 | 5·4 0·1 | 11·4 0·3 | 17·4 0·4 |
| 55 | 0 28·8 | 0 28·8 | 0 27·4 | 5·5 0·1 | 11·5 0·3 | 17·5 0·4 |
| 56 | 0 29·0 | 0 29·1 | 0 27·7 | 5·6 0·1 | 11·6 0·3 | 17·6 0·4 |
| 57 | 0 29·3 | 0 29·3 | 0 27·9 | 5·7 0·1 | 11·7 0·3 | 17·7 0·4 |
| 58 | 0 29·5 | 0 29·6 | 0 28·2 | 5·8 0·1 | 11·8 0·3 | 17·8 0·4 |
| 59 | 0 29·8 | 0 29·8 | 0 28·4 | 5·9 0·1 | 11·9 0·3 | 17·9 0·4 |
| 60 | 0 30·0 | 0 30·1 | 0 28·6 | 6·0 0·2 | 12·0 0·3 | 18·0 0·5 |

## INCREMENTS AND CORRECTIONS

### 10ᵐ

| 10 | SUN PLANETS | ARIES | MOON | v or d / Corrⁿ | v or d / Corrⁿ | v or d / Corrⁿ |
|---|---|---|---|---|---|---|
| s | ° ′ | ° ′ | ° ′ | ′ ′ | ′ ′ | ′ ′ |
| 00 | 2 30·0 | 2 30·4 | 2 23·2 | 0·0 0·0 | 6·0 1·1 | 12·0 2·1 |
| 01 | 2 30·3 | 2 30·7 | 2 23·4 | 0·1 0·0 | 6·1 1·1 | 12·1 2·1 |
| 02 | 2 30·5 | 2 30·9 | 2 23·6 | 0·2 0·0 | 6·2 1·1 | 12·2 2·1 |
| 03 | 2 30·8 | 2 31·2 | 2 23·9 | 0·3 0·1 | 6·3 1·1 | 12·3 2·2 |
| 04 | 2 31·0 | 2 31·4 | 2 24·1 | 0·4 0·1 | 6·4 1·1 | 12·4 2·2 |
| 05 | 2 31·3 | 2 31·7 | 2 24·4 | 0·5 0·1 | 6·5 1·1 | 12·5 2·2 |
| 06 | 2 31·5 | 2 31·9 | 2 24·6 | 0·6 0·1 | 6·6 1·2 | 12·6 2·2 |
| 07 | 2 31·8 | 2 32·2 | 2 24·8 | 0·7 0·1 | 6·7 1·2 | 12·7 2·2 |
| 08 | 2 32·0 | 2 32·4 | 2 25·1 | 0·8 0·1 | 6·8 1·2 | 12·8 2·2 |
| 09 | 2 32·3 | 2 32·7 | 2 25·3 | 0·9 0·2 | 6·9 1·2 | 12·9 2·3 |
| 10 | 2 32·5 | 2 32·9 | 2 25·6 | 1·0 0·2 | 7·0 1·2 | 13·0 2·3 |
| 11 | 2 32·8 | 2 33·2 | 2 25·8 | 1·1 0·2 | 7·1 1·2 | 13·1 2·3 |
| 12 | 2 33·0 | 2 33·4 | 2 26·0 | 1·2 0·2 | 7·2 1·3 | 13·2 2·3 |
| 13 | 2 33·3 | 2 33·7 | 2 26·3 | 1·3 0·2 | 7·3 1·3 | 13·3 2·3 |
| 14 | 2 33·5 | 2 33·9 | 2 26·5 | 1·4 0·2 | 7·4 1·3 | 13·4 2·3 |
| 15 | 2 33·8 | 2 34·2 | 2 26·7 | 1·5 0·3 | 7·5 1·3 | 13·5 2·4 |
| 16 | 2 34·0 | 2 34·4 | 2 27·0 | 1·6 0·3 | 7·6 1·3 | 13·6 2·4 |
| 17 | 2 34·3 | 2 34·7 | 2 27·2 | 1·7 0·3 | 7·7 1·3 | 13·7 2·4 |
| 18 | 2 34·5 | 2 34·9 | 2 27·5 | 1·8 0·3 | 7·8 1·4 | 13·8 2·4 |
| 19 | 2 34·8 | 2 35·2 | 2 27·7 | 1·9 0·3 | 7·9 1·4 | 13·9 2·4 |
| 20 | 2 35·0 | 2 35·4 | 2 27·9 | 2·0 0·4 | 8·0 1·4 | 14·0 2·5 |
| 21 | 2 35·3 | 2 35·7 | 2 28·2 | 2·1 0·4 | 8·1 1·4 | 14·1 2·5 |
| 22 | 2 35·5 | 2 35·9 | 2 28·4 | 2·2 0·4 | 8·2 1·4 | 14·2 2·5 |
| 23 | 2 35·8 | 2 36·2 | 2 28·7 | 2·3 0·4 | 8·3 1·5 | 14·3 2·5 |
| 24 | 2 36·0 | 2 36·4 | 2 28·9 | 2·4 0·4 | 8·4 1·5 | 14·4 2·5 |
| 25 | 2 36·3 | 2 36·7 | 2 29·1 | 2·5 0·4 | 8·5 1·5 | 14·5 2·5 |
| 26 | 2 36·5 | 2 36·9 | 2 29·4 | 2·6 0·5 | 8·6 1·5 | 14·6 2·6 |
| 27 | 2 36·8 | 2 37·2 | 2 29·6 | 2·7 0·5 | 8·7 1·5 | 14·7 2·6 |
| 28 | 2 37·0 | 2 37·4 | 2 29·8 | 2·8 0·5 | 8·8 1·5 | 14·8 2·6 |
| 29 | 2 37·3 | 2 37·7 | 2 30·1 | 2·9 0·5 | 8·9 1·6 | 14·9 2·6 |
| 30 | 2 37·5 | 2 37·9 | 2 30·3 | 3·0 0·5 | 9·0 1·6 | 15·0 2·6 |
| 31 | 2 37·8 | 2 38·2 | 2 30·6 | 3·1 0·5 | 9·1 1·6 | 15·1 2·6 |
| 32 | 2 38·0 | 2 38·4 | 2 30·8 | 3·2 0·6 | 9·2 1·6 | 15·2 2·7 |
| 33 | 2 38·3 | 2 38·7 | 2 31·0 | 3·3 0·6 | 9·3 1·6 | 15·3 2·7 |
| 34 | 2 38·5 | 2 38·9 | 2 31·3 | 3·4 0·6 | 9·4 1·6 | 15·4 2·7 |
| 35 | 2 38·8 | 2 39·2 | 2 31·5 | 3·5 0·6 | 9·5 1·7 | 15·5 2·7 |
| 36 | 2 39·0 | 2 39·4 | 2 31·8 | 3·6 0·6 | 9·6 1·7 | 15·6 2·7 |
| 37 | 2 39·3 | 2 39·7 | 2 32·0 | 3·7 0·6 | 9·7 1·7 | 15·7 2·7 |
| 38 | 2 39·5 | 2 39·9 | 2 32·2 | 3·8 0·7 | 9·8 1·7 | 15·8 2·8 |
| 39 | 2 39·8 | 2 40·2 | 2 32·5 | 3·9 0·7 | 9·9 1·7 | 15·9 2·8 |
| 40 | 2 40·0 | 2 40·4 | 2 32·7 | 4·0 0·7 | 10·0 1·8 | 16·0 2·8 |
| 41 | 2 40·3 | 2 40·7 | 2 32·9 | 4·1 0·7 | 10·1 1·8 | 16·1 2·8 |
| 42 | 2 40·5 | 2 40·9 | 2 33·2 | 4·2 0·7 | 10·2 1·8 | 16·2 2·8 |
| 43 | 2 40·8 | 2 41·2 | 2 33·4 | 4·3 0·8 | 10·3 1·8 | 16·3 2·9 |
| 44 | 2 41·0 | 2 41·4 | 2 33·7 | 4·4 0·8 | 10·4 1·8 | 16·4 2·9 |
| 45 | 2 41·3 | 2 41·7 | 2 33·9 | 4·5 0·8 | 10·5 1·8 | 16·5 2·9 |
| 46 | 2 41·5 | 2 41·9 | 2 34·1 | 4·6 0·8 | 10·6 1·9 | 16·6 2·9 |
| 47 | 2 41·8 | 2 42·2 | 2 34·4 | 4·7 0·8 | 10·7 1·9 | 16·7 2·9 |
| 48 | 2 42·0 | 2 42·4 | 2 34·6 | 4·8 0·8 | 10·8 1·9 | 16·8 2·9 |
| 49 | 2 42·3 | 2 42·7 | 2 34·9 | 4·9 0·9 | 10·9 1·9 | 16·9 3·0 |
| 50 | 2 42·5 | 2 42·9 | 2 35·1 | 5·0 0·9 | 11·0 1·9 | 17·0 3·0 |
| 51 | 2 42·8 | 2 43·2 | 2 35·3 | 5·1 0·9 | 11·1 1·9 | 17·1 3·0 |
| 52 | 2 43·0 | 2 43·4 | 2 35·6 | 5·2 0·9 | 11·2 2·0 | 17·2 3·0 |
| 53 | 2 43·3 | 2 43·7 | 2 35·8 | 5·3 0·9 | 11·3 2·0 | 17·3 3·0 |
| 54 | 2 43·5 | 2 43·9 | 2 36·1 | 5·4 0·9 | 11·4 2·0 | 17·4 3·0 |
| 55 | 2 43·8 | 2 44·2 | 2 36·3 | 5·5 1·0 | 11·5 2·0 | 17·5 3·1 |
| 56 | 2 44·0 | 2 44·4 | 2 36·5 | 5·6 1·0 | 11·6 2·0 | 17·6 3·1 |
| 57 | 2 44·3 | 2 44·7 | 2 36·8 | 5·7 1·0 | 11·7 2·0 | 17·7 3·1 |
| 58 | 2 44·5 | 2 45·0 | 2 37·0 | 5·8 1·0 | 11·8 2·1 | 17·8 3·1 |
| 59 | 2 44·8 | 2 45·2 | 2 37·2 | 5·9 1·0 | 11·9 2·1 | 17·9 3·1 |
| 60 | 2 45·0 | 2 45·5 | 2 37·5 | 6·0 1·1 | 12·0 2·1 | 18·0 3·2 |

### 11ᵐ

| 11 | SUN PLANETS | ARIES | MOON | v or d / Corrⁿ | v or d / Corrⁿ | v or d / Corrⁿ |
|---|---|---|---|---|---|---|
| s | ° ′ | ° ′ | ° ′ | ′ ′ | ′ ′ | ′ ′ |
| 00 | 2 45·0 | 2 45·5 | 2 37·5 | 0·0 0·0 | 6·0 1·2 | 12·0 2·3 |
| 01 | 2 45·3 | 2 45·7 | 2 37·7 | 0·1 0·0 | 6·1 1·2 | 12·1 2·3 |
| 02 | 2 45·5 | 2 46·0 | 2 38·0 | 0·2 0·0 | 6·2 1·2 | 12·2 2·3 |
| 03 | 2 45·8 | 2 46·2 | 2 38·2 | 0·3 0·1 | 6·3 1·2 | 12·3 2·4 |
| 04 | 2 46·0 | 2 46·5 | 2 38·4 | 0·4 0·1 | 6·4 1·2 | 12·4 2·4 |
| 05 | 2 46·3 | 2 46·7 | 2 38·7 | 0·5 0·1 | 6·5 1·2 | 12·5 2·4 |
| 06 | 2 46·5 | 2 47·0 | 2 38·9 | 0·6 0·1 | 6·6 1·3 | 12·6 2·4 |
| 07 | 2 46·8 | 2 47·2 | 2 39·2 | 0·7 0·1 | 6·7 1·3 | 12·7 2·4 |
| 08 | 2 47·0 | 2 47·5 | 2 39·4 | 0·8 0·2 | 6·8 1·3 | 12·8 2·5 |
| 09 | 2 47·3 | 2 47·7 | 2 39·6 | 0·9 0·2 | 6·9 1·3 | 12·9 2·5 |
| 10 | 2 47·5 | 2 48·0 | 2 39·9 | 1·0 0·2 | 7·0 1·3 | 13·0 2·5 |
| 11 | 2 47·8 | 2 48·2 | 2 40·1 | 1·1 0·2 | 7·1 1·4 | 13·1 2·5 |
| 12 | 2 48·0 | 2 48·5 | 2 40·3 | 1·2 0·2 | 7·2 1·4 | 13·2 2·5 |
| 13 | 2 48·3 | 2 48·7 | 2 40·6 | 1·3 0·2 | 7·3 1·4 | 13·3 2·5 |
| 14 | 2 48·5 | 2 49·0 | 2 40·8 | 1·4 0·3 | 7·4 1·4 | 13·4 2·6 |
| 15 | 2 48·8 | 2 49·2 | 2 41·1 | 1·5 0·3 | 7·5 1·4 | 13·5 2·6 |
| 16 | 2 49·0 | 2 49·5 | 2 41·3 | 1·6 0·3 | 7·6 1·5 | 13·6 2·6 |
| 17 | 2 49·3 | 2 49·7 | 2 41·5 | 1·7 0·3 | 7·7 1·5 | 13·7 2·6 |
| 18 | 2 49·5 | 2 50·0 | 2 41·8 | 1·8 0·3 | 7·8 1·5 | 13·8 2·6 |
| 19 | 2 49·8 | 2 50·2 | 2 42·0 | 1·9 0·4 | 7·9 1·5 | 13·9 2·7 |
| 20 | 2 50·0 | 2 50·5 | 2 42·3 | 2·0 0·4 | 8·0 1·5 | 14·0 2·7 |
| 21 | 2 50·3 | 2 50·7 | 2 42·5 | 2·1 0·4 | 8·1 1·6 | 14·1 2·7 |
| 22 | 2 50·5 | 2 51·0 | 2 42·7 | 2·2 0·4 | 8·2 1·6 | 14·2 2·7 |
| 23 | 2 50·8 | 2 51·2 | 2 43·0 | 2·3 0·4 | 8·3 1·6 | 14·3 2·7 |
| 24 | 2 51·0 | 2 51·5 | 2 43·2 | 2·4 0·5 | 8·4 1·6 | 14·4 2·8 |
| 25 | 2 51·3 | 2 51·7 | 2 43·4 | 2·5 0·5 | 8·5 1·6 | 14·5 2·8 |
| 26 | 2 51·5 | 2 52·0 | 2 43·7 | 2·6 0·5 | 8·6 1·6 | 14·6 2·8 |
| 27 | 2 51·8 | 2 52·2 | 2 43·9 | 2·7 0·5 | 8·7 1·7 | 14·7 2·8 |
| 28 | 2 52·0 | 2 52·5 | 2 44·2 | 2·8 0·5 | 8·8 1·7 | 14·8 2·8 |
| 29 | 2 52·3 | 2 52·7 | 2 44·4 | 2·9 0·6 | 8·9 1·7 | 14·9 2·9 |
| 30 | 2 52·5 | 2 53·0 | 2 44·6 | 3·0 0·6 | 9·0 1·7 | 15·0 2·9 |
| 31 | 2 52·8 | 2 53·2 | 2 44·9 | 3·1 0·6 | 9·1 1·7 | 15·1 2·9 |
| 32 | 2 53·0 | 2 53·5 | 2 45·1 | 3·2 0·6 | 9·2 1·8 | 15·2 2·9 |
| 33 | 2 53·3 | 2 53·7 | 2 45·4 | 3·3 0·6 | 9·3 1·8 | 15·3 2·9 |
| 34 | 2 53·5 | 2 54·0 | 2 45·6 | 3·4 0·7 | 9·4 1·8 | 15·4 3·0 |
| 35 | 2 53·8 | 2 54·2 | 2 45·8 | 3·5 0·7 | 9·5 1·8 | 15·5 3·0 |
| 36 | 2 54·0 | 2 54·5 | 2 46·1 | 3·6 0·7 | 9·6 1·8 | 15·6 3·0 |
| 37 | 2 54·3 | 2 54·7 | 2 46·3 | 3·7 0·7 | 9·7 1·9 | 15·7 3·0 |
| 38 | 2 54·5 | 2 55·0 | 2 46·6 | 3·8 0·7 | 9·8 1·9 | 15·8 3·0 |
| 39 | 2 54·8 | 2 55·2 | 2 46·8 | 3·9 0·7 | 9·9 1·9 | 15·9 3·0 |
| 40 | 2 55·0 | 2 55·5 | 2 47·0 | 4·0 0·8 | 10·0 1·9 | 16·0 3·1 |
| 41 | 2 55·3 | 2 55·7 | 2 47·3 | 4·1 0·8 | 10·1 1·9 | 16·1 3·1 |
| 42 | 2 55·5 | 2 56·0 | 2 47·5 | 4·2 0·8 | 10·2 2·0 | 16·2 3·1 |
| 43 | 2 55·8 | 2 56·2 | 2 47·7 | 4·3 0·8 | 10·3 2·0 | 16·3 3·1 |
| 44 | 2 56·0 | 2 56·5 | 2 48·0 | 4·4 0·8 | 10·4 2·0 | 16·4 3·1 |
| 45 | 2 56·3 | 2 56·7 | 2 48·2 | 4·5 0·9 | 10·5 2·0 | 16·5 3·2 |
| 46 | 2 56·5 | 2 57·0 | 2 48·5 | 4·6 0·9 | 10·6 2·0 | 16·6 3·2 |
| 47 | 2 56·8 | 2 57·2 | 2 48·7 | 4·7 0·9 | 10·7 2·1 | 16·7 3·2 |
| 48 | 2 57·0 | 2 57·5 | 2 48·9 | 4·8 0·9 | 10·8 2·1 | 16·8 3·2 |
| 49 | 2 57·3 | 2 57·7 | 2 49·2 | 4·9 0·9 | 10·9 2·1 | 16·9 3·2 |
| 50 | 2 57·5 | 2 58·0 | 2 49·4 | 5·0 1·0 | 11·0 2·1 | 17·0 3·3 |
| 51 | 2 57·8 | 2 58·2 | 2 49·7 | 5·1 1·0 | 11·1 2·1 | 17·1 3·3 |
| 52 | 2 58·0 | 2 58·5 | 2 49·9 | 5·2 1·0 | 11·2 2·1 | 17·2 3·3 |
| 53 | 2 58·3 | 2 58·7 | 2 50·1 | 5·3 1·0 | 11·3 2·2 | 17·3 3·3 |
| 54 | 2 58·5 | 2 59·0 | 2 50·4 | 5·4 1·0 | 11·4 2·2 | 17·4 3·3 |
| 55 | 2 58·8 | 2 59·2 | 2 50·6 | 5·5 1·1 | 11·5 2·2 | 17·5 3·4 |
| 56 | 2 59·0 | 2 59·5 | 2 50·8 | 5·6 1·1 | 11·6 2·2 | 17·6 3·4 |
| 57 | 2 59·3 | 2 59·7 | 2 51·1 | 5·7 1·1 | 11·7 2·2 | 17·7 3·4 |
| 58 | 2 59·5 | 3 00·0 | 2 51·3 | 5·8 1·1 | 11·8 2·3 | 17·8 3·4 |
| 59 | 2 59·8 | 3 00·2 | 2 51·6 | 5·9 1·1 | 11·9 2·3 | 17·9 3·4 |
| 60 | 3 00·0 | 3 00·5 | 2 51·8 | 6·0 1·2 | 12·0 2·3 | 18·0 3·5 |

## 14ᵐ INCREMENTS AND CORRECTIONS 15ᵐ

| 14ᵐ | SUN PLANETS | ARIES | MOON | v or d | Corrⁿ | v or d | Corrⁿ | v or d | Corrⁿ | 15ᵐ | SUN PLANETS | ARIES | MOON | v or d | Corrⁿ | v or d | Corrⁿ | v or d | Corrⁿ |
|---|---|---|---|---|---|---|---|---|---|---|---|---|---|---|---|---|---|---|---|
| s | ° ′ | ° ′ | ° ′ | ′ | ′ | ′ | ′ | ′ | ′ | s | ° ′ | ° ′ | ° ′ | ′ | ′ | ′ | ′ | ′ | ′ |
| 00 | 3 30·0 | 3 30·6 | 3 20·4 | 0·0 | 0·0 | 6·0 | 1·5 | 12·0 | 2·9 | 00 | 3 45·0 | 3 45·6 | 3 34·8 | 0·0 | 0·0 | 6·0 | 1·6 | 12·0 | 3·1 |
| 01 | 3 30·3 | 3 30·8 | 3 20·7 | 0·1 | 0·0 | 6·1 | 1·5 | 12·1 | 2·9 | 01 | 3 45·3 | 3 45·9 | 3 35·0 | 0·1 | 0·0 | 6·1 | 1·6 | 12·1 | 3·1 |
| 02 | 3 30·5 | 3 31·1 | 3 20·9 | 0·2 | 0·0 | 6·2 | 1·5 | 12·2 | 2·9 | 02 | 3 45·5 | 3 46·1 | 3 35·2 | 0·2 | 0·1 | 6·2 | 1·6 | 12·2 | 3·2 |
| 03 | 3 30·8 | 3 31·3 | 3 21·1 | 0·3 | 0·1 | 6·3 | 1·5 | 12·3 | 3·0 | 03 | 3 45·8 | 3 46·4 | 3 35·5 | 0·3 | 0·1 | 6·3 | 1·6 | 12·3 | 3·2 |
| 04 | 3 31·0 | 3 31·6 | 3 21·4 | 0·4 | 0·1 | 6·4 | 1·5 | 12·4 | 3·0 | 04 | 3 46·0 | 3 46·6 | 3 35·7 | 0·4 | 0·1 | 6·4 | 1·7 | 12·4 | 3·2 |
| 05 | 3 31·3 | 3 31·8 | 3 21·6 | 0·5 | 0·1 | 6·5 | 1·6 | 12·5 | 3·0 | 05 | 3 46·3 | 3 46·9 | 3 35·9 | 0·5 | 0·1 | 6·5 | 1·7 | 12·5 | 3·2 |
| 06 | 3 31·5 | 3 32·1 | 3 21·9 | 0·6 | 0·1 | 6·6 | 1·6 | 12·6 | 3·0 | 06 | 3 46·5 | 3 47·1 | 3 36·2 | 0·6 | 0·2 | 6·6 | 1·7 | 12·6 | 3·3 |
| 07 | 3 31·8 | 3 32·3 | 3 22·1 | 0·7 | 0·2 | 6·7 | 1·6 | 12·7 | 3·1 | 07 | 3 46·8 | 3 47·4 | 3 36·4 | 0·7 | 0·2 | 6·7 | 1·7 | 12·7 | 3·3 |
| 08 | 3 32·0 | 3 32·6 | 3 22·3 | 0·8 | 0·2 | 6·8 | 1·6 | 12·8 | 3·1 | 08 | 3 47·0 | 3 47·6 | 3 36·7 | 0·8 | 0·2 | 6·8 | 1·8 | 12·8 | 3·3 |
| 09 | 3 32·3 | 3 32·8 | 3 22·6 | 0·9 | 0·2 | 6·9 | 1·7 | 12·9 | 3·1 | 09 | 3 47·3 | 3 47·9 | 3 36·9 | 0·9 | 0·2 | 6·9 | 1·8 | 12·9 | 3·3 |
| 10 | 3 32·5 | 3 33·1 | 3 22·8 | 1·0 | 0·2 | 7·0 | 1·7 | 13·0 | 3·1 | 10 | 3 47·5 | 3 48·1 | 3 37·1 | 1·0 | 0·3 | 7·0 | 1·8 | 13·0 | 3·4 |
| 11 | 3 32·8 | 3 33·3 | 3 23·1 | 1·1 | 0·3 | 7·1 | 1·7 | 13·1 | 3·2 | 11 | 3 47·8 | 3 48·4 | 3 37·4 | 1·1 | 0·3 | 7·1 | 1·8 | 13·1 | 3·4 |
| 12 | 3 33·0 | 3 33·6 | 3 23·3 | 1·2 | 0·3 | 7·2 | 1·7 | 13·2 | 3·2 | 12 | 3 48·0 | 3 48·6 | 3 37·6 | 1·2 | 0·3 | 7·2 | 1·9 | 13·2 | 3·4 |
| 13 | 3 33·3 | 3 33·8 | 3 23·5 | 1·3 | 0·3 | 7·3 | 1·8 | 13·3 | 3·2 | 13 | 3 48·3 | 3 48·9 | 3 37·9 | 1·3 | 0·3 | 7·3 | 1·9 | 13·3 | 3·4 |
| 14 | 3 33·5 | 3 34·1 | 3 23·8 | 1·4 | 0·3 | 7·4 | 1·8 | 13·4 | 3·2 | 14 | 3 48·5 | 3 49·1 | 3 38·1 | 1·4 | 0·4 | 7·4 | 1·9 | 13·4 | 3·5 |
| 15 | 3 33·8 | 3 34·3 | 3 24·0 | 1·5 | 0·4 | 7·5 | 1·8 | 13·5 | 3·3 | 15 | 3 48·8 | 3 49·4 | 3 38·3 | 1·5 | 0·4 | 7·5 | 1·9 | 13·5 | 3·5 |
| 16 | 3 34·0 | 3 34·6 | 3 24·3 | 1·6 | 0·4 | 7·6 | 1·8 | 13·6 | 3·3 | 16 | 3 49·0 | 3 49·6 | 3 38·6 | 1·6 | 0·4 | 7·6 | 2·0 | 13·6 | 3·5 |
| 17 | 3 34·3 | 3 34·8 | 3 24·5 | 1·7 | 0·4 | 7·7 | 1·9 | 13·7 | 3·3 | 17 | 3 49·3 | 3 49·9 | 3 38·8 | 1·7 | 0·4 | 7·7 | 2·0 | 13·7 | 3·5 |
| 18 | 3 34·5 | 3 35·1 | 3 24·7 | 1·8 | 0·4 | 7·8 | 1·9 | 13·8 | 3·3 | 18 | 3 49·5 | 3 50·1 | 3 39·0 | 1·8 | 0·5 | 7·8 | 2·0 | 13·8 | 3·6 |
| 19 | 3 34·8 | 3 35·3 | 3 25·0 | 1·9 | 0·5 | 7·9 | 1·9 | 13·9 | 3·4 | 19 | 3 49·8 | 3 50·4 | 3 39·3 | 1·9 | 0·5 | 7·9 | 2·0 | 13·9 | 3·6 |
| 20 | 3 35·0 | 3 35·6 | 3 25·2 | 2·0 | 0·5 | 8·0 | 1·9 | 14·0 | 3·4 | 20 | 3 50·0 | 3 50·6 | 3 39·5 | 2·0 | 0·5 | 8·0 | 2·1 | 14·0 | 3·6 |
| 21 | 3 35·3 | 3 35·8 | 3 25·4 | 2·1 | 0·5 | 8·1 | 2·0 | 14·1 | 3·4 | 21 | 3 50·3 | 3 50·9 | 3 39·8 | 2·1 | 0·5 | 8·1 | 2·1 | 14·1 | 3·6 |
| 22 | 3 35·5 | 3 36·1 | 3 25·7 | 2·2 | 0·5 | 8·2 | 2·0 | 14·2 | 3·4 | 22 | 3 50·5 | 3 51·1 | 3 40·0 | 2·2 | 0·6 | 8·2 | 2·1 | 14·2 | 3·7 |
| 23 | 3 35·8 | 3 36·3 | 3 25·9 | 2·3 | 0·6 | 8·3 | 2·0 | 14·3 | 3·5 | 23 | 3 50·8 | 3 51·4 | 3 40·2 | 2·3 | 0·6 | 8·3 | 2·1 | 14·3 | 3·7 |
| 24 | 3 36·0 | 3 36·6 | 3 26·2 | 2·4 | 0·6 | 8·4 | 2·0 | 14·4 | 3·5 | 24 | 3 51·0 | 3 51·6 | 3 40·5 | 2·4 | 0·6 | 8·4 | 2·2 | 14·4 | 3·7 |
| 25 | 3 36·3 | 3 36·8 | 3 26·4 | 2·5 | 0·6 | 8·5 | 2·1 | 14·5 | 3·5 | 25 | 3 51·3 | 3 51·9 | 3 40·7 | 2·5 | 0·6 | 8·5 | 2·2 | 14·5 | 3·7 |
| 26 | 3 36·5 | 3 37·1 | 3 26·6 | 2·6 | 0·6 | 8·6 | 2·1 | 14·6 | 3·5 | 26 | 3 51·5 | 3 52·1 | 3 41·0 | 2·6 | 0·7 | 8·6 | 2·2 | 14·6 | 3·8 |
| 27 | 3 36·8 | 3 37·3 | 3 26·9 | 2·7 | 0·7 | 8·7 | 2·1 | 14·7 | 3·6 | 27 | 3 51·8 | 3 52·4 | 3 41·2 | 2·7 | 0·7 | 8·7 | 2·2 | 14·7 | 3·8 |
| 28 | 3 37·0 | 3 37·6 | 3 27·1 | 2·8 | 0·7 | 8·8 | 2·1 | 14·8 | 3·6 | 28 | 3 52·0 | 3 52·6 | 3 41·4 | 2·8 | 0·7 | 8·8 | 2·3 | 14·8 | 3·8 |
| 29 | 3 37·3 | 3 37·8 | 3 27·4 | 2·9 | 0·7 | 8·9 | 2·2 | 14·9 | 3·6 | 29 | 3 52·3 | 3 52·9 | 3 41·7 | 2·9 | 0·7 | 8·9 | 2·3 | 14·9 | 3·8 |
| 30 | 3 37·5 | 3 38·1 | 3 27·6 | 3·0 | 0·7 | 9·0 | 2·2 | 15·0 | 3·6 | 30 | 3 52·5 | 3 53·1 | 3 41·9 | 3·0 | 0·8 | 9·0 | 2·3 | 15·0 | 3·9 |
| 31 | 3 37·8 | 3 38·3 | 3 27·8 | 3·1 | 0·7 | 9·1 | 2·2 | 15·1 | 3·6 | 31 | 3 52·8 | 3 53·4 | 3 42·1 | 3·1 | 0·8 | 9·1 | 2·4 | 15·1 | 3·9 |
| 32 | 3 38·0 | 3 38·6 | 3 28·1 | 3·2 | 0·8 | 9·2 | 2·2 | 15·2 | 3·7 | 32 | 3 53·0 | 3 53·6 | 3 42·4 | 3·2 | 0·8 | 9·2 | 2·4 | 15·2 | 3·9 |
| 33 | 3 38·3 | 3 38·8 | 3 28·3 | 3·3 | 0·8 | 9·3 | 2·2 | 15·3 | 3·7 | 33 | 3 53·3 | 3 53·9 | 3 42·6 | 3·3 | 0·9 | 9·3 | 2·4 | 15·3 | 4·0 |
| 34 | 3 38·5 | 3 39·1 | 3 28·5 | 3·4 | 0·8 | 9·4 | 2·3 | 15·4 | 3·7 | 34 | 3 53·5 | 3 54·1 | 3 42·9 | 3·4 | 0·9 | 9·4 | 2·4 | 15·4 | 4·0 |
| 35 | 3 38·8 | 3 39·3 | 3 28·8 | 3·5 | 0·8 | 9·5 | 2·3 | 15·5 | 3·7 | 35 | 3 53·8 | 3 54·4 | 3 43·1 | 3·5 | 0·9 | 9·5 | 2·5 | 15·5 | 4·0 |
| 36 | 3 39·0 | 3 39·6 | 3 29·0 | 3·6 | 0·9 | 9·6 | 2·3 | 15·6 | 3·8 | 36 | 3 54·0 | 3 54·6 | 3 43·3 | 3·6 | 0·9 | 9·6 | 2·5 | 15·6 | 4·0 |
| 37 | 3 39·3 | 3 39·9 | 3 29·3 | 3·7 | 0·9 | 9·7 | 2·3 | 15·7 | 3·8 | 37 | 3 54·3 | 3 54·9 | 3 43·6 | 3·7 | 1·0 | 9·7 | 2·5 | 15·7 | 4·1 |
| 38 | 3 39·5 | 3 40·1 | 3 29·5 | 3·8 | 0·9 | 9·8 | 2·4 | 15·8 | 3·8 | 38 | 3 54·5 | 3 55·1 | 3 43·8 | 3·8 | 1·0 | 9·8 | 2·5 | 15·8 | 4·1 |
| 39 | 3 39·8 | 3 40·4 | 3 29·7 | 3·9 | 0·9 | 9·9 | 2·4 | 15·9 | 3·8 | 39 | 3 54·8 | 3 55·4 | 3 44·1 | 3·9 | 1·0 | 9·9 | 2·6 | 15·9 | 4·1 |
| 40 | 3 40·0 | 3 40·6 | 3 30·0 | 4·0 | 1·0 | 10·0 | 2·4 | 16·0 | 3·9 | 40 | 3 55·0 | 3 55·6 | 3 44·3 | 4·0 | 1·0 | 10·0 | 2·6 | 16·0 | 4·1 |
| 41 | 3 40·3 | 3 40·9 | 3 30·2 | 4·1 | 1·0 | 10·1 | 2·4 | 16·1 | 3·9 | 41 | 3 55·3 | 3 55·9 | 3 44·5 | 4·1 | 1·1 | 10·1 | 2·6 | 16·1 | 4·2 |
| 42 | 3 40·5 | 3 41·1 | 3 30·5 | 4·2 | 1·0 | 10·2 | 2·5 | 16·2 | 3·9 | 42 | 3 55·5 | 3 56·1 | 3 44·8 | 4·2 | 1·1 | 10·2 | 2·6 | 16·2 | 4·2 |
| 43 | 3 40·8 | 3 41·4 | 3 30·7 | 4·3 | 1·0 | 10·3 | 2·5 | 16·3 | 3·9 | 43 | 3 55·8 | 3 56·4 | 3 45·0 | 4·3 | 1·1 | 10·3 | 2·7 | 16·3 | 4·2 |
| 44 | 3 41·0 | 3 41·6 | 3 30·9 | 4·4 | 1·1 | 10·4 | 2·5 | 16·4 | 4·0 | 44 | 3 56·0 | 3 56·6 | 3 45·2 | 4·4 | 1·1 | 10·4 | 2·7 | 16·4 | 4·2 |
| 45 | 3 41·3 | 3 41·9 | 3 31·2 | 4·5 | 1·1 | 10·5 | 2·5 | 16·5 | 4·0 | 45 | 3 56·3 | 3 56·9 | 3 45·5 | 4·5 | 1·2 | 10·5 | 2·7 | 16·5 | 4·3 |
| 46 | 3 41·5 | 3 42·1 | 3 31·4 | 4·6 | 1·1 | 10·6 | 2·6 | 16·6 | 4·0 | 46 | 3 56·5 | 3 57·1 | 3 45·7 | 4·6 | 1·2 | 10·6 | 2·7 | 16·6 | 4·3 |
| 47 | 3 41·8 | 3 42·4 | 3 31·6 | 4·7 | 1·1 | 10·7 | 2·6 | 16·7 | 4·0 | 47 | 3 56·8 | 3 57·4 | 3 46·0 | 4·7 | 1·2 | 10·7 | 2·8 | 16·7 | 4·3 |
| 48 | 3 42·0 | 3 42·6 | 3 31·9 | 4·8 | 1·2 | 10·8 | 2·6 | 16·8 | 4·1 | 48 | 3 57·0 | 3 57·6 | 3 46·2 | 4·8 | 1·2 | 10·8 | 2·8 | 16·8 | 4·3 |
| 49 | 3 42·3 | 3 42·9 | 3 32·1 | 4·9 | 1·2 | 10·9 | 2·6 | 16·9 | 4·1 | 49 | 3 57·3 | 3 57·9 | 3 46·4 | 4·9 | 1·3 | 10·9 | 2·8 | 16·9 | 4·4 |
| 50 | 3 42·5 | 3 43·1 | 3 32·4 | 5·0 | 1·2 | 11·0 | 2·7 | 17·0 | 4·1 | 50 | 3 57·5 | 3 58·2 | 3 46·7 | 5·0 | 1·3 | 11·0 | 2·8 | 17·0 | 4·4 |
| 51 | 3 42·8 | 3 43·4 | 3 32·6 | 5·1 | 1·2 | 11·1 | 2·7 | 17·1 | 4·1 | 51 | 3 57·8 | 3 58·4 | 3 46·9 | 5·1 | 1·3 | 11·1 | 2·9 | 17·1 | 4·4 |
| 52 | 3 43·0 | 3 43·6 | 3 32·8 | 5·2 | 1·3 | 11·2 | 2·7 | 17·2 | 4·2 | 52 | 3 58·0 | 3 58·7 | 3 47·2 | 5·2 | 1·3 | 11·2 | 2·9 | 17·2 | 4·4 |
| 53 | 3 43·3 | 3 43·9 | 3 33·1 | 5·3 | 1·3 | 11·3 | 2·7 | 17·3 | 4·2 | 53 | 3 58·3 | 3 58·9 | 3 47·4 | 5·3 | 1·4 | 11·3 | 2·9 | 17·3 | 4·5 |
| 54 | 3 43·5 | 3 44·1 | 3 33·3 | 5·4 | 1·3 | 11·4 | 2·8 | 17·4 | 4·2 | 54 | 3 58·5 | 3 59·2 | 3 47·6 | 5·4 | 1·4 | 11·4 | 2·9 | 17·4 | 4·5 |
| 55 | 3 43·8 | 3 44·4 | 3 33·6 | 5·5 | 1·3 | 11·5 | 2·8 | 17·5 | 4·2 | 55 | 3 58·8 | 3 59·4 | 3 47·9 | 5·5 | 1·4 | 11·5 | 3·0 | 17·5 | 4·5 |
| 56 | 3 44·0 | 3 44·6 | 3 33·8 | 5·6 | 1·4 | 11·6 | 2·8 | 17·6 | 4·3 | 56 | 3 59·0 | 3 59·7 | 3 48·1 | 5·6 | 1·4 | 11·6 | 3·0 | 17·6 | 4·5 |
| 57 | 3 44·3 | 3 44·9 | 3 34·0 | 5·7 | 1·4 | 11·7 | 2·8 | 17·7 | 4·3 | 57 | 3 59·3 | 3 59·9 | 3 48·4 | 5·7 | 1·5 | 11·7 | 3·0 | 17·7 | 4·6 |
| 58 | 3 44·5 | 3 45·1 | 3 34·3 | 5·8 | 1·4 | 11·8 | 2·9 | 17·8 | 4·3 | 58 | 3 59·5 | 4 00·2 | 3 48·6 | 5·8 | 1·5 | 11·8 | 3·0 | 17·8 | 4·6 |
| 59 | 3 44·8 | 3 45·4 | 3 34·5 | 5·9 | 1·4 | 11·9 | 2·9 | 17·9 | 4·3 | 59 | 3 59·8 | 4 00·4 | 3 48·8 | 5·9 | 1·5 | 11·9 | 3·1 | 17·9 | 4·6 |
| 60 | 3 45·0 | 3 45·6 | 3 34·8 | 6·0 | 1·5 | 12·0 | 2·9 | 18·0 | 4·4 | 60 | 4 00·0 | 4 00·7 | 3 49·1 | 6·0 | 1·6 | 12·0 | 3·1 | 18·0 | 4·7 |

# INCREMENTS AND CORRECTIONS

## 26m

| 26 | SUN PLANETS | ARIES | MOON | v or d | Corrⁿ | v or d | Corrⁿ | v or d | Corrⁿ |
|---|---|---|---|---|---|---|---|---|---|
| s | ° ′ | ° ′ | ° ′ | ′ | ′ | ′ | ′ | ′ | ′ |
| 00 | 6 30·0 | 6 31·1 | 6 12·2 | 0·0 | 0·0 | 6·0 | 2·7 | 12·0 | 5·3 |
| 01 | 6 30·3 | 6 31·3 | 6 12·5 | 0·1 | 0·0 | 6·1 | 2·7 | 12·1 | 5·3 |
| 02 | 6 30·5 | 6 31·6 | 6 12·7 | 0·2 | 0·1 | 6·2 | 2·7 | 12·2 | 5·4 |
| 03 | 6 30·8 | 6 31·8 | 6 12·9 | 0·3 | 0·1 | 6·3 | 2·8 | 12·3 | 5·4 |
| 04 | 6 31·0 | 6 32·1 | 6 13·2 | 0·4 | 0·2 | 6·4 | 2·8 | 12·4 | 5·5 |
| 05 | 6 31·3 | 6 32·3 | 6 13·4 | 0·5 | 0·2 | 6·5 | 2·9 | 12·5 | 5·5 |
| 06 | 6 31·5 | 6 32·6 | 6 13·7 | 0·6 | 0·3 | 6·6 | 2·9 | 12·6 | 5·6 |
| 07 | 6 31·8 | 6 32·8 | 6 13·9 | 0·7 | 0·3 | 6·7 | 3·0 | 12·7 | 5·6 |
| 08 | 6 32·0 | 6 33·1 | 6 14·1 | 0·8 | 0·4 | 6·8 | 3·0 | 12·8 | 5·7 |
| 09 | 6 32·3 | 6 33·3 | 6 14·4 | 0·9 | 0·4 | 6·9 | 3·0 | 12·9 | 5·7 |
| 10 | 6 32·5 | 6 33·6 | 6 14·6 | 1·0 | 0·4 | 7·0 | 3·1 | 13·0 | 5·7 |
| 11 | 6 32·8 | 6 33·8 | 6 14·9 | 1·1 | 0·5 | 7·1 | 3·1 | 13·1 | 5·8 |
| 12 | 6 33·0 | 6 34·1 | 6 15·1 | 1·2 | 0·5 | 7·2 | 3·2 | 13·2 | 5·8 |
| 13 | 6 33·3 | 6 34·3 | 6 15·3 | 1·3 | 0·6 | 7·3 | 3·2 | 13·3 | 5·9 |
| 14 | 6 33·5 | 6 34·6 | 6 15·6 | 1·4 | 0·6 | 7·4 | 3·3 | 13·4 | 5·9 |
| 15 | 6 33·8 | 6 34·8 | 6 15·8 | 1·5 | 0·7 | 7·5 | 3·3 | 13·5 | 6·0 |
| 16 | 6 34·0 | 6 35·1 | 6 16·1 | 1·6 | 0·7 | 7·6 | 3·4 | 13·6 | 6·0 |
| 17 | 6 34·3 | 6 35·3 | 6 16·3 | 1·7 | 0·8 | 7·7 | 3·4 | 13·7 | 6·1 |
| 18 | 6 34·5 | 6 35·6 | 6 16·5 | 1·8 | 0·8 | 7·8 | 3·4 | 13·8 | 6·1 |
| 19 | 6 34·8 | 6 35·8 | 6 16·8 | 1·9 | 0·8 | 7·9 | 3·5 | 13·9 | 6·1 |
| 20 | 6 35·0 | 6 36·1 | 6 17·0 | 2·0 | 0·9 | 8·0 | 3·5 | 14·0 | 6·2 |
| 21 | 6 35·3 | 6 36·3 | 6 17·2 | 2·1 | 0·9 | 8·1 | 3·6 | 14·1 | 6·2 |
| 22 | 6 35·5 | 6 36·6 | 6 17·5 | 2·2 | 1·0 | 8·2 | 3·6 | 14·2 | 6·3 |
| 23 | 6 35·8 | 6 36·8 | 6 17·7 | 2·3 | 1·0 | 8·3 | 3·7 | 14·3 | 6·3 |
| 24 | 6 36·0 | 6 37·1 | 6 18·0 | 2·4 | 1·1 | 8·4 | 3·7 | 14·4 | 6·4 |
| 25 | 6 36·3 | 6 37·3 | 6 18·2 | 2·5 | 1·1 | 8·5 | 3·8 | 14·5 | 6·4 |
| 26 | 6 36·5 | 6 37·6 | 6 18·4 | 2·6 | 1·1 | 8·6 | 3·8 | 14·6 | 6·4 |
| 27 | 6 36·8 | 6 37·8 | 6 18·7 | 2·7 | 1·2 | 8·7 | 3·8 | 14·7 | 6·5 |
| 28 | 6 37·0 | 6 38·1 | 6 18·9 | 2·8 | 1·2 | 8·8 | 3·9 | 14·8 | 6·5 |
| 29 | 6 37·3 | 6 38·3 | 6 19·2 | 2·9 | 1·3 | 8·9 | 3·9 | 14·9 | 6·6 |
| 30 | 6 37·5 | 6 38·6 | 6 19·4 | 3·0 | 1·3 | 9·0 | 4·0 | 15·0 | 6·6 |
| 31 | 6 37·8 | 6 38·8 | 6 19·6 | 3·1 | 1·4 | 9·1 | 4·0 | 15·1 | 6·7 |
| 32 | 6 38·0 | 6 39·1 | 6 19·9 | 3·2 | 1·4 | 9·2 | 4·1 | 15·2 | 6·7 |
| 33 | 6 38·3 | 6 39·3 | 6 20·1 | 3·3 | 1·5 | 9·3 | 4·1 | 15·3 | 6·8 |
| 34 | 6 38·5 | 6 39·6 | 6 20·3 | 3·4 | 1·5 | 9·4 | 4·2 | 15·4 | 6·8 |
| 35 | 6 38·8 | 6 39·8 | 6 20·6 | 3·5 | 1·5 | 9·5 | 4·2 | 15·5 | 6·8 |
| 36 | 6 39·0 | 6 40·1 | 6 20·8 | 3·6 | 1·6 | 9·6 | 4·2 | 15·6 | 6·9 |
| 37 | 6 39·3 | 6 40·3 | 6 21·1 | 3·7 | 1·6 | 9·7 | 4·3 | 15·7 | 6·9 |
| 38 | 6 39·5 | 6 40·6 | 6 21·3 | 3·8 | 1·7 | 9·8 | 4·3 | 15·8 | 7·0 |
| 39 | 6 39·8 | 6 40·8 | 6 21·5 | 3·9 | 1·7 | 9·9 | 4·4 | 15·9 | 7·0 |
| 40 | 6 40·0 | 6 41·1 | 6 21·8 | 4·0 | 1·8 | 10·0 | 4·4 | 16·0 | 7·1 |
| 41 | 6 40·3 | 6 41·3 | 6 22·0 | 4·1 | 1·8 | 10·1 | 4·5 | 16·1 | 7·1 |
| 42 | 6 40·5 | 6 41·6 | 6 22·3 | 4·2 | 1·9 | 10·2 | 4·5 | 16·2 | 7·2 |
| 43 | 6 40·8 | 6 41·8 | 6 22·5 | 4·3 | 1·9 | 10·3 | 4·5 | 16·3 | 7·2 |
| 44 | 6 41·0 | 6 42·1 | 6 22·7 | 4·4 | 1·9 | 10·4 | 4·6 | 16·4 | 7·2 |
| 45 | 6 41·3 | 6 42·3 | 6 23·0 | 4·5 | 2·0 | 10·5 | 4·6 | 16·5 | 7·3 |
| 46 | 6 41·5 | 6 42·6 | 6 23·2 | 4·6 | 2·0 | 10·6 | 4·7 | 16·6 | 7·3 |
| 47 | 6 41·8 | 6 42·8 | 6 23·4 | 4·7 | 2·1 | 10·7 | 4·7 | 16·7 | 7·4 |
| 48 | 6 42·0 | 6 43·1 | 6 23·7 | 4·8 | 2·1 | 10·8 | 4·8 | 16·8 | 7·4 |
| 49 | 6 42·3 | 6 43·3 | 6 23·9 | 4·9 | 2·2 | 10·9 | 4·8 | 16·9 | 7·5 |
| 50 | 6 42·5 | 6 43·6 | 6 24·2 | 5·0 | 2·2 | 11·0 | 4·9 | 17·0 | 7·5 |
| 51 | 6 42·8 | 6 43·9 | 6 24·4 | 5·1 | 2·3 | 11·1 | 4·9 | 17·1 | 7·6 |
| 52 | 6 43·0 | 6 44·1 | 6 24·6 | 5·2 | 2·3 | 11·2 | 4·9 | 17·2 | 7·6 |
| 53 | 6 43·3 | 6 44·4 | 6 24·9 | 5·3 | 2·3 | 11·3 | 5·0 | 17·3 | 7·6 |
| 54 | 6 43·5 | 6 44·6 | 6 25·1 | 5·4 | 2·4 | 11·4 | 5·0 | 17·4 | 7·7 |
| 55 | 6 43·8 | 6 44·9 | 6 25·4 | 5·5 | 2·4 | 11·5 | 5·1 | 17·5 | 7·7 |
| 56 | 6 44·0 | 6 45·1 | 6 25·6 | 5·6 | 2·5 | 11·6 | 5·1 | 17·6 | 7·8 |
| 57 | 6 44·3 | 6 45·4 | 6 25·8 | 5·7 | 2·5 | 11·7 | 5·2 | 17·7 | 7·8 |
| 58 | 6 44·5 | 6 45·6 | 6 26·1 | 5·8 | 2·6 | 11·8 | 5·2 | 17·8 | 7·9 |
| 59 | 6 44·8 | 6 45·9 | 6 26·3 | 5·9 | 2·6 | 11·9 | 5·3 | 17·9 | 7·9 |
| 60 | 6 45·0 | 6 46·1 | 6 26·6 | 6·0 | 2·7 | 12·0 | 5·3 | 18·0 | 8·0 |

## 27m

| 27 | SUN PLANETS | ARIES | MOON | v or d | Corrⁿ | v or d | Corrⁿ | v or d | Corrⁿ |
|---|---|---|---|---|---|---|---|---|---|
| s | ° ′ | ° ′ | ° ′ | ′ | ′ | ′ | ′ | ′ | ′ |
| 00 | 6 45·0 | 6 46·1 | 6 26·6 | 0·0 | 0·0 | 6·0 | 2·8 | 12·0 | 5·5 |
| 01 | 6 45·3 | 6 46·4 | 6 26·8 | 0·1 | 0·0 | 6·1 | 2·8 | 12·1 | 5·5 |
| 02 | 6 45·5 | 6 46·6 | 6 27·0 | 0·2 | 0·1 | 6·2 | 2·8 | 12·2 | 5·6 |
| 03 | 6 45·8 | 6 46·9 | 6 27·3 | 0·3 | 0·1 | 6·3 | 2·9 | 12·3 | 5·6 |
| 04 | 6 46·0 | 6 47·1 | 6 27·5 | 0·4 | 0·2 | 6·4 | 2·9 | 12·4 | 5·7 |
| 05 | 6 46·3 | 6 47·4 | 6 27·7 | 0·5 | 0·2 | 6·5 | 3·0 | 12·5 | 5·7 |
| 06 | 6 46·5 | 6 47·6 | 6 28·0 | 0·6 | 0·3 | 6·6 | 3·0 | 12·6 | 5·8 |
| 07 | 6 46·8 | 6 47·9 | 6 28·2 | 0·7 | 0·3 | 6·7 | 3·1 | 12·7 | 5·8 |
| 08 | 6 47·0 | 6 48·1 | 6 28·5 | 0·8 | 0·4 | 6·8 | 3·1 | 12·8 | 5·9 |
| 09 | 6 47·3 | 6 48·4 | 6 28·7 | 0·9 | 0·4 | 6·9 | 3·2 | 12·9 | 5·9 |
| 10 | 6 47·5 | 6 48·6 | 6 28·9 | 1·0 | 0·5 | 7·0 | 3·2 | 13·0 | 6·0 |
| 11 | 6 47·8 | 6 48·9 | 6 29·2 | 1·1 | 0·5 | 7·1 | 3·3 | 13·1 | 6·0 |
| 12 | 6 48·0 | 6 49·1 | 6 29·4 | 1·2 | 0·6 | 7·2 | 3·3 | 13·2 | 6·1 |
| 13 | 6 48·3 | 6 49·4 | 6 29·7 | 1·3 | 0·6 | 7·3 | 3·3 | 13·3 | 6·1 |
| 14 | 6 48·5 | 6 49·6 | 6 29·9 | 1·4 | 0·6 | 7·4 | 3·4 | 13·4 | 6·1 |
| 15 | 6 48·8 | 6 49·9 | 6 30·1 | 1·5 | 0·7 | 7·5 | 3·4 | 13·5 | 6·2 |
| 16 | 6 49·0 | 6 50·1 | 6 30·4 | 1·6 | 0·7 | 7·6 | 3·5 | 13·6 | 6·2 |
| 17 | 6 49·3 | 6 50·4 | 6 30·6 | 1·7 | 0·8 | 7·7 | 3·5 | 13·7 | 6·3 |
| 18 | 6 49·5 | 6 50·6 | 6 30·8 | 1·8 | 0·8 | 7·8 | 3·6 | 13·8 | 6·3 |
| 19 | 6 49·8 | 6 50·9 | 6 31·1 | 1·9 | 0·9 | 7·9 | 3·6 | 13·9 | 6·4 |
| 20 | 6 50·0 | 6 51·1 | 6 31·3 | 2·0 | 0·9 | 8·0 | 3·7 | 14·0 | 6·4 |
| 21 | 6 50·3 | 6 51·4 | 6 31·6 | 2·1 | 1·0 | 8·1 | 3·7 | 14·1 | 6·5 |
| 22 | 6 50·5 | 6 51·6 | 6 31·8 | 2·2 | 1·0 | 8·2 | 3·8 | 14·2 | 6·5 |
| 23 | 6 50·8 | 6 51·9 | 6 32·0 | 2·3 | 1·1 | 8·3 | 3·8 | 14·3 | 6·6 |
| 24 | 6 51·0 | 6 52·1 | 6 32·3 | 2·4 | 1·1 | 8·4 | 3·9 | 14·4 | 6·6 |
| 25 | 6 51·3 | 6 52·4 | 6 32·5 | 2·5 | 1·1 | 8·5 | 3·9 | 14·5 | 6·6 |
| 26 | 6 51·5 | 6 52·6 | 6 32·8 | 2·6 | 1·2 | 8·6 | 3·9 | 14·6 | 6·7 |
| 27 | 6 51·8 | 6 52·9 | 6 33·0 | 2·7 | 1·2 | 8·7 | 4·0 | 14·7 | 6·7 |
| 28 | 6 52·0 | 6 53·1 | 6 33·2 | 2·8 | 1·3 | 8·8 | 4·0 | 14·8 | 6·8 |
| 29 | 6 52·3 | 6 53·4 | 6 33·5 | 2·9 | 1·3 | 8·9 | 4·1 | 14·9 | 6·8 |
| 30 | 6 52·5 | 6 53·6 | 6 33·7 | 3·0 | 1·4 | 9·0 | 4·1 | 15·0 | 6·9 |
| 31 | 6 52·8 | 6 53·9 | 6 33·9 | 3·1 | 1·4 | 9·1 | 4·2 | 15·1 | 6·9 |
| 32 | 6 53·0 | 6 54·1 | 6 34·2 | 3·2 | 1·5 | 9·2 | 4·2 | 15·2 | 7·0 |
| 33 | 6 53·3 | 6 54·4 | 6 34·4 | 3·3 | 1·5 | 9·3 | 4·3 | 15·3 | 7·0 |
| 34 | 6 53·5 | 6 54·6 | 6 34·7 | 3·4 | 1·6 | 9·4 | 4·3 | 15·4 | 7·1 |
| 35 | 6 53·8 | 6 54·9 | 6 34·9 | 3·5 | 1·6 | 9·5 | 4·4 | 15·5 | 7·1 |
| 36 | 6 54·0 | 6 55·1 | 6 35·1 | 3·6 | 1·7 | 9·6 | 4·4 | 15·6 | 7·2 |
| 37 | 6 54·3 | 6 55·4 | 6 35·4 | 3·7 | 1·7 | 9·7 | 4·4 | 15·7 | 7·2 |
| 38 | 6 54·5 | 6 55·6 | 6 35·6 | 3·8 | 1·7 | 9·8 | 4·5 | 15·8 | 7·2 |
| 39 | 6 54·8 | 6 55·9 | 6 35·9 | 3·9 | 1·8 | 9·9 | 4·5 | 15·9 | 7·3 |
| 40 | 6 55·0 | 6 56·1 | 6 36·1 | 4·0 | 1·8 | 10·0 | 4·6 | 16·0 | 7·3 |
| 41 | 6 55·3 | 6 56·4 | 6 36·3 | 4·1 | 1·9 | 10·1 | 4·6 | 16·1 | 7·4 |
| 42 | 6 55·5 | 6 56·6 | 6 36·6 | 4·2 | 1·9 | 10·2 | 4·7 | 16·2 | 7·4 |
| 43 | 6 55·8 | 6 56·9 | 6 36·8 | 4·3 | 2·0 | 10·3 | 4·7 | 16·3 | 7·5 |
| 44 | 6 56·0 | 6 57·1 | 6 37·0 | 4·4 | 2·0 | 10·4 | 4·8 | 16·4 | 7·5 |
| 45 | 6 56·3 | 6 57·4 | 6 37·3 | 4·5 | 2·1 | 10·5 | 4·8 | 16·5 | 7·6 |
| 46 | 6 56·5 | 6 57·6 | 6 37·5 | 4·6 | 2·1 | 10·6 | 4·9 | 16·6 | 7·6 |
| 47 | 6 56·8 | 6 57·9 | 6 37·8 | 4·7 | 2·2 | 10·7 | 4·9 | 16·7 | 7·7 |
| 48 | 6 57·0 | 6 58·1 | 6 38·0 | 4·8 | 2·2 | 10·8 | 5·0 | 16·8 | 7·7 |
| 49 | 6 57·3 | 6 58·4 | 6 38·2 | 4·9 | 2·2 | 10·9 | 5·0 | 16·9 | 7·7 |
| 50 | 6 57·5 | 6 58·6 | 6 38·5 | 5·0 | 2·3 | 11·0 | 5·0 | 17·0 | 7·8 |
| 51 | 6 57·8 | 6 58·9 | 6 38·7 | 5·1 | 2·3 | 11·1 | 5·1 | 17·1 | 7·8 |
| 52 | 6 58·0 | 6 59·1 | 6 39·0 | 5·2 | 2·4 | 11·2 | 5·1 | 17·2 | 7·9 |
| 53 | 6 58·3 | 6 59·4 | 6 39·2 | 5·3 | 2·4 | 11·3 | 5·2 | 17·3 | 7·9 |
| 54 | 6 58·5 | 6 59·6 | 6 39·4 | 5·4 | 2·5 | 11·4 | 5·2 | 17·4 | 8·0 |
| 55 | 6 58·8 | 6 59·9 | 6 39·7 | 5·5 | 2·5 | 11·5 | 5·3 | 17·5 | 8·0 |
| 56 | 6 59·0 | 7 00·1 | 6 39·9 | 5·6 | 2·6 | 11·6 | 5·3 | 17·6 | 8·1 |
| 57 | 6 59·3 | 7 00·4 | 6 40·2 | 5·7 | 2·6 | 11·7 | 5·4 | 17·7 | 8·1 |
| 58 | 6 59·5 | 7 00·6 | 6 40·4 | 5·8 | 2·7 | 11·8 | 5·4 | 17·8 | 8·2 |
| 59 | 6 59·8 | 7 00·9 | 6 40·6 | 5·9 | 2·7 | 11·9 | 5·5 | 17·9 | 8·2 |
| 60 | 7 00·0 | 7 01·1 | 6 40·9 | 6·0 | 2·8 | 12·0 | 5·5 | 18·0 | 8·3 |

S

## INCREMENTS AND CORRECTIONS

### 30ᵐ

| 30 | SUN PLANETS | ARIES | MOON | v or d | Corrⁿ | v or d | Corrⁿ | v or d | Corrⁿ |
|---|---|---|---|---|---|---|---|---|---|
| s | ° ′ | ° ′ | ° ′ | ′ | ′ | ′ | ′ | ′ | ′ |
| 00 | 7 30·0 | 7 31·2 | 7 09·5 | 0·0 | 0·0 | 6·0 | 3·1 | 12·0 | 6·1 |
| 01 | 7 30·3 | 7 31·5 | 7 09·7 | 0·1 | 0·1 | 6·1 | 3·1 | 12·1 | 6·2 |
| 02 | 7 30·5 | 7 31·7 | 7 10·0 | 0·2 | 0·1 | 6·2 | 3·2 | 12·2 | 6·2 |
| 03 | 7 30·8 | 7 32·0 | 7 10·2 | 0·3 | 0·2 | 6·3 | 3·2 | 12·3 | 6·3 |
| 04 | 7 31·0 | 7 32·2 | 7 10·5 | 0·4 | 0·2 | 6·4 | 3·3 | 12·4 | 6·3 |
| 05 | 7 31·3 | 7 32·5 | 7 10·7 | 0·5 | 0·3 | 6·5 | 3·3 | 12·5 | 6·4 |
| 06 | 7 31·5 | 7 32·7 | 7 10·9 | 0·6 | 0·3 | 6·6 | 3·4 | 12·6 | 6·4 |
| 07 | 7 31·8 | 7 33·0 | 7 11·2 | 0·7 | 0·4 | 6·7 | 3·4 | 12·7 | 6·5 |
| 08 | 7 32·0 | 7 33·2 | 7 11·4 | 0·8 | 0·4 | 6·8 | 3·5 | 12·8 | 6·5 |
| 09 | 7 32·3 | 7 33·5 | 7 11·6 | 0·9 | 0·5 | 6·9 | 3·5 | 12·9 | 6·6 |
| 10 | 7 32·5 | 7 33·7 | 7 11·9 | 1·0 | 0·5 | 7·0 | 3·6 | 13·0 | 6·6 |
| 11 | 7 32·8 | 7 34·0 | 7 12·1 | 1·1 | 0·6 | 7·1 | 3·6 | 13·1 | 6·7 |
| 12 | 7 33·0 | 7 34·2 | 7 12·4 | 1·2 | 0·6 | 7·2 | 3·7 | 13·2 | 6·7 |
| 13 | 7 33·3 | 7 34·5 | 7 12·6 | 1·3 | 0·7 | 7·3 | 3·7 | 13·3 | 6·8 |
| 14 | 7 33·5 | 7 34·7 | 7 12·8 | 1·4 | 0·7 | 7·4 | 3·8 | 13·4 | 6·8 |
| 15 | 7 33·8 | 7 35·0 | 7 13·1 | 1·5 | 0·8 | 7·5 | 3·8 | 13·5 | 6·9 |
| 16 | 7 34·0 | 7 35·2 | 7 13·3 | 1·6 | 0·8 | 7·6 | 3·9 | 13·6 | 6·9 |
| 17 | 7 34·3 | 7 35·5 | 7 13·6 | 1·7 | 0·9 | 7·7 | 3·9 | 13·7 | 7·0 |
| 18 | 7 34·5 | 7 35·7 | 7 13·8 | 1·8 | 0·9 | 7·8 | 4·0 | 13·8 | 7·0 |
| 19 | 7 34·8 | 7 36·0 | 7 14·0 | 1·9 | 1·0 | 7·9 | 4·0 | 13·9 | 7·1 |
| 20 | 7 35·0 | 7 36·2 | 7 14·3 | 2·0 | 1·0 | 8·0 | 4·1 | 14·0 | 7·1 |
| 21 | 7 35·3 | 7 36·5 | 7 14·5 | 2·1 | 1·1 | 8·1 | 4·1 | 14·1 | 7·2 |
| 22 | 7 35·5 | 7 36·7 | 7 14·7 | 2·2 | 1·1 | 8·2 | 4·2 | 14·2 | 7·2 |
| 23 | 7 35·8 | 7 37·0 | 7 15·0 | 2·3 | 1·2 | 8·3 | 4·2 | 14·3 | 7·3 |
| 24 | 7 36·0 | 7 37·2 | 7 15·2 | 2·4 | 1·2 | 8·4 | 4·3 | 14·4 | 7·3 |
| 25 | 7 36·3 | 7 37·5 | 7 15·5 | 2·5 | 1·3 | 8·5 | 4·3 | 14·5 | 7·4 |
| 26 | 7 36·5 | 7 37·7 | 7 15·7 | 2·6 | 1·3 | 8·6 | 4·4 | 14·6 | 7·4 |
| 27 | 7 36·8 | 7 38·0 | 7 15·9 | 2·7 | 1·4 | 8·7 | 4·4 | 14·7 | 7·5 |
| 28 | 7 37·0 | 7 38·3 | 7 16·2 | 2·8 | 1·4 | 8·8 | 4·5 | 14·8 | 7·5 |
| 29 | 7 37·3 | 7 38·5 | 7 16·4 | 2·9 | 1·5 | 8·9 | 4·5 | 14·9 | 7·6 |
| 30 | 7 37·5 | 7 38·8 | 7 16·7 | 3·0 | 1·5 | 9·0 | 4·6 | 15·0 | 7·6 |
| 31 | 7 37·8 | 7 39·0 | 7 16·9 | 3·1 | 1·6 | 9·1 | 4·6 | 15·1 | 7·7 |
| 32 | 7 38·0 | 7 39·3 | 7 17·1 | 3·2 | 1·6 | 9·2 | 4·7 | 15·2 | 7·7 |
| 33 | 7 38·3 | 7 39·5 | 7 17·4 | 3·3 | 1·7 | 9·3 | 4·7 | 15·3 | 7·8 |
| 34 | 7 38·5 | 7 39·8 | 7 17·6 | 3·4 | 1·7 | 9·4 | 4·8 | 15·4 | 7·8 |
| 35 | 7 38·8 | 7 40·0 | 7 17·9 | 3·5 | 1·8 | 9·5 | 4·8 | 15·5 | 7·9 |
| 36 | 7 39·0 | 7 40·3 | 7 18·1 | 3·6 | 1·8 | 9·6 | 4·9 | 15·6 | 7·9 |
| 37 | 7 39·3 | 7 40·5 | 7 18·3 | 3·7 | 1·9 | 9·7 | 4·9 | 15·7 | 8·0 |
| 38 | 7 39·5 | 7 40·8 | 7 18·6 | 3·8 | 1·9 | 9·8 | 5·0 | 15·8 | 8·0 |
| 39 | 7 39·8 | 7 41·0 | 7 18·8 | 3·9 | 2·0 | 9·9 | 5·0 | 15·9 | 8·1 |
| 40 | 7 40·0 | 7 41·3 | 7 19·0 | 4·0 | 2·0 | 10·0 | 5·1 | 16·0 | 8·1 |
| 41 | 7 40·3 | 7 41·5 | 7 19·3 | 4·1 | 2·1 | 10·1 | 5·1 | 16·1 | 8·2 |
| 42 | 7 40·5 | 7 41·8 | 7 19·5 | 4·2 | 2·1 | 10·2 | 5·2 | 16·2 | 8·2 |
| 43 | 7 40·8 | 7 42·0 | 7 19·8 | 4·3 | 2·2 | 10·3 | 5·2 | 16·3 | 8·3 |
| 44 | 7 41·0 | 7 42·3 | 7 20·0 | 4·4 | 2·2 | 10·4 | 5·3 | 16·4 | 8·3 |
| 45 | 7 41·3 | 7 42·5 | 7 20·2 | 4·5 | 2·3 | 10·5 | 5·3 | 16·5 | 8·4 |
| 46 | 7 41·5 | 7 42·8 | 7 20·5 | 4·6 | 2·3 | 10·6 | 5·4 | 16·6 | 8·4 |
| 47 | 7 41·8 | 7 43·0 | 7 20·7 | 4·7 | 2·4 | 10·7 | 5·4 | 16·7 | 8·5 |
| 48 | 7 42·0 | 7 43·3 | 7 21·0 | 4·8 | 2·4 | 10·8 | 5·5 | 16·8 | 8·5 |
| 49 | 7 42·3 | 7 43·5 | 7 21·2 | 4·9 | 2·5 | 10·9 | 5·5 | 16·9 | 8·6 |
| 50 | 7 42·5 | 7 43·8 | 7 21·4 | 5·0 | 2·5 | 11·0 | 5·6 | 17·0 | 8·6 |
| 51 | 7 42·8 | 7 44·0 | 7 21·7 | 5·1 | 2·6 | 11·1 | 5·6 | 17·1 | 8·7 |
| 52 | 7 43·0 | 7 44·3 | 7 21·9 | 5·2 | 2·6 | 11·2 | 5·7 | 17·2 | 8·7 |
| 53 | 7 43·3 | 7 44·5 | 7 22·1 | 5·3 | 2·7 | 11·3 | 5·7 | 17·3 | 8·8 |
| 54 | 7 43·5 | 7 44·8 | 7 22·4 | 5·4 | 2·7 | 11·4 | 5·8 | 17·4 | 8·8 |
| 55 | 7 43·8 | 7 45·0 | 7 22·6 | 5·5 | 2·8 | 11·5 | 5·8 | 17·5 | 8·9 |
| 56 | 7 44·0 | 7 45·3 | 7 22·9 | 5·6 | 2·8 | 11·6 | 5·9 | 17·6 | 8·9 |
| 57 | 7 44·3 | 7 45·5 | 7 23·1 | 5·7 | 2·9 | 11·7 | 5·9 | 17·7 | 9·0 |
| 58 | 7 44·5 | 7 45·8 | 7 23·3 | 5·8 | 2·9 | 11·8 | 6·0 | 17·8 | 9·0 |
| 59 | 7 44·8 | 7 46·0 | 7 23·6 | 5·9 | 3·0 | 11·9 | 6·0 | 17·9 | 9·1 |
| 60 | 7 45·0 | 7 46·3 | 7 23·8 | 6·0 | 3·1 | 12·0 | 6·1 | 18·0 | 9·2 |

### 31ᵐ

| 31 | SUN PLANETS | ARIES | MOON | v or d | Corrⁿ | v or d | Corrⁿ | v or d | Corrⁿ |
|---|---|---|---|---|---|---|---|---|---|
| s | ° ′ | ° ′ | ° ′ | ′ | ′ | ′ | ′ | ′ | ′ |
| 00 | 7 45·0 | 7 46·3 | 7 23·8 | 0·0 | 0·0 | 6·0 | 3·2 | 12·0 | 6·3 |
| 01 | 7 45·3 | 7 46·5 | 7 24·1 | 0·1 | 0·1 | 6·1 | 3·2 | 12·1 | 6·4 |
| 02 | 7 45·5 | 7 46·8 | 7 24·3 | 0·2 | 0·1 | 6·2 | 3·3 | 12·2 | 6·4 |
| 03 | 7 45·8 | 7 47·0 | 7 24·5 | 0·3 | 0·2 | 6·3 | 3·3 | 12·3 | 6·5 |
| 04 | 7 46·0 | 7 47·3 | 7 24·8 | 0·4 | 0·2 | 6·4 | 3·4 | 12·4 | 6·5 |
| 05 | 7 46·3 | 7 47·5 | 7 25·0 | 0·5 | 0·3 | 6·5 | 3·4 | 12·5 | 6·6 |
| 06 | 7 46·5 | 7 47·8 | 7 25·2 | 0·6 | 0·3 | 6·6 | 3·5 | 12·6 | 6·6 |
| 07 | 7 46·8 | 7 48·0 | 7 25·5 | 0·7 | 0·4 | 6·7 | 3·5 | 12·7 | 6·7 |
| 08 | 7 47·0 | 7 48·3 | 7 25·7 | 0·8 | 0·4 | 6·8 | 3·6 | 12·8 | 6·7 |
| 09 | 7 47·3 | 7 48·5 | 7 26·0 | 0·9 | 0·5 | 6·9 | 3·6 | 12·9 | 6·8 |
| 10 | 7 47·5 | 7 48·8 | 7 26·2 | 1·0 | 0·5 | 7·0 | 3·7 | 13·0 | 6·8 |
| 11 | 7 47·8 | 7 49·0 | 7 26·4 | 1·1 | 0·6 | 7·1 | 3·7 | 13·1 | 6·9 |
| 12 | 7 48·0 | 7 49·3 | 7 26·7 | 1·2 | 0·6 | 7·2 | 3·8 | 13·2 | 6·9 |
| 13 | 7 48·3 | 7 49·5 | 7 26·9 | 1·3 | 0·7 | 7·3 | 3·8 | 13·3 | 7·0 |
| 14 | 7 48·5 | 7 49·8 | 7 27·2 | 1·4 | 0·7 | 7·4 | 3·9 | 13·4 | 7·0 |
| 15 | 7 48·8 | 7 50·0 | 7 27·4 | 1·5 | 0·8 | 7·5 | 3·9 | 13·5 | 7·1 |
| 16 | 7 49·0 | 7 50·3 | 7 27·6 | 1·6 | 0·8 | 7·6 | 4·0 | 13·6 | 7·1 |
| 17 | 7 49·3 | 7 50·5 | 7 27·9 | 1·7 | 0·9 | 7·7 | 4·0 | 13·7 | 7·2 |
| 18 | 7 49·5 | 7 50·8 | 7 28·1 | 1·8 | 0·9 | 7·8 | 4·1 | 13·8 | 7·2 |
| 19 | 7 49·8 | 7 51·0 | 7 28·4 | 1·9 | 1·0 | 7·9 | 4·1 | 13·9 | 7·3 |
| 20 | 7 50·0 | 7 51·3 | 7 28·6 | 2·0 | 1·1 | 8·0 | 4·2 | 14·0 | 7·4 |
| 21 | 7 50·3 | 7 51·5 | 7 28·8 | 2·1 | 1·1 | 8·1 | 4·3 | 14·1 | 7·4 |
| 22 | 7 50·5 | 7 51·8 | 7 29·1 | 2·2 | 1·2 | 8·2 | 4·3 | 14·2 | 7·5 |
| 23 | 7 50·8 | 7 52·0 | 7 29·3 | 2·3 | 1·2 | 8·3 | 4·4 | 14·3 | 7·5 |
| 24 | 7 51·0 | 7 52·3 | 7 29·5 | 2·4 | 1·3 | 8·4 | 4·4 | 14·4 | 7·6 |
| 25 | 7 51·3 | 7 52·5 | 7 29·8 | 2·5 | 1·3 | 8·5 | 4·5 | 14·5 | 7·6 |
| 26 | 7 51·5 | 7 52·8 | 7 30·0 | 2·6 | 1·4 | 8·6 | 4·5 | 14·6 | 7·7 |
| 27 | 7 51·8 | 7 53·0 | 7 30·3 | 2·7 | 1·4 | 8·7 | 4·6 | 14·7 | 7·7 |
| 28 | 7 52·0 | 7 53·3 | 7 30·5 | 2·8 | 1·5 | 8·8 | 4·6 | 14·8 | 7·8 |
| 29 | 7 52·3 | 7 53·5 | 7 30·7 | 2·9 | 1·5 | 8·9 | 4·7 | 14·9 | 7·8 |
| 30 | 7 52·5 | 7 53·8 | 7 31·0 | 3·0 | 1·6 | 9·0 | 4·7 | 15·0 | 7·9 |
| 31 | 7 52·8 | 7 54·0 | 7 31·2 | 3·1 | 1·6 | 9·1 | 4·8 | 15·1 | 7·9 |
| 32 | 7 53·0 | 7 54·3 | 7 31·5 | 3·2 | 1·7 | 9·2 | 4·8 | 15·2 | 8·0 |
| 33 | 7 53·3 | 7 54·5 | 7 31·7 | 3·3 | 1·7 | 9·3 | 4·9 | 15·3 | 8·0 |
| 34 | 7 53·5 | 7 54·8 | 7 31·9 | 3·4 | 1·8 | 9·4 | 4·9 | 15·4 | 8·1 |
| 35 | 7 53·8 | 7 55·0 | 7 32·2 | 3·5 | 1·8 | 9·5 | 5·0 | 15·5 | 8·1 |
| 36 | 7 54·0 | 7 55·3 | 7 32·4 | 3·6 | 1·9 | 9·6 | 5·0 | 15·6 | 8·2 |
| 37 | 7 54·3 | 7 55·5 | 7 32·6 | 3·7 | 1·9 | 9·7 | 5·1 | 15·7 | 8·2 |
| 38 | 7 54·5 | 7 55·8 | 7 32·9 | 3·8 | 2·0 | 9·8 | 5·1 | 15·8 | 8·3 |
| 39 | 7 54·8 | 7 56·0 | 7 33·1 | 3·9 | 2·0 | 9·9 | 5·2 | 15·9 | 8·3 |
| 40 | 7 55·0 | 7 56·3 | 7 33·4 | 4·0 | 2·1 | 10·0 | 5·3 | 16·0 | 8·4 |
| 41 | 7 55·3 | 7 56·6 | 7 33·6 | 4·1 | 2·2 | 10·1 | 5·3 | 16·1 | 8·5 |
| 42 | 7 55·5 | 7 56·8 | 7 33·8 | 4·2 | 2·2 | 10·2 | 5·4 | 16·2 | 8·5 |
| 43 | 7 55·8 | 7 57·1 | 7 34·1 | 4·3 | 2·3 | 10·3 | 5·4 | 16·3 | 8·6 |
| 44 | 7 56·0 | 7 57·3 | 7 34·3 | 4·4 | 2·3 | 10·4 | 5·5 | 16·4 | 8·6 |
| 45 | 7 56·3 | 7 57·6 | 7 34·6 | 4·5 | 2·4 | 10·5 | 5·5 | 16·5 | 8·7 |
| 46 | 7 56·5 | 7 57·8 | 7 34·8 | 4·6 | 2·4 | 10·6 | 5·6 | 16·6 | 8·7 |
| 47 | 7 56·8 | 7 58·1 | 7 35·0 | 4·7 | 2·5 | 10·7 | 5·6 | 16·7 | 8·8 |
| 48 | 7 57·0 | 7 58·3 | 7 35·3 | 4·8 | 2·5 | 10·8 | 5·7 | 16·8 | 8·8 |
| 49 | 7 57·3 | 7 58·6 | 7 35·5 | 4·9 | 2·6 | 10·9 | 5·7 | 16·9 | 8·9 |
| 50 | 7 57·5 | 7 58·8 | 7 35·7 | 5·0 | 2·6 | 11·0 | 5·8 | 17·0 | 8·9 |
| 51 | 7 57·8 | 7 59·1 | 7 36·0 | 5·1 | 2·7 | 11·1 | 5·8 | 17·1 | 9·0 |
| 52 | 7 58·0 | 7 59·3 | 7 36·2 | 5·2 | 2·7 | 11·2 | 5·9 | 17·2 | 9·0 |
| 53 | 7 58·3 | 7 59·6 | 7 36·5 | 5·3 | 2·8 | 11·3 | 5·9 | 17·3 | 9·1 |
| 54 | 7 58·5 | 7 59·8 | 7 36·7 | 5·4 | 2·8 | 11·4 | 6·0 | 17·4 | 9·1 |
| 55 | 7 58·8 | 8 00·1 | 7 36·9 | 5·5 | 2·9 | 11·5 | 6·0 | 17·5 | 9·2 |
| 56 | 7 59·0 | 8 00·3 | 7 37·2 | 5·6 | 2·9 | 11·6 | 6·1 | 17·6 | 9·2 |
| 57 | 7 59·3 | 8 00·6 | 7 37·4 | 5·7 | 3·0 | 11·7 | 6·1 | 17·7 | 9·3 |
| 58 | 7 59·5 | 8 00·8 | 7 37·7 | 5·8 | 3·0 | 11·8 | 6·2 | 17·8 | 9·3 |
| 59 | 7 59·8 | 8 01·1 | 7 37·9 | 5·9 | 3·1 | 11·9 | 6·2 | 17·9 | 9·4 |
| 60 | 8 00·0 | 8 01·3 | 7 38·1 | 6·0 | 3·2 | 12·0 | 6·3 | 18·0 | 9·5 |

## INCREMENTS AND CORRECTIONS

### 34m

| 34s | SUN PLANETS | ARIES | MOON | v or d | Corrn | v or d | Corrn | v or d | Corrn |
|---|---|---|---|---|---|---|---|---|---|
| s | ° ′ | ° ′ | ° ′ | ′ | ′ | ′ | ′ | ′ | ′ |
| 00 | 8 30.0 | 8 31.4 | 8 06.8 | 0.0 | 0.0 | 6.0 | 3.5 | 12.0 | 6.9 |
| 01 | 8 30.3 | 8 31.6 | 8 07.0 | 0.1 | 0.1 | 6.1 | 3.5 | 12.1 | 7.0 |
| 02 | 8 30.5 | 8 31.9 | 8 07.2 | 0.2 | 0.1 | 6.2 | 3.6 | 12.2 | 7.0 |
| 03 | 8 30.8 | 8 32.1 | 8 07.5 | 0.3 | 0.2 | 6.3 | 3.6 | 12.3 | 7.1 |
| 04 | 8 31.0 | 8 32.4 | 8 07.7 | 0.4 | 0.2 | 6.4 | 3.7 | 12.4 | 7.1 |
| 05 | 8 31.3 | 8 32.6 | 8 08.0 | 0.5 | 0.3 | 6.5 | 3.7 | 12.5 | 7.2 |
| 06 | 8 31.5 | 8 32.9 | 8 08.2 | 0.6 | 0.3 | 6.6 | 3.8 | 12.6 | 7.2 |
| 07 | 8 31.8 | 8 33.2 | 8 08.4 | 0.7 | 0.4 | 6.7 | 3.9 | 12.7 | 7.3 |
| 08 | 8 32.0 | 8 33.4 | 8 08.7 | 0.8 | 0.5 | 6.8 | 3.9 | 12.8 | 7.4 |
| 09 | 8 32.3 | 8 33.7 | 8 08.9 | 0.9 | 0.5 | 6.9 | 4.0 | 12.9 | 7.4 |
| 10 | 8 32.5 | 8 33.9 | 8 09.2 | 1.0 | 0.6 | 7.0 | 4.0 | 13.0 | 7.5 |
| 11 | 8 32.8 | 8 34.2 | 8 09.4 | 1.1 | 0.6 | 7.1 | 4.1 | 13.1 | 7.5 |
| 12 | 8 33.0 | 8 34.4 | 8 09.6 | 1.2 | 0.7 | 7.2 | 4.1 | 13.2 | 7.6 |
| 13 | 8 33.3 | 8 34.7 | 8 09.9 | 1.3 | 0.7 | 7.3 | 4.2 | 13.3 | 7.6 |
| 14 | 8 33.5 | 8 34.9 | 8 10.1 | 1.4 | 0.8 | 7.4 | 4.3 | 13.4 | 7.7 |
| 15 | 8 33.8 | 8 35.2 | 8 10.3 | 1.5 | 0.9 | 7.5 | 4.3 | 13.5 | 7.8 |
| 16 | 8 34.0 | 8 35.4 | 8 10.6 | 1.6 | 0.9 | 7.6 | 4.4 | 13.6 | 7.8 |
| 17 | 8 34.3 | 8 35.7 | 8 10.8 | 1.7 | 1.0 | 7.7 | 4.4 | 13.7 | 7.9 |
| 18 | 8 34.5 | 8 35.9 | 8 11.1 | 1.8 | 1.0 | 7.8 | 4.5 | 13.8 | 7.9 |
| 19 | 8 34.8 | 8 36.2 | 8 11.3 | 1.9 | 1.1 | 7.9 | 4.5 | 13.9 | 8.0 |
| 20 | 8 35.0 | 8 36.4 | 8 11.5 | 2.0 | 1.2 | 8.0 | 4.6 | 14.0 | 8.1 |
| 21 | 8 35.3 | 8 36.7 | 8 11.8 | 2.1 | 1.2 | 8.1 | 4.7 | 14.1 | 8.1 |
| 22 | 8 35.5 | 8 36.9 | 8 12.0 | 2.2 | 1.3 | 8.2 | 4.7 | 14.2 | 8.2 |
| 23 | 8 35.8 | 8 37.2 | 8 12.3 | 2.3 | 1.3 | 8.3 | 4.8 | 14.3 | 8.2 |
| 24 | 8 36.0 | 8 37.4 | 8 12.5 | 2.4 | 1.4 | 8.4 | 4.8 | 14.4 | 8.3 |
| 25 | 8 36.3 | 8 37.7 | 8 12.7 | 2.5 | 1.4 | 8.5 | 4.9 | 14.5 | 8.3 |
| 26 | 8 36.5 | 8 37.9 | 8 13.0 | 2.6 | 1.5 | 8.6 | 4.9 | 14.6 | 8.4 |
| 27 | 8 36.8 | 8 38.2 | 8 13.2 | 2.7 | 1.6 | 8.7 | 5.0 | 14.7 | 8.5 |
| 28 | 8 37.0 | 8 38.4 | 8 13.4 | 2.8 | 1.6 | 8.8 | 5.1 | 14.8 | 8.5 |
| 29 | 8 37.3 | 8 38.7 | 8 13.7 | 2.9 | 1.7 | 8.9 | 5.1 | 14.9 | 8.6 |
| 30 | 8 37.5 | 8 38.9 | 8 13.9 | 3.0 | 1.7 | 9.0 | 5.2 | 15.0 | 8.6 |
| 31 | 8 37.8 | 8 39.2 | 8 14.2 | 3.1 | 1.8 | 9.1 | 5.2 | 15.1 | 8.7 |
| 32 | 8 38.0 | 8 39.4 | 8 14.4 | 3.2 | 1.8 | 9.2 | 5.3 | 15.2 | 8.7 |
| 33 | 8 38.3 | 8 39.7 | 8 14.6 | 3.3 | 1.9 | 9.3 | 5.3 | 15.3 | 8.8 |
| 34 | 8 38.5 | 8 39.9 | 8 14.9 | 3.4 | 2.0 | 9.4 | 5.4 | 15.4 | 8.9 |
| 35 | 8 38.8 | 8 40.2 | 8 15.1 | 3.5 | 2.0 | 9.5 | 5.5 | 15.5 | 8.9 |
| 36 | 8 39.0 | 8 40.4 | 8 15.4 | 3.6 | 2.1 | 9.6 | 5.5 | 15.6 | 9.0 |
| 37 | 8 39.3 | 8 40.7 | 8 15.6 | 3.7 | 2.1 | 9.7 | 5.6 | 15.7 | 9.0 |
| 38 | 8 39.5 | 8 40.9 | 8 15.8 | 3.8 | 2.2 | 9.8 | 5.6 | 15.8 | 9.1 |
| 39 | 8 39.8 | 8 41.2 | 8 16.1 | 3.9 | 2.2 | 9.9 | 5.7 | 15.9 | 9.1 |
| 40 | 8 40.0 | 8 41.4 | 8 16.3 | 4.0 | 2.3 | 10.0 | 5.8 | 16.0 | 9.2 |
| 41 | 8 40.3 | 8 41.7 | 8 16.5 | 4.1 | 2.4 | 10.1 | 5.8 | 16.1 | 9.3 |
| 42 | 8 40.5 | 8 41.9 | 8 16.8 | 4.2 | 2.4 | 10.2 | 5.9 | 16.2 | 9.3 |
| 43 | 8 40.8 | 8 42.2 | 8 17.0 | 4.3 | 2.5 | 10.3 | 5.9 | 16.3 | 9.4 |
| 44 | 8 41.0 | 8 42.4 | 8 17.3 | 4.4 | 2.5 | 10.4 | 6.0 | 16.4 | 9.4 |
| 45 | 8 41.3 | 8 42.7 | 8 17.5 | 4.5 | 2.6 | 10.5 | 6.0 | 16.5 | 9.5 |
| 46 | 8 41.5 | 8 42.9 | 8 17.7 | 4.6 | 2.6 | 10.6 | 6.1 | 16.6 | 9.5 |
| 47 | 8 41.8 | 8 43.2 | 8 18.0 | 4.7 | 2.7 | 10.7 | 6.2 | 16.7 | 9.6 |
| 48 | 8 42.0 | 8 43.4 | 8 18.2 | 4.8 | 2.8 | 10.8 | 6.2 | 16.8 | 9.7 |
| 49 | 8 42.3 | 8 43.7 | 8 18.5 | 4.9 | 2.8 | 10.9 | 6.3 | 16.9 | 9.7 |
| 50 | 8 42.5 | 8 43.9 | 8 18.7 | 5.0 | 2.9 | 11.0 | 6.3 | 17.0 | 9.8 |
| 51 | 8 42.8 | 8 44.2 | 8 18.9 | 5.1 | 2.9 | 11.1 | 6.4 | 17.1 | 9.8 |
| 52 | 8 43.0 | 8 44.4 | 8 19.2 | 5.2 | 3.0 | 11.2 | 6.4 | 17.2 | 9.9 |
| 53 | 8 43.3 | 8 44.7 | 8 19.4 | 5.3 | 3.0 | 11.3 | 6.5 | 17.3 | 9.9 |
| 54 | 8 43.5 | 8 44.9 | 8 19.7 | 5.4 | 3.1 | 11.4 | 6.6 | 17.4 | 10.0 |
| 55 | 8 43.8 | 8 45.2 | 8 19.9 | 5.5 | 3.2 | 11.5 | 6.6 | 17.5 | 10.1 |
| 56 | 8 44.0 | 8 45.4 | 8 20.1 | 5.6 | 3.2 | 11.6 | 6.7 | 17.6 | 10.1 |
| 57 | 8 44.3 | 8 45.7 | 8 20.4 | 5.7 | 3.3 | 11.7 | 6.7 | 17.7 | 10.2 |
| 58 | 8 44.5 | 8 45.9 | 8 20.6 | 5.8 | 3.3 | 11.8 | 6.8 | 17.8 | 10.2 |
| 59 | 8 44.8 | 8 46.2 | 8 20.8 | 5.9 | 3.4 | 11.9 | 6.8 | 17.9 | 10.3 |
| 60 | 8 45.0 | 8 46.4 | 8 21.1 | 6.0 | 3.5 | 12.0 | 6.9 | 18.0 | 10.4 |

### 35m

| 35s | SUN PLANETS | ARIES | MOON | v or d | Corrn | v or d | Corrn | v or d | Corrn |
|---|---|---|---|---|---|---|---|---|---|
| s | ° ′ | ° ′ | ° ′ | ′ | ′ | ′ | ′ | ′ | ′ |
| 00 | 8 45.0 | 8 46.4 | 8 21.1 | 0.0 | 0.0 | 6.0 | 3.6 | 12.0 | 7.1 |
| 01 | 8 45.3 | 8 46.7 | 8 21.3 | 0.1 | 0.1 | 6.1 | 3.6 | 12.1 | 7.2 |
| 02 | 8 45.5 | 8 46.9 | 8 21.6 | 0.2 | 0.1 | 6.2 | 3.7 | 12.2 | 7.2 |
| 03 | 8 45.8 | 8 47.2 | 8 21.8 | 0.3 | 0.2 | 6.3 | 3.7 | 12.3 | 7.3 |
| 04 | 8 46.0 | 8 47.4 | 8 22.0 | 0.4 | 0.2 | 6.4 | 3.8 | 12.4 | 7.3 |
| 05 | 8 46.3 | 8 47.7 | 8 22.3 | 0.5 | 0.3 | 6.5 | 3.8 | 12.5 | 7.4 |
| 06 | 8 46.5 | 8 47.9 | 8 22.5 | 0.6 | 0.4 | 6.6 | 3.9 | 12.6 | 7.5 |
| 07 | 8 46.8 | 8 48.2 | 8 22.8 | 0.7 | 0.4 | 6.7 | 4.0 | 12.7 | 7.5 |
| 08 | 8 47.0 | 8 48.4 | 8 23.0 | 0.8 | 0.5 | 6.8 | 4.0 | 12.8 | 7.6 |
| 09 | 8 47.3 | 8 48.7 | 8 23.2 | 0.9 | 0.5 | 6.9 | 4.1 | 12.9 | 7.6 |
| 10 | 8 47.5 | 8 48.9 | 8 23.5 | 1.0 | 0.6 | 7.0 | 4.1 | 13.0 | 7.7 |
| 11 | 8 47.8 | 8 49.2 | 8 23.7 | 1.1 | 0.7 | 7.1 | 4.2 | 13.1 | 7.8 |
| 12 | 8 48.0 | 8 49.4 | 8 23.9 | 1.2 | 0.7 | 7.2 | 4.3 | 13.2 | 7.8 |
| 13 | 8 48.3 | 8 49.7 | 8 24.2 | 1.3 | 0.8 | 7.3 | 4.3 | 13.3 | 7.9 |
| 14 | 8 48.5 | 8 49.9 | 8 24.4 | 1.4 | 0.8 | 7.4 | 4.4 | 13.4 | 7.9 |
| 15 | 8 48.8 | 8 50.2 | 8 24.7 | 1.5 | 0.9 | 7.5 | 4.4 | 13.5 | 8.0 |
| 16 | 8 49.0 | 8 50.4 | 8 24.9 | 1.6 | 0.9 | 7.6 | 4.5 | 13.6 | 8.0 |
| 17 | 8 49.3 | 8 50.7 | 8 25.1 | 1.7 | 1.0 | 7.7 | 4.6 | 13.7 | 8.1 |
| 18 | 8 49.5 | 8 50.9 | 8 25.4 | 1.8 | 1.1 | 7.8 | 4.6 | 13.8 | 8.2 |
| 19 | 8 49.8 | 8 51.2 | 8 25.6 | 1.9 | 1.1 | 7.9 | 4.7 | 13.9 | 8.2 |
| 20 | 8 50.0 | 8 51.5 | 8 25.9 | 2.0 | 1.2 | 8.0 | 4.7 | 14.0 | 8.3 |
| 21 | 8 50.3 | 8 51.7 | 8 26.1 | 2.1 | 1.2 | 8.1 | 4.8 | 14.1 | 8.3 |
| 22 | 8 50.5 | 8 52.0 | 8 26.3 | 2.2 | 1.3 | 8.2 | 4.9 | 14.2 | 8.4 |
| 23 | 8 50.8 | 8 52.2 | 8 26.6 | 2.3 | 1.4 | 8.3 | 4.9 | 14.3 | 8.5 |
| 24 | 8 51.0 | 8 52.5 | 8 26.8 | 2.4 | 1.4 | 8.4 | 5.0 | 14.4 | 8.5 |
| 25 | 8 51.3 | 8 52.7 | 8 27.0 | 2.5 | 1.5 | 8.5 | 5.0 | 14.5 | 8.6 |
| 26 | 8 51.5 | 8 53.0 | 8 27.3 | 2.6 | 1.5 | 8.6 | 5.1 | 14.6 | 8.6 |
| 27 | 8 51.8 | 8 53.2 | 8 27.5 | 2.7 | 1.6 | 8.7 | 5.1 | 14.7 | 8.7 |
| 28 | 8 52.0 | 8 53.5 | 8 27.8 | 2.8 | 1.7 | 8.8 | 5.2 | 14.8 | 8.8 |
| 29 | 8 52.3 | 8 53.7 | 8 28.0 | 2.9 | 1.7 | 8.9 | 5.3 | 14.9 | 8.8 |
| 30 | 8 52.5 | 8 54.0 | 8 28.2 | 3.0 | 1.8 | 9.0 | 5.3 | 15.0 | 8.9 |
| 31 | 8 52.8 | 8 54.2 | 8 28.5 | 3.1 | 1.8 | 9.1 | 5.4 | 15.1 | 8.9 |
| 32 | 8 53.0 | 8 54.5 | 8 28.7 | 3.2 | 1.9 | 9.2 | 5.4 | 15.2 | 9.0 |
| 33 | 8 53.3 | 8 54.7 | 8 29.0 | 3.3 | 2.0 | 9.3 | 5.5 | 15.3 | 9.1 |
| 34 | 8 53.5 | 8 55.0 | 8 29.2 | 3.4 | 2.0 | 9.4 | 5.6 | 15.4 | 9.1 |
| 35 | 8 53.8 | 8 55.2 | 8 29.4 | 3.5 | 2.1 | 9.5 | 5.6 | 15.5 | 9.2 |
| 36 | 8 54.0 | 8 55.5 | 8 29.7 | 3.6 | 2.1 | 9.6 | 5.7 | 15.6 | 9.2 |
| 37 | 8 54.3 | 8 55.7 | 8 29.9 | 3.7 | 2.2 | 9.7 | 5.7 | 15.7 | 9.3 |
| 38 | 8 54.5 | 8 56.0 | 8 30.2 | 3.8 | 2.2 | 9.8 | 5.8 | 15.8 | 9.3 |
| 39 | 8 54.8 | 8 56.2 | 8 30.4 | 3.9 | 2.3 | 9.9 | 5.9 | 15.9 | 9.4 |
| 40 | 8 55.0 | 8 56.5 | 8 30.6 | 4.0 | 2.4 | 10.0 | 5.9 | 16.0 | 9.5 |
| 41 | 8 55.3 | 8 56.7 | 8 30.9 | 4.1 | 2.4 | 10.1 | 6.0 | 16.1 | 9.5 |
| 42 | 8 55.5 | 8 57.0 | 8 31.1 | 4.2 | 2.5 | 10.2 | 6.0 | 16.2 | 9.6 |
| 43 | 8 55.8 | 8 57.2 | 8 31.3 | 4.3 | 2.5 | 10.3 | 6.1 | 16.3 | 9.6 |
| 44 | 8 56.0 | 8 57.5 | 8 31.6 | 4.4 | 2.6 | 10.4 | 6.2 | 16.4 | 9.7 |
| 45 | 8 56.3 | 8 57.7 | 8 31.8 | 4.5 | 2.7 | 10.5 | 6.2 | 16.5 | 9.8 |
| 46 | 8 56.5 | 8 58.0 | 8 32.1 | 4.6 | 2.7 | 10.6 | 6.3 | 16.6 | 9.8 |
| 47 | 8 56.8 | 8 58.2 | 8 32.3 | 4.7 | 2.8 | 10.7 | 6.3 | 16.7 | 9.9 |
| 48 | 8 57.0 | 8 58.5 | 8 32.5 | 4.8 | 2.8 | 10.8 | 6.4 | 16.8 | 9.9 |
| 49 | 8 57.3 | 8 58.7 | 8 32.8 | 4.9 | 2.9 | 10.9 | 6.4 | 16.9 | 10.0 |
| 50 | 8 57.5 | 8 59.0 | 8 33.0 | 5.0 | 3.0 | 11.0 | 6.5 | 17.0 | 10.1 |
| 51 | 8 57.8 | 8 59.2 | 8 33.3 | 5.1 | 3.0 | 11.1 | 6.6 | 17.1 | 10.1 |
| 52 | 8 58.0 | 8 59.5 | 8 33.5 | 5.2 | 3.1 | 11.2 | 6.6 | 17.2 | 10.2 |
| 53 | 8 58.3 | 8 59.7 | 8 33.7 | 5.3 | 3.1 | 11.3 | 6.7 | 17.3 | 10.2 |
| 54 | 8 58.5 | 9 00.0 | 8 34.0 | 5.4 | 3.2 | 11.4 | 6.7 | 17.4 | 10.3 |
| 55 | 8 58.8 | 9 00.2 | 8 34.2 | 5.5 | 3.3 | 11.5 | 6.8 | 17.5 | 10.4 |
| 56 | 8 59.0 | 9 00.5 | 8 34.4 | 5.6 | 3.3 | 11.6 | 6.9 | 17.6 | 10.4 |
| 57 | 8 59.3 | 9 00.7 | 8 34.7 | 5.7 | 3.4 | 11.7 | 6.9 | 17.7 | 10.5 |
| 58 | 8 59.5 | 9 01.0 | 8 34.9 | 5.8 | 3.4 | 11.8 | 7.0 | 17.8 | 10.5 |
| 59 | 8 59.8 | 9 01.2 | 8 35.2 | 5.9 | 3.5 | 11.9 | 7.0 | 17.9 | 10.6 |
| 60 | 9 00.0 | 9 01.5 | 8 35.4 | 6.0 | 3.6 | 12.0 | 7.1 | 18.0 | 10.7 |

## INCREMENTS AND CORRECTIONS

| 40ᵐ | SUN PLANETS | ARIES | MOON | v or d | Corrⁿ | v or d | Corrⁿ | v or d | Corrⁿ |
|---|---|---|---|---|---|---|---|---|---|
| s | ° ′ | ° ′ | ° ′ | ′ | ′ | ′ | ′ | ′ | ′ |
| 00 | 10 00·0 | 10 01·6 | 9 32·7 | 0·0 | 0·0 | 6·0 | 4·1 | 12·0 | 8·1 |
| 01 | 10 00·3 | 10 01·9 | 9 32·9 | 0·1 | 0·1 | 6·1 | 4·1 | 12·1 | 8·2 |
| 02 | 10 00·5 | 10 02·1 | 9 33·1 | 0·2 | 0·1 | 6·2 | 4·2 | 12·2 | 8·2 |
| 03 | 10 00·8 | 10 02·4 | 9 33·4 | 0·3 | 0·2 | 6·3 | 4·3 | 12·3 | 8·3 |
| 04 | 10 01·0 | 10 02·6 | 9 33·6 | 0·4 | 0·3 | 6·4 | 4·3 | 12·4 | 8·4 |
| 05 | 10 01·3 | 10 02·9 | 9 33·9 | 0·5 | 0·3 | 6·5 | 4·4 | 12·5 | 8·4 |
| 06 | 10 01·5 | 10 03·1 | 9 34·1 | 0·6 | 0·4 | 6·6 | 4·5 | 12·6 | 8·5 |
| 07 | 10 01·8 | 10 03·4 | 9 34·3 | 0·7 | 0·5 | 6·7 | 4·5 | 12·7 | 8·6 |
| 08 | 10 02·0 | 10 03·6 | 9 34·6 | 0·8 | 0·5 | 6·8 | 4·6 | 12·8 | 8·6 |
| 09 | 10 02·3 | 10 03·9 | 9 34·8 | 0·9 | 0·6 | 6·9 | 4·7 | 12·9 | 8·7 |
| 10 | 10 02·5 | 10 04·1 | 9 35·1 | 1·0 | 0·7 | 7·0 | 4·8 | 13·0 | 8·8 |
| 11 | 10 02·8 | 10 04·4 | 9 35·3 | 1·1 | 0·7 | 7·1 | 4·8 | 13·1 | 8·8 |
| 12 | 10 03·0 | 10 04·7 | 9 35·5 | 1·2 | 0·8 | 7·2 | 4·9 | 13·2 | 8·9 |
| 13 | 10 03·3 | 10 04·9 | 9 35·8 | 1·3 | 0·9 | 7·3 | 5·0 | 13·3 | 9·0 |
| 14 | 10 03·5 | 10 05·2 | 9 36·0 | 1·4 | 0·9 | 7·4 | 5·0 | 13·4 | 9·0 |
| 15 | 10 03·8 | 10 05·4 | 9 36·2 | 1·5 | 1·0 | 7·5 | 5·1 | 13·5 | 9·1 |
| 16 | 10 04·0 | 10 05·7 | 9 36·5 | 1·6 | 1·1 | 7·6 | 5·1 | 13·6 | 9·2 |
| 17 | 10 04·3 | 10 05·9 | 9 36·7 | 1·7 | 1·1 | 7·7 | 5·2 | 13·7 | 9·2 |
| 18 | 10 04·5 | 10 06·2 | 9 37·0 | 1·8 | 1·2 | 7·8 | 5·3 | 13·8 | 9·3 |
| 19 | 10 04·8 | 10 06·4 | 9 37·2 | 1·9 | 1·3 | 7·9 | 5·3 | 13·9 | 9·4 |
| 20 | 10 05·0 | 10 06·7 | 9 37·4 | 2·0 | 1·4 | 8·0 | 5·4 | 14·0 | 9·5 |
| 21 | 10 05·3 | 10 06·9 | 9 37·7 | 2·1 | 1·4 | 8·1 | 5·5 | 14·1 | 9·5 |
| 22 | 10 05·5 | 10 07·2 | 9 37·9 | 2·2 | 1·5 | 8·2 | 5·5 | 14·2 | 9·6 |
| 23 | 10 05·8 | 10 07·4 | 9 38·2 | 2·3 | 1·6 | 8·3 | 5·6 | 14·3 | 9·7 |
| 24 | 10 06·0 | 10 07·7 | 9 38·4 | 2·4 | 1·6 | 8·4 | 5·7 | 14·4 | 9·7 |
| 25 | 10 06·3 | 10 07·9 | 9 38·6 | 2·5 | 1·7 | 8·5 | 5·7 | 14·5 | 9·8 |
| 26 | 10 06·5 | 10 08·2 | 9 38·9 | 2·6 | 1·8 | 8·6 | 5·8 | 14·6 | 9·9 |
| 27 | 10 06·8 | 10 08·4 | 9 39·1 | 2·7 | 1·8 | 8·7 | 5·9 | 14·7 | 9·9 |
| 28 | 10 07·0 | 10 08·7 | 9 39·3 | 2·8 | 1·9 | 8·8 | 5·9 | 14·8 | 10·0 |
| 29 | 10 07·3 | 10 08·9 | 9 39·6 | 2·9 | 2·0 | 8·9 | 6·0 | 14·9 | 10·1 |
| 30 | 10 07·5 | 10 09·2 | 9 39·8 | 3·0 | 2·0 | 9·0 | 6·1 | 15·0 | 10·1 |
| 31 | 10 07·8 | 10 09·4 | 9 40·1 | 3·1 | 2·1 | 9·1 | 6·1 | 15·1 | 10·2 |
| 32 | 10 08·0 | 10 09·7 | 9 40·3 | 3·2 | 2·2 | 9·2 | 6·2 | 15·2 | 10·3 |
| 33 | 10 08·3 | 10 09·9 | 9 40·5 | 3·3 | 2·2 | 9·3 | 6·3 | 15·3 | 10·3 |
| 34 | 10 08·5 | 10 10·2 | 9 40·8 | 3·4 | 2·3 | 9·4 | 6·3 | 15·4 | 10·4 |
| 35 | 10 08·8 | 10 10·4 | 9 41·0 | 3·5 | 2·4 | 9·5 | 6·4 | 15·5 | 10·5 |
| 36 | 10 09·0 | 10 10·7 | 9 41·3 | 3·6 | 2·4 | 9·6 | 6·5 | 15·6 | 10·5 |
| 37 | 10 09·3 | 10 10·9 | 9 41·5 | 3·7 | 2·5 | 9·7 | 6·5 | 15·7 | 10·6 |
| 38 | 10 09·5 | 10 11·2 | 9 41·7 | 3·8 | 2·6 | 9·8 | 6·6 | 15·8 | 10·7 |
| 39 | 10 09·8 | 10 11·4 | 9 42·0 | 3·9 | 2·6 | 9·9 | 6·7 | 15·9 | 10·7 |
| 40 | 10 10·0 | 10 11·7 | 9 42·2 | 4·0 | 2·7 | 10·0 | 6·8 | 16·0 | 10·8 |
| 41 | 10 10·3 | 10 11·9 | 9 42·4 | 4·1 | 2·8 | 10·1 | 6·8 | 16·1 | 10·9 |
| 42 | 10 10·5 | 10 12·2 | 9 42·7 | 4·2 | 2·8 | 10·2 | 6·9 | 16·2 | 10·9 |
| 43 | 10 10·8 | 10 12·4 | 9 42·9 | 4·3 | 2·9 | 10·3 | 7·0 | 16·3 | 11·0 |
| 44 | 10 11·0 | 10 12·7 | 9 43·2 | 4·4 | 3·0 | 10·4 | 7·0 | 16·4 | 11·1 |
| 45 | 10 11·3 | 10 12·9 | 9 43·4 | 4·5 | 3·0 | 10·5 | 7·1 | 16·5 | 11·1 |
| 46 | 10 11·5 | 10 13·2 | 9 43·6 | 4·6 | 3·1 | 10·6 | 7·2 | 16·6 | 11·2 |
| 47 | 10 11·8 | 10 13·4 | 9 43·9 | 4·7 | 3·2 | 10·7 | 7·2 | 16·7 | 11·3 |
| 48 | 10 12·0 | 10 13·7 | 9 44·1 | 4·8 | 3·2 | 10·8 | 7·3 | 16·8 | 11·3 |
| 49 | 10 12·3 | 10 13·9 | 9 44·4 | 4·9 | 3·3 | 10·9 | 7·4 | 16·9 | 11·4 |
| 50 | 10 12·5 | 10 14·2 | 9 44·6 | 5·0 | 3·4 | 11·0 | 7·4 | 17·0 | 11·5 |
| 51 | 10 12·8 | 10 14·4 | 9 44·8 | 5·1 | 3·4 | 11·1 | 7·5 | 17·1 | 11·5 |
| 52 | 10 13·0 | 10 14·7 | 9 45·1 | 5·2 | 3·5 | 11·2 | 7·6 | 17·2 | 11·6 |
| 53 | 10 13·3 | 10 14·9 | 9 45·3 | 5·3 | 3·6 | 11·3 | 7·6 | 17·3 | 11·7 |
| 54 | 10 13·5 | 10 15·2 | 9 45·6 | 5·4 | 3·6 | 11·4 | 7·7 | 17·4 | 11·7 |
| 55 | 10 13·8 | 10 15·4 | 9 45·8 | 5·5 | 3·7 | 11·5 | 7·8 | 17·5 | 11·8 |
| 56 | 10 14·0 | 10 15·7 | 9 46·0 | 5·6 | 3·8 | 11·6 | 7·8 | 17·6 | 11·9 |
| 57 | 10 14·3 | 10 15·9 | 9 46·3 | 5·7 | 3·8 | 11·7 | 7·9 | 17·7 | 11·9 |
| 58 | 10 14·5 | 10 16·2 | 9 46·5 | 5·8 | 3·9 | 11·8 | 8·0 | 17·8 | 12·0 |
| 59 | 10 14·8 | 10 16·4 | 9 46·7 | 5·9 | 4·0 | 11·9 | 8·0 | 17·9 | 12·1 |
| 60 | 10 15·0 | 10 16·7 | 9 47·0 | 6·0 | 4·1 | 12·0 | 8·1 | 18·0 | 12·2 |

| 41ᵐ | SUN PLANETS | ARIES | MOON | v or d | Corrⁿ | v or d | Corrⁿ | v or d | Corrⁿ |
|---|---|---|---|---|---|---|---|---|---|
| s | ° ′ | ° ′ | ° ′ | ′ | ′ | ′ | ′ | ′ | ′ |
| 00 | 10 15·0 | 10 16·7 | 9 47·0 | 0·0 | 0·0 | 6·0 | 4·2 | 12·0 | 8·3 |
| 01 | 10 15·3 | 10 16·9 | 9 47·2 | 0·1 | 0·1 | 6·1 | 4·2 | 12·1 | 8·4 |
| 02 | 10 15·5 | 10 17·2 | 9 47·5 | 0·2 | 0·1 | 6·2 | 4·3 | 12·2 | 8·4 |
| 03 | 10 15·8 | 10 17·4 | 9 47·7 | 0·3 | 0·2 | 6·3 | 4·4 | 12·3 | 8·5 |
| 04 | 10 16·0 | 10 17·7 | 9 47·9 | 0·4 | 0·3 | 6·4 | 4·4 | 12·4 | 8·6 |
| 05 | 10 16·3 | 10 17·9 | 9 48·2 | 0·5 | 0·3 | 6·5 | 4·5 | 12·5 | 8·6 |
| 06 | 10 16·5 | 10 18·2 | 9 48·4 | 0·6 | 0·4 | 6·6 | 4·6 | 12·6 | 8·7 |
| 07 | 10 16·8 | 10 18·4 | 9 48·7 | 0·7 | 0·5 | 6·7 | 4·6 | 12·7 | 8·8 |
| 08 | 10 17·0 | 10 18·7 | 9 48·9 | 0·8 | 0·6 | 6·8 | 4·7 | 12·8 | 8·9 |
| 09 | 10 17·3 | 10 18·9 | 9 49·1 | 0·9 | 0·6 | 6·9 | 4·8 | 12·9 | 8·9 |
| 10 | 10 17·5 | 10 19·2 | 9 49·4 | 1·0 | 0·7 | 7·0 | 4·8 | 13·0 | 9·0 |
| 11 | 10 17·8 | 10 19·4 | 9 49·6 | 1·1 | 0·8 | 7·1 | 4·9 | 13·1 | 9·1 |
| 12 | 10 18·0 | 10 19·7 | 9 49·8 | 1·2 | 0·8 | 7·2 | 5·0 | 13·2 | 9·1 |
| 13 | 10 18·3 | 10 19·9 | 9 50·1 | 1·3 | 0·9 | 7·3 | 5·0 | 13·3 | 9·2 |
| 14 | 10 18·5 | 10 20·2 | 9 50·3 | 1·4 | 1·0 | 7·4 | 5·1 | 13·4 | 9·3 |
| 15 | 10 18·8 | 10 20·4 | 9 50·6 | 1·5 | 1·0 | 7·5 | 5·2 | 13·5 | 9·3 |
| 16 | 10 19·0 | 10 20·7 | 9 50·8 | 1·6 | 1·1 | 7·6 | 5·3 | 13·6 | 9·4 |
| 17 | 10 19·3 | 10 20·9 | 9 51·0 | 1·7 | 1·2 | 7·7 | 5·3 | 13·7 | 9·5 |
| 18 | 10 19·5 | 10 21·2 | 9 51·3 | 1·8 | 1·2 | 7·8 | 5·4 | 13·8 | 9·5 |
| 19 | 10 19·8 | 10 21·4 | 9 51·5 | 1·9 | 1·3 | 7·9 | 5·5 | 13·9 | 9·6 |
| 20 | 10 20·0 | 10 21·7 | 9 51·8 | 2·0 | 1·4 | 8·0 | 5·5 | 14·0 | 9·7 |
| 21 | 10 20·3 | 10 21·9 | 9 52·0 | 2·1 | 1·5 | 8·1 | 5·6 | 14·1 | 9·8 |
| 22 | 10 20·5 | 10 22·2 | 9 52·2 | 2·2 | 1·5 | 8·2 | 5·7 | 14·2 | 9·8 |
| 23 | 10 20·8 | 10 22·4 | 9 52·5 | 2·3 | 1·6 | 8·3 | 5·7 | 14·3 | 9·9 |
| 24 | 10 21·0 | 10 22·7 | 9 52·7 | 2·4 | 1·7 | 8·4 | 5·8 | 14·4 | 10·0 |
| 25 | 10 21·3 | 10 23·0 | 9 52·9 | 2·5 | 1·7 | 8·5 | 5·9 | 14·5 | 10·0 |
| 26 | 10 21·5 | 10 23·2 | 9 53·2 | 2·6 | 1·8 | 8·6 | 5·9 | 14·6 | 10·1 |
| 27 | 10 21·8 | 10 23·5 | 9 53·4 | 2·7 | 1·9 | 8·7 | 6·0 | 14·7 | 10·2 |
| 28 | 10 22·0 | 10 23·7 | 9 53·7 | 2·8 | 1·9 | 8·8 | 6·1 | 14·8 | 10·2 |
| 29 | 10 22·3 | 10 24·0 | 9 53·9 | 2·9 | 2·0 | 8·9 | 6·2 | 14·9 | 10·3 |
| 30 | 10 22·5 | 10 24·2 | 9 54·1 | 3·0 | 2·1 | 9·0 | 6·2 | 15·0 | 10·4 |
| 31 | 10 22·8 | 10 24·5 | 9 54·4 | 3·1 | 2·1 | 9·1 | 6·3 | 15·1 | 10·4 |
| 32 | 10 23·0 | 10 24·7 | 9 54·6 | 3·2 | 2·2 | 9·2 | 6·4 | 15·2 | 10·5 |
| 33 | 10 23·3 | 10 25·0 | 9 54·9 | 3·3 | 2·3 | 9·3 | 6·4 | 15·3 | 10·6 |
| 34 | 10 23·5 | 10 25·2 | 9 55·1 | 3·4 | 2·4 | 9·4 | 6·5 | 15·4 | 10·7 |
| 35 | 10 23·8 | 10 25·5 | 9 55·3 | 3·5 | 2·4 | 9·5 | 6·6 | 15·5 | 10·7 |
| 36 | 10 24·0 | 10 25·7 | 9 55·6 | 3·6 | 2·5 | 9·6 | 6·6 | 15·6 | 10·8 |
| 37 | 10 24·3 | 10 26·0 | 9 55·8 | 3·7 | 2·6 | 9·7 | 6·7 | 15·7 | 10·9 |
| 38 | 10 24·5 | 10 26·2 | 9 56·1 | 3·8 | 2·6 | 9·8 | 6·8 | 15·8 | 10·9 |
| 39 | 10 24·8 | 10 26·5 | 9 56·3 | 3·9 | 2·7 | 9·9 | 6·8 | 15·9 | 11·0 |
| 40 | 10 25·0 | 10 26·7 | 9 56·5 | 4·0 | 2·8 | 10·0 | 6·9 | 16·0 | 11·1 |
| 41 | 10 25·3 | 10 27·0 | 9 56·8 | 4·1 | 2·8 | 10·1 | 7·0 | 16·1 | 11·1 |
| 42 | 10 25·5 | 10 27·2 | 9 57·0 | 4·2 | 2·9 | 10·2 | 7·1 | 16·2 | 11·2 |
| 43 | 10 25·8 | 10 27·5 | 9 57·2 | 4·3 | 3·0 | 10·3 | 7·1 | 16·3 | 11·3 |
| 44 | 10 26·0 | 10 27·7 | 9 57·5 | 4·4 | 3·0 | 10·4 | 7·2 | 16·4 | 11·3 |
| 45 | 10 26·3 | 10 28·0 | 9 57·7 | 4·5 | 3·1 | 10·5 | 7·3 | 16·5 | 11·4 |
| 46 | 10 26·5 | 10 28·2 | 9 58·0 | 4·6 | 3·2 | 10·6 | 7·3 | 16·6 | 11·5 |
| 47 | 10 26·8 | 10 28·5 | 9 58·2 | 4·7 | 3·3 | 10·7 | 7·4 | 16·7 | 11·6 |
| 48 | 10 27·0 | 10 28·7 | 9 58·4 | 4·8 | 3·3 | 10·8 | 7·5 | 16·8 | 11·6 |
| 49 | 10 27·3 | 10 29·0 | 9 58·7 | 4·9 | 3·4 | 10·9 | 7·5 | 16·9 | 11·7 |
| 50 | 10 27·5 | 10 29·2 | 9 58·9 | 5·0 | 3·5 | 11·0 | 7·6 | 17·0 | 11·8 |
| 51 | 10 27·8 | 10 29·5 | 9 59·2 | 5·1 | 3·5 | 11·1 | 7·7 | 17·1 | 11·8 |
| 52 | 10 28·0 | 10 29·7 | 9 59·4 | 5·2 | 3·6 | 11·2 | 7·7 | 17·2 | 11·9 |
| 53 | 10 28·3 | 10 30·0 | 9 59·6 | 5·3 | 3·7 | 11·3 | 7·8 | 17·3 | 12·0 |
| 54 | 10 28·5 | 10 30·2 | 9 59·9 | 5·4 | 3·7 | 11·4 | 7·9 | 17·4 | 12·0 |
| 55 | 10 28·8 | 10 30·5 | 10 00·1 | 5·5 | 3·8 | 11·5 | 8·0 | 17·5 | 12·1 |
| 56 | 10 29·0 | 10 30·7 | 10 00·3 | 5·6 | 3·9 | 11·6 | 8·0 | 17·6 | 12·2 |
| 57 | 10 29·3 | 10 31·0 | 10 00·6 | 5·7 | 3·9 | 11·7 | 8·1 | 17·7 | 12·2 |
| 58 | 10 29·5 | 10 31·2 | 10 00·8 | 5·8 | 4·0 | 11·8 | 8·2 | 17·8 | 12·3 |
| 59 | 10 29·8 | 10 31·5 | 10 01·1 | 5·9 | 4·1 | 11·9 | 8·2 | 17·9 | 12·4 |
| 60 | 10 30·0 | 10 31·7 | 10 01·3 | 6·0 | 4·2 | 12·0 | 8·3 | 18·0 | 12·5 |

## INCREMENTS AND CORRECTIONS

### 44ᵐ

| 44 | SUN PLANETS | ARIES | MOON | v or d | Corrⁿ | v or d | Corrⁿ | v or d | Corrⁿ |
|---|---|---|---|---|---|---|---|---|---|
| s | ° ′ | ° ′ | ° ′ | ′ | ′ | ′ | ′ | ′ | ′ |
| 00 | 11 00·0 | 11 01·8 | 10 29·9 | 0·0 | 0·0 | 6·0 | 4·5 | 12·0 | 8·9 |
| 01 | 11 00·3 | 11 02·1 | 10 30·2 | 0·1 | 0·1 | 6·1 | 4·5 | 12·1 | 9·0 |
| 02 | 11 00·5 | 11 02·3 | 10 30·4 | 0·2 | 0·1 | 6·2 | 4·6 | 12·2 | 9·0 |
| 03 | 11 00·8 | 11 02·6 | 10 30·6 | 0·3 | 0·2 | 6·3 | 4·7 | 12·3 | 9·1 |
| 04 | 11 01·0 | 11 02·8 | 10 30·9 | 0·4 | 0·3 | 6·4 | 4·7 | 12·4 | 9·2 |
| 05 | 11 01·3 | 11 03·1 | 10 31·1 | 0·5 | 0·4 | 6·5 | 4·8 | 12·5 | 9·3 |
| 06 | 11 01·5 | 11 03·3 | 10 31·4 | 0·6 | 0·4 | 6·6 | 4·9 | 12·6 | 9·3 |
| 07 | 11 01·8 | 11 03·6 | 10 31·6 | 0·7 | 0·5 | 6·7 | 5·0 | 12·7 | 9·4 |
| 08 | 11 02·0 | 11 03·8 | 10 31·8 | 0·8 | 0·6 | 6·8 | 5·0 | 12·8 | 9·5 |
| 09 | 11 02·3 | 11 04·1 | 10 32·1 | 0·9 | 0·7 | 6·9 | 5·1 | 12·9 | 9·6 |
| 10 | 11 02·5 | 11 04·3 | 10 32·3 | 1·0 | 0·7 | 7·0 | 5·2 | 13·0 | 9·6 |
| 11 | 11 02·8 | 11 04·6 | 10 32·6 | 1·1 | 0·8 | 7·1 | 5·3 | 13·1 | 9·7 |
| 12 | 11 03·0 | 11 04·8 | 10 32·8 | 1·2 | 0·9 | 7·2 | 5·3 | 13·2 | 9·8 |
| 13 | 11 03·3 | 11 05·1 | 10 33·0 | 1·3 | 1·0 | 7·3 | 5·4 | 13·3 | 9·9 |
| 14 | 11 03·5 | 11 05·3 | 10 33·3 | 1·4 | 1·0 | 7·4 | 5·5 | 13·4 | 9·9 |
| 15 | 11 03·8 | 11 05·6 | 10 33·5 | 1·5 | 1·1 | 7·5 | 5·6 | 13·5 | 10·0 |
| 16 | 11 04·0 | 11 05·8 | 10 33·8 | 1·6 | 1·2 | 7·6 | 5·6 | 13·6 | 10·1 |
| 17 | 11 04·3 | 11 06·1 | 10 34·0 | 1·7 | 1·3 | 7·7 | 5·7 | 13·7 | 10·2 |
| 18 | 11 04·5 | 11 06·3 | 10 34·2 | 1·8 | 1·3 | 7·8 | 5·8 | 13·8 | 10·2 |
| 19 | 11 04·8 | 11 06·6 | 10 34·5 | 1·9 | 1·4 | 7·9 | 5·9 | 13·9 | 10·3 |
| 20 | 11 05·0 | 11 06·8 | 10 34·7 | 2·0 | 1·5 | 8·0 | 5·9 | 14·0 | 10·4 |
| 21 | 11 05·3 | 11 07·1 | 10 34·9 | 2·1 | 1·6 | 8·1 | 6·0 | 14·1 | 10·5 |
| 22 | 11 05·5 | 11 07·3 | 10 35·2 | 2·2 | 1·6 | 8·2 | 6·1 | 14·2 | 10·5 |
| 23 | 11 05·8 | 11 07·6 | 10 35·4 | 2·3 | 1·7 | 8·3 | 6·2 | 14·3 | 10·6 |
| 24 | 11 06·0 | 11 07·8 | 10 35·7 | 2·4 | 1·8 | 8·4 | 6·2 | 14·4 | 10·7 |
| 25 | 11 06·3 | 11 08·1 | 10 35·9 | 2·5 | 1·9 | 8·5 | 6·3 | 14·5 | 10·8 |
| 26 | 11 06·5 | 11 08·3 | 10 36·1 | 2·6 | 1·9 | 8·6 | 6·4 | 14·6 | 10·8 |
| 27 | 11 06·8 | 11 08·6 | 10 36·4 | 2·7 | 2·0 | 8·7 | 6·5 | 14·7 | 10·9 |
| 28 | 11 07·0 | 11 08·8 | 10 36·6 | 2·8 | 2·1 | 8·8 | 6·5 | 14·8 | 11·0 |
| 29 | 11 07·3 | 11 09·1 | 10 36·9 | 2·9 | 2·2 | 8·9 | 6·6 | 14·9 | 11·1 |
| 30 | 11 07·5 | 11 09·3 | 10 37·1 | 3·0 | 2·2 | 9·0 | 6·7 | 15·0 | 11·1 |
| 31 | 11 07·8 | 11 09·6 | 10 37·3 | 3·1 | 2·3 | 9·1 | 6·7 | 15·1 | 11·2 |
| 32 | 11 08·0 | 11 09·8 | 10 37·6 | 3·2 | 2·4 | 9·2 | 6·8 | 15·2 | 11·3 |
| 33 | 11 08·3 | 11 10·1 | 10 37·8 | 3·3 | 2·4 | 9·3 | 6·9 | 15·3 | 11·3 |
| 34 | 11 08·5 | 11 10·3 | 10 38·0 | 3·4 | 2·5 | 9·4 | 7·0 | 15·4 | 11·4 |
| 35 | 11 08·8 | 11 10·6 | 10 38·3 | 3·5 | 2·6 | 9·5 | 7·0 | 15·5 | 11·5 |
| 36 | 11 09·0 | 11 10·8 | 10 38·5 | 3·6 | 2·7 | 9·6 | 7·1 | 15·6 | 11·6 |
| 37 | 11 09·3 | 11 11·1 | 10 38·8 | 3·7 | 2·7 | 9·7 | 7·2 | 15·7 | 11·6 |
| 38 | 11 09·5 | 11 11·3 | 10 39·0 | 3·8 | 2·8 | 9·8 | 7·3 | 15·8 | 11·7 |
| 39 | 11 09·8 | 11 11·6 | 10 39·2 | 3·9 | 2·9 | 9·9 | 7·3 | 15·9 | 11·8 |
| 40 | 11 10·0 | 11 11·8 | 10 39·5 | 4·0 | 3·0 | 10·0 | 7·4 | 16·0 | 11·9 |
| 41 | 11 10·3 | 11 12·1 | 10 39·7 | 4·1 | 3·0 | 10·1 | 7·5 | 16·1 | 11·9 |
| 42 | 11 10·5 | 11 12·3 | 10 40·0 | 4·2 | 3·1 | 10·2 | 7·6 | 16·2 | 12·0 |
| 43 | 11 10·8 | 11 12·6 | 10 40·2 | 4·3 | 3·2 | 10·3 | 7·6 | 16·3 | 12·1 |
| 44 | 11 11·0 | 11 12·8 | 10 40·4 | 4·4 | 3·3 | 10·4 | 7·7 | 16·4 | 12·2 |
| 45 | 11 11·3 | 11 13·1 | 10 40·7 | 4·5 | 3·3 | 10·5 | 7·8 | 16·5 | 12·2 |
| 46 | 11 11·5 | 11 13·3 | 10 40·9 | 4·6 | 3·4 | 10·6 | 7·9 | 16·6 | 12·3 |
| 47 | 11 11·8 | 11 13·6 | 10 41·1 | 4·7 | 3·5 | 10·7 | 7·9 | 16·7 | 12·4 |
| 48 | 11 12·0 | 11 13·8 | 10 41·4 | 4·8 | 3·6 | 10·8 | 8·0 | 16·8 | 12·5 |
| 49 | 11 12·3 | 11 14·1 | 10 41·6 | 4·9 | 3·6 | 10·9 | 8·1 | 16·9 | 12·5 |
| 50 | 11 12·5 | 11 14·3 | 10 41·9 | 5·0 | 3·7 | 11·0 | 8·2 | 17·0 | 12·6 |
| 51 | 11 12·8 | 11 14·6 | 10 42·1 | 5·1 | 3·8 | 11·1 | 8·2 | 17·1 | 12·7 |
| 52 | 11 13·0 | 11 14·8 | 10 42·3 | 5·2 | 3·9 | 11·2 | 8·3 | 17·2 | 12·8 |
| 53 | 11 13·3 | 11 15·1 | 10 42·6 | 5·3 | 3·9 | 11·3 | 8·4 | 17·3 | 12·8 |
| 54 | 11 13·5 | 11 15·3 | 10 42·8 | 5·4 | 4·0 | 11·4 | 8·5 | 17·4 | 12·9 |
| 55 | 11 13·8 | 11 15·6 | 10 43·1 | 5·5 | 4·1 | 11·5 | 8·5 | 17·5 | 13·0 |
| 56 | 11 14·0 | 11 15·8 | 10 43·3 | 5·6 | 4·2 | 11·6 | 8·6 | 17·6 | 13·1 |
| 57 | 11 14·3 | 11 16·1 | 10 43·5 | 5·7 | 4·2 | 11·7 | 8·7 | 17·7 | 13·1 |
| 58 | 11 14·5 | 11 16·3 | 10 43·8 | 5·8 | 4·3 | 11·8 | 8·8 | 17·8 | 13·2 |
| 59 | 11 14·8 | 11 16·6 | 10 44·0 | 5·9 | 4·4 | 11·9 | 8·8 | 17·9 | 13·3 |
| 60 | 11 15·0 | 11 16·8 | 10 44·3 | 6·0 | 4·5 | 12·0 | 8·9 | 18·0 | 13·4 |

### 45ᵐ

| 45 | SUN PLANETS | ARIES | MOON | v or d | Corrⁿ | v or d | Corrⁿ | v or d | Corrⁿ |
|---|---|---|---|---|---|---|---|---|---|
| s | ° ′ | ° ′ | ° ′ | ′ | ′ | ′ | ′ | ′ | ′ |
| 00 | 11 15·0 | 11 16·8 | 10 44·3 | 0·0 | 0·0 | 6·0 | 4·6 | 12·0 | 9·1 |
| 01 | 11 15·3 | 11 17·1 | 10 44·5 | 0·1 | 0·1 | 6·1 | 4·6 | 12·1 | 9·2 |
| 02 | 11 15·5 | 11 17·3 | 10 44·7 | 0·2 | 0·2 | 6·2 | 4·7 | 12·2 | 9·3 |
| 03 | 11 15·8 | 11 17·6 | 10 45·0 | 0·3 | 0·2 | 6·3 | 4·8 | 12·3 | 9·3 |
| 04 | 11 16·0 | 11 17·9 | 10 45·2 | 0·4 | 0·3 | 6·4 | 4·9 | 12·4 | 9·4 |
| 05 | 11 16·3 | 11 18·1 | 10 45·4 | 0·5 | 0·4 | 6·5 | 4·9 | 12·5 | 9·5 |
| 06 | 11 16·5 | 11 18·4 | 10 45·7 | 0·6 | 0·5 | 6·6 | 5·0 | 12·6 | 9·6 |
| 07 | 11 16·8 | 11 18·6 | 10 45·9 | 0·7 | 0·5 | 6·7 | 5·1 | 12·7 | 9·6 |
| 08 | 11 17·0 | 11 18·9 | 10 46·2 | 0·8 | 0·6 | 6·8 | 5·2 | 12·8 | 9·7 |
| 09 | 11 17·3 | 11 19·1 | 10 46·4 | 0·9 | 0·7 | 6·9 | 5·2 | 12·9 | 9·8 |
| 10 | 11 17·5 | 11 19·4 | 10 46·6 | 1·0 | 0·8 | 7·0 | 5·3 | 13·0 | 9·9 |
| 11 | 11 17·8 | 11 19·6 | 10 46·9 | 1·1 | 0·8 | 7·1 | 5·4 | 13·1 | 9·9 |
| 12 | 11 18·0 | 11 19·9 | 10 47·1 | 1·2 | 0·9 | 7·2 | 5·5 | 13·2 | 10·0 |
| 13 | 11 18·3 | 11 20·1 | 10 47·4 | 1·3 | 1·0 | 7·3 | 5·5 | 13·3 | 10·1 |
| 14 | 11 18·5 | 11 20·4 | 10 47·6 | 1·4 | 1·1 | 7·4 | 5·6 | 13·4 | 10·2 |
| 15 | 11 18·8 | 11 20·6 | 10 47·8 | 1·5 | 1·1 | 7·5 | 5·7 | 13·5 | 10·2 |
| 16 | 11 19·0 | 11 20·9 | 10 48·1 | 1·6 | 1·2 | 7·6 | 5·8 | 13·6 | 10·3 |
| 17 | 11 19·3 | 11 21·1 | 10 48·3 | 1·7 | 1·3 | 7·7 | 5·8 | 13·7 | 10·4 |
| 18 | 11 19·5 | 11 21·4 | 10 48·5 | 1·8 | 1·4 | 7·8 | 5·9 | 13·8 | 10·5 |
| 19 | 11 19·8 | 11 21·6 | 10 48·8 | 1·9 | 1·4 | 7·9 | 6·0 | 13·9 | 10·5 |
| 20 | 11 20·0 | 11 21·9 | 10 49·0 | 2·0 | 1·5 | 8·0 | 6·1 | 14·0 | 10·6 |
| 21 | 11 20·3 | 11 22·1 | 10 49·3 | 2·1 | 1·6 | 8·1 | 6·1 | 14·1 | 10·7 |
| 22 | 11 20·5 | 11 22·4 | 10 49·5 | 2·2 | 1·7 | 8·2 | 6·2 | 14·2 | 10·8 |
| 23 | 11 20·8 | 11 22·6 | 10 49·7 | 2·3 | 1·7 | 8·3 | 6·3 | 14·3 | 10·8 |
| 24 | 11 21·0 | 11 22·9 | 10 50·0 | 2·4 | 1·8 | 8·4 | 6·4 | 14·4 | 10·9 |
| 25 | 11 21·3 | 11 23·1 | 10 50·2 | 2·5 | 1·9 | 8·5 | 6·4 | 14·5 | 11·0 |
| 26 | 11 21·5 | 11 23·4 | 10 50·5 | 2·6 | 2·0 | 8·6 | 6·5 | 14·6 | 11·1 |
| 27 | 11 21·8 | 11 23·6 | 10 50·7 | 2·7 | 2·0 | 8·7 | 6·6 | 14·7 | 11·1 |
| 28 | 11 22·0 | 11 23·9 | 10 50·9 | 2·8 | 2·1 | 8·8 | 6·7 | 14·8 | 11·2 |
| 29 | 11 22·3 | 11 24·1 | 10 51·2 | 2·9 | 2·2 | 8·9 | 6·7 | 14·9 | 11·3 |
| 30 | 11 22·5 | 11 24·4 | 10 51·4 | 3·0 | 2·3 | 9·0 | 6·8 | 15·0 | 11·4 |
| 31 | 11 22·8 | 11 24·6 | 10 51·6 | 3·1 | 2·4 | 9·1 | 6·9 | 15·1 | 11·5 |
| 32 | 11 23·0 | 11 24·9 | 10 51·9 | 3·2 | 2·4 | 9·2 | 7·0 | 15·2 | 11·5 |
| 33 | 11 23·3 | 11 25·1 | 10 52·1 | 3·3 | 2·5 | 9·3 | 7·1 | 15·3 | 11·6 |
| 34 | 11 23·5 | 11 25·4 | 10 52·4 | 3·4 | 2·6 | 9·4 | 7·1 | 15·4 | 11·7 |
| 35 | 11 23·8 | 11 25·6 | 10 52·6 | 3·5 | 2·7 | 9·5 | 7·2 | 15·5 | 11·8 |
| 36 | 11 24·0 | 11 25·9 | 10 52·8 | 3·6 | 2·7 | 9·6 | 7·3 | 15·6 | 11·8 |
| 37 | 11 24·3 | 11 26·1 | 10 53·1 | 3·7 | 2·8 | 9·7 | 7·4 | 15·7 | 11·9 |
| 38 | 11 24·5 | 11 26·4 | 10 53·3 | 3·8 | 2·9 | 9·8 | 7·4 | 15·8 | 12·0 |
| 39 | 11 24·8 | 11 26·6 | 10 53·6 | 3·9 | 3·0 | 9·9 | 7·5 | 15·9 | 12·1 |
| 40 | 11 25·0 | 11 26·9 | 10 53·8 | 4·0 | 3·0 | 10·0 | 7·6 | 16·0 | 12·1 |
| 41 | 11 25·3 | 11 27·1 | 10 54·0 | 4·1 | 3·1 | 10·1 | 7·7 | 16·1 | 12·2 |
| 42 | 11 25·5 | 11 27·4 | 10 54·3 | 4·2 | 3·2 | 10·2 | 7·7 | 16·2 | 12·3 |
| 43 | 11 25·8 | 11 27·6 | 10 54·5 | 4·3 | 3·3 | 10·3 | 7·8 | 16·3 | 12·4 |
| 44 | 11 26·0 | 11 27·9 | 10 54·7 | 4·4 | 3·3 | 10·4 | 7·9 | 16·4 | 12·4 |
| 45 | 11 26·3 | 11 28·1 | 10 55·0 | 4·5 | 3·4 | 10·5 | 8·0 | 16·5 | 12·5 |
| 46 | 11 26·5 | 11 28·4 | 10 55·2 | 4·6 | 3·5 | 10·6 | 8·0 | 16·6 | 12·6 |
| 47 | 11 26·8 | 11 28·6 | 10 55·5 | 4·7 | 3·6 | 10·7 | 8·1 | 16·7 | 12·7 |
| 48 | 11 27·0 | 11 28·9 | 10 55·7 | 4·8 | 3·6 | 10·8 | 8·2 | 16·8 | 12·7 |
| 49 | 11 27·3 | 11 29·1 | 10 55·9 | 4·9 | 3·7 | 10·9 | 8·3 | 16·9 | 12·8 |
| 50 | 11 27·5 | 11 29·4 | 10 56·2 | 5·0 | 3·8 | 11·0 | 8·3 | 17·0 | 12·9 |
| 51 | 11 27·8 | 11 29·6 | 10 56·4 | 5·1 | 3·9 | 11·1 | 8·4 | 17·1 | 13·0 |
| 52 | 11 28·0 | 11 29·9 | 10 56·7 | 5·2 | 3·9 | 11·2 | 8·5 | 17·2 | 13·0 |
| 53 | 11 28·3 | 11 30·1 | 10 56·9 | 5·3 | 4·0 | 11·3 | 8·6 | 17·3 | 13·1 |
| 54 | 11 28·5 | 11 30·4 | 10 57·1 | 5·4 | 4·1 | 11·4 | 8·6 | 17·4 | 13·2 |
| 55 | 11 28·8 | 11 30·6 | 10 57·4 | 5·5 | 4·2 | 11·5 | 8·7 | 17·5 | 13·3 |
| 56 | 11 29·0 | 11 30·9 | 10 57·6 | 5·6 | 4·2 | 11·6 | 8·8 | 17·6 | 13·3 |
| 57 | 11 29·3 | 11 31·1 | 10 57·9 | 5·7 | 4·3 | 11·7 | 8·9 | 17·7 | 13·4 |
| 58 | 11 29·5 | 11 31·4 | 10 58·1 | 5·8 | 4·4 | 11·8 | 8·9 | 17·8 | 13·5 |
| 59 | 11 29·8 | 11 31·6 | 10 58·3 | 5·9 | 4·5 | 11·9 | 9·0 | 17·9 | 13·6 |
| 60 | 11 30·0 | 11 31·9 | 10 58·6 | 6·0 | 4·6 | 12·0 | 9·1 | 18·0 | 13·7 |

## 50ᵐ INCREMENTS AND CORRECTIONS 51ᵐ

| 50 | SUN PLANETS | ARIES | MOON | v or Corrⁿ d | v or Corrⁿ d | v or Corrⁿ d | 51 | SUN PLANETS | ARIES | MOON | v or Corrⁿ d | v or Corrⁿ d | v or Corrⁿ d |
|---|---|---|---|---|---|---|---|---|---|---|---|---|---|
| s | ° ′ | ° ′ | ° ′ | ′ ′ | ′ ′ | ′ ′ | s | ° ′ | ° ′ | ° ′ | ′ ′ | ′ ′ | ′ ′ |
| 00 | 12 30·0 | 12 32·1 | 11 55·8 | 0·0 0·0 | 6·0 5·1 | 12·0 10·1 | 00 | 12 45·0 | 12 47·1 | 12 10·2 | 0·0 0·0 | 6·0 5·2 | 12·0 10·3 |
| 01 | 12 30·3 | 12 32·3 | 11 56·1 | 0·1 0·1 | 6·1 5·1 | 12·1 10·2 | 01 | 12 45·3 | 12 47·3 | 12 10·4 | 0·1 0·1 | 6·1 5·2 | 12·1 10·4 |
| 02 | 12 30·5 | 12 32·6 | 11 56·3 | 0·2 0·2 | 6·2 5·2 | 12·2 10·3 | 02 | 12 45·5 | 12 47·6 | 12 10·6 | 0·2 0·2 | 6·2 5·3 | 12·2 10·5 |
| 03 | 12 30·8 | 12 32·8 | 11 56·5 | 0·3 0·3 | 6·3 5·3 | 12·3 10·4 | 03 | 12 45·8 | 12 47·8 | 12 10·9 | 0·3 0·3 | 6·3 5·4 | 12·3 10·6 |
| 04 | 12 31·0 | 12 33·1 | 11 56·8 | 0·4 0·3 | 6·4 5·4 | 12·4 10·4 | 04 | 12 46·0 | 12 48·1 | 12 11·1 | 0·4 0·3 | 6·4 5·5 | 12·4 10·6 |
| 05 | 12 31·3 | 12 33·3 | 11 57·0 | 0·5 0·4 | 6·5 5·5 | 12·5 10·5 | 05 | 12 46·3 | 12 48·3 | 12 11·3 | 0·5 0·4 | 6·5 5·6 | 12·5 10·7 |
| 06 | 12 31·5 | 12 33·6 | 11 57·3 | 0·6 0·5 | 6·6 5·6 | 12·6 10·6 | 06 | 12 46·5 | 12 48·6 | 12 11·6 | 0·6 0·5 | 6·6 5·7 | 12·6 10·8 |
| 07 | 12 31·8 | 12 33·8 | 11 57·5 | 0·7 0·6 | 6·7 5·6 | 12·7 10·7 | 07 | 12 46·8 | 12 48·8 | 12 11·8 | 0·7 0·6 | 6·7 5·8 | 12·7 10·9 |
| 08 | 12 32·0 | 12 34·1 | 11 57·7 | 0·8 0·7 | 6·8 5·7 | 12·8 10·8 | 08 | 12 47·0 | 12 49·1 | 12 12·1 | 0·8 0·7 | 6·8 5·8 | 12·8 11·0 |
| 09 | 12 32·3 | 12 34·3 | 11 58·0 | 0·9 0·8 | 6·9 5·8 | 12·9 10·9 | 09 | 12 47·3 | 12 49·4 | 12 12·3 | 0·9 0·8 | 6·9 5·9 | 12·9 11·1 |
| 10 | 12 32·5 | 12 34·6 | 11 58·2 | 1·0 0·8 | 7·0 5·9 | 13·0 10·9 | 10 | 12 47·5 | 12 49·6 | 12 12·5 | 1·0 0·9 | 7·0 6·0 | 13·0 11·2 |
| 11 | 12 32·8 | 12 34·8 | 11 58·5 | 1·1 0·9 | 7·1 6·0 | 13·1 11·0 | 11 | 12 47·8 | 12 49·9 | 12 12·8 | 1·1 0·9 | 7·1 6·1 | 13·1 11·2 |
| 12 | 12 33·0 | 12 35·1 | 11 58·7 | 1·2 1·0 | 7·2 6·1 | 13·2 11·1 | 12 | 12 48·0 | 12 50·1 | 12 13·0 | 1·2 1·0 | 7·2 6·2 | 13·2 11·3 |
| 13 | 12 33·3 | 12 35·3 | 11 58·9 | 1·3 1·1 | 7·3 6·1 | 13·3 11·2 | 13 | 12 48·3 | 12 50·4 | 12 13·3 | 1·3 1·1 | 7·3 6·3 | 13·3 11·4 |
| 14 | 12 33·5 | 12 35·6 | 11 59·2 | 1·4 1·2 | 7·4 6·2 | 13·4 11·3 | 14 | 12 48·5 | 12 50·6 | 12 13·5 | 1·4 1·2 | 7·4 6·4 | 13·4 11·5 |
| 15 | 12 33·8 | 12 35·8 | 11 59·4 | 1·5 1·3 | 7·5 6·3 | 13·5 11·4 | 15 | 12 48·8 | 12 50·9 | 12 13·7 | 1·5 1·3 | 7·5 6·4 | 13·5 11·6 |
| 16 | 12 34·0 | 12 36·1 | 11 59·7 | 1·6 1·3 | 7·6 6·4 | 13·6 11·4 | 16 | 12 49·0 | 12 51·1 | 12 14·0 | 1·6 1·4 | 7·6 6·5 | 13·6 11·7 |
| 17 | 12 34·3 | 12 36·3 | 11 59·9 | 1·7 1·4 | 7·7 6·5 | 13·7 11·5 | 17 | 12 49·3 | 12 51·4 | 12 14·2 | 1·7 1·5 | 7·7 6·6 | 13·7 11·8 |
| 18 | 12 34·5 | 12 36·6 | 12 00·1 | 1·8 1·5 | 7·8 6·6 | 13·8 11·6 | 18 | 12 49·5 | 12 51·6 | 12 14·4 | 1·8 1·5 | 7·8 6·7 | 13·8 11·8 |
| 19 | 12 34·8 | 12 36·8 | 12 00·4 | 1·9 1·6 | 7·9 6·6 | 13·9 11·7 | 19 | 12 49·8 | 12 51·9 | 12 14·7 | 1·9 1·6 | 7·9 6·8 | 13·9 11·9 |
| 20 | 12 35·0 | 12 37·1 | 12 00·6 | 2·0 1·7 | 8·0 6·7 | 14·0 11·8 | 20 | 12 50·0 | 12 52·1 | 12 14·9 | 2·0 1·7 | 8·0 6·9 | 14·0 12·0 |
| 21 | 12 35·3 | 12 37·3 | 12 00·8 | 2·1 1·8 | 8·1 6·8 | 14·1 11·9 | 21 | 12 50·3 | 12 52·4 | 12 15·2 | 2·1 1·8 | 8·1 7·0 | 14·1 12·1 |
| 22 | 12 35·5 | 12 37·6 | 12 01·1 | 2·2 1·9 | 8·2 6·9 | 14·2 12·0 | 22 | 12 50·5 | 12 52·6 | 12 15·4 | 2·2 1·9 | 8·2 7·0 | 14·2 12·2 |
| 23 | 12 35·8 | 12 37·8 | 12 01·3 | 2·3 1·9 | 8·3 7·0 | 14·3 12·0 | 23 | 12 50·8 | 12 52·9 | 12 15·6 | 2·3 2·0 | 8·3 7·1 | 14·3 12·3 |
| 24 | 12 36·0 | 12 38·1 | 12 01·6 | 2·4 2·0 | 8·4 7·1 | 14·4 12·1 | 24 | 12 51·0 | 12 53·1 | 12 15·9 | 2·4 2·1 | 8·4 7·2 | 14·4 12·4 |
| 25 | 12 36·3 | 12 38·3 | 12 01·8 | 2·5 2·1 | 8·5 7·2 | 14·5 12·2 | 25 | 12 51·3 | 12 53·4 | 12 16·1 | 2·5 2·1 | 8·5 7·3 | 14·5 12·4 |
| 26 | 12 36·5 | 12 38·6 | 12 02·0 | 2·6 2·2 | 8·6 7·2 | 14·6 12·3 | 26 | 12 51·5 | 12 53·6 | 12 16·4 | 2·6 2·2 | 8·6 7·4 | 14·6 12·5 |
| 27 | 12 36·8 | 12 38·8 | 12 02·3 | 2·7 2·3 | 8·7 7·3 | 14·7 12·4 | 27 | 12 51·8 | 12 53·9 | 12 16·6 | 2·7 2·3 | 8·7 7·5 | 14·7 12·6 |
| 28 | 12 37·0 | 12 39·1 | 12 02·5 | 2·8 2·4 | 8·8 7·4 | 14·8 12·5 | 28 | 12 52·0 | 12 54·1 | 12 16·8 | 2·8 2·4 | 8·8 7·6 | 14·8 12·7 |
| 29 | 12 37·3 | 12 39·3 | 12 02·8 | 2·9 2·4 | 8·9 7·5 | 14·9 12·5 | 29 | 12 52·3 | 12 54·4 | 12 17·1 | 2·9 2·5 | 8·9 7·6 | 14·9 12·8 |
| 30 | 12 37·5 | 12 39·6 | 12 03·0 | 3·0 2·5 | 9·0 7·6 | 15·0 12·6 | 30 | 12 52·5 | 12 54·6 | 12 17·3 | 3·0 2·6 | 9·0 7·7 | 15·0 12·9 |
| 31 | 12 37·8 | 12 39·8 | 12 03·2 | 3·1 2·6 | 9·1 7·7 | 15·1 12·7 | 31 | 12 52·8 | 12 54·9 | 12 17·5 | 3·1 2·7 | 9·1 7·8 | 15·1 13·0 |
| 32 | 12 38·0 | 12 40·1 | 12 03·5 | 3·2 2·7 | 9·2 7·7 | 15·2 12·8 | 32 | 12 53·0 | 12 55·1 | 12 17·8 | 3·2 2·7 | 9·2 7·9 | 15·2 13·0 |
| 33 | 12 38·3 | 12 40·3 | 12 03·7 | 3·3 2·8 | 9·3 7·8 | 15·3 12·9 | 33 | 12 53·3 | 12 55·4 | 12 18·0 | 3·3 2·8 | 9·3 8·0 | 15·3 13·1 |
| 34 | 12 38·5 | 12 40·6 | 12 03·9 | 3·4 2·9 | 9·4 7·9 | 15·4 13·0 | 34 | 12 53·5 | 12 55·6 | 12 18·3 | 3·4 2·9 | 9·4 8·1 | 15·4 13·2 |
| 35 | 12 38·8 | 12 40·8 | 12 04·2 | 3·5 2·9 | 9·5 8·0 | 15·5 13·0 | 35 | 12 53·8 | 12 55·9 | 12 18·5 | 3·5 3·0 | 9·5 8·2 | 15·5 13·3 |
| 36 | 12 39·0 | 12 41·1 | 12 04·4 | 3·6 3·0 | 9·6 8·1 | 15·6 13·1 | 36 | 12 54·0 | 12 56·1 | 12 18·7 | 3·6 3·1 | 9·6 8·2 | 15·6 13·4 |
| 37 | 12 39·3 | 12 41·3 | 12 04·7 | 3·7 3·1 | 9·7 8·2 | 15·7 13·2 | 37 | 12 54·3 | 12 56·4 | 12 19·0 | 3·7 3·2 | 9·7 8·3 | 15·7 13·5 |
| 38 | 12 39·5 | 12 41·6 | 12 04·9 | 3·8 3·2 | 9·8 8·2 | 15·8 13·3 | 38 | 12 54·5 | 12 56·6 | 12 19·2 | 3·8 3·3 | 9·8 8·4 | 15·8 13·6 |
| 39 | 12 39·8 | 12 41·8 | 12 05·1 | 3·9 3·3 | 9·9 8·3 | 15·9 13·4 | 39 | 12 54·8 | 12 56·9 | 12 19·5 | 3·9 3·3 | 9·9 8·5 | 15·9 13·6 |
| 40 | 12 40·0 | 12 42·1 | 12 05·4 | 4·0 3·4 | 10·0 8·4 | 16·0 13·5 | 40 | 12 55·0 | 12 57·1 | 12 19·7 | 4·0 3·4 | 10·0 8·6 | 16·0 13·7 |
| 41 | 12 40·3 | 12 42·3 | 12 05·6 | 4·1 3·5 | 10·1 8·5 | 16·1 13·6 | 41 | 12 55·3 | 12 57·4 | 12 19·9 | 4·1 3·5 | 10·1 8·7 | 16·1 13·8 |
| 42 | 12 40·5 | 12 42·6 | 12 05·9 | 4·2 3·5 | 10·2 8·6 | 16·2 13·6 | 42 | 12 55·5 | 12 57·6 | 12 20·2 | 4·2 3·6 | 10·2 8·8 | 16·2 13·9 |
| 43 | 12 40·8 | 12 42·8 | 12 06·1 | 4·3 3·6 | 10·3 8·7 | 16·3 13·7 | 43 | 12 55·8 | 12 57·9 | 12 20·4 | 4·3 3·7 | 10·3 8·8 | 16·3 14·0 |
| 44 | 12 41·0 | 12 43·1 | 12 06·3 | 4·4 3·7 | 10·4 8·8 | 16·4 13·8 | 44 | 12 56·0 | 12 58·1 | 12 20·6 | 4·4 3·8 | 10·4 8·9 | 16·4 14·1 |
| 45 | 12 41·3 | 12 43·3 | 12 06·6 | 4·5 3·8 | 10·5 8·8 | 16·5 13·9 | 45 | 12 56·3 | 12 58·4 | 12 20·9 | 4·5 3·9 | 10·5 9·0 | 16·5 14·2 |
| 46 | 12 41·5 | 12 43·6 | 12 06·8 | 4·6 3·9 | 10·6 8·9 | 16·6 14·0 | 46 | 12 56·5 | 12 58·6 | 12 21·1 | 4·6 3·9 | 10·6 9·1 | 16·6 14·2 |
| 47 | 12 41·8 | 12 43·8 | 12 07·0 | 4·7 4·0 | 10·7 9·0 | 16·7 14·1 | 47 | 12 56·8 | 12 58·9 | 12 21·4 | 4·7 4·0 | 10·7 9·2 | 16·7 14·3 |
| 48 | 12 42·0 | 12 44·1 | 12 07·3 | 4·8 4·0 | 10·8 9·1 | 16·8 14·1 | 48 | 12 57·0 | 12 59·1 | 12 21·6 | 4·8 4·1 | 10·8 9·3 | 16·8 14·4 |
| 49 | 12 42·3 | 12 44·3 | 12 07·5 | 4·9 4·1 | 10·9 9·2 | 16·9 14·2 | 49 | 12 57·3 | 12 59·4 | 12 21·8 | 4·9 4·2 | 10·9 9·4 | 16·9 14·5 |
| 50 | 12 42·5 | 12 44·6 | 12 07·8 | 5·0 4·2 | 11·0 9·3 | 17·0 14·3 | 50 | 12 57·5 | 12 59·6 | 12 22·1 | 5·0 4·3 | 11·0 9·4 | 17·0 14·6 |
| 51 | 12 42·8 | 12 44·8 | 12 08·0 | 5·1 4·3 | 11·1 9·3 | 17·1 14·4 | 51 | 12 57·8 | 12 59·9 | 12 22·3 | 5·1 4·4 | 11·1 9·5 | 17·1 14·7 |
| 52 | 12 43·0 | 12 45·1 | 12 08·2 | 5·2 4·4 | 11·2 9·4 | 17·2 14·5 | 52 | 12 58·0 | 13 00·1 | 12 22·6 | 5·2 4·5 | 11·2 9·6 | 17·2 14·8 |
| 53 | 12 43·3 | 12 45·3 | 12 08·5 | 5·3 4·5 | 11·3 9·5 | 17·3 14·6 | 53 | 12 58·3 | 13 00·4 | 12 22·8 | 5·3 4·5 | 11·3 9·7 | 17·3 14·8 |
| 54 | 12 43·5 | 12 45·6 | 12 08·7 | 5·4 4·5 | 11·4 9·6 | 17·4 14·6 | 54 | 12 58·5 | 13 00·6 | 12 23·0 | 5·4 4·6 | 11·4 9·8 | 17·4 14·9 |
| 55 | 12 43·8 | 12 45·8 | 12 09·0 | 5·5 4·6 | 11·5 9·7 | 17·5 14·7 | 55 | 12 58·8 | 13 00·9 | 12 23·3 | 5·5 4·7 | 11·5 9·9 | 17·5 15·0 |
| 56 | 12 44·0 | 12 46·1 | 12 09·2 | 5·6 4·7 | 11·6 9·8 | 17·6 14·8 | 56 | 12 59·0 | 13 01·1 | 12 23·5 | 5·6 4·8 | 11·6 10·0 | 17·6 15·1 |
| 57 | 12 44·3 | 12 46·3 | 12 09·4 | 5·7 4·8 | 11·7 9·8 | 17·7 14·9 | 57 | 12 59·3 | 13 01·4 | 12 23·8 | 5·7 4·9 | 11·7 10·0 | 17·7 15·2 |
| 58 | 12 44·5 | 12 46·6 | 12 09·7 | 5·8 4·9 | 11·8 9·9 | 17·8 15·0 | 58 | 12 59·5 | 13 01·6 | 12 24·0 | 5·8 5·0 | 11·8 10·1 | 17·8 15·3 |
| 59 | 12 44·8 | 12 46·8 | 12 09·9 | 5·9 5·0 | 11·9 10·0 | 17·9 15·1 | 59 | 12 59·8 | 13 01·9 | 12 24·2 | 5·9 5·1 | 11·9 10·2 | 17·9 15·4 |
| 60 | 12 45·0 | 12 47·1 | 12 10·2 | 6·0 5·1 | 12·0 10·1 | 18·0 15·2 | 60 | 13 00·0 | 13 02·1 | 12 24·5 | 6·0 5·2 | 12·0 10·3 | 18·0 15·5 |

## INCREMENTS AND CORRECTIONS

### 58m

| 58 | SUN PLANETS | ARIES | MOON | v or Corrn d | v or Corrn d | v or Corrn d |
|---|---|---|---|---|---|---|
| s | ° ' | ° ' | ° ' | ' ' | ' ' | ' ' |
| 00 | 14 30·0 | 14 32·4 | 13 50·4 | 0·0 0·0 | 6·0 5·9 | 12·0 11·7 |
| 01 | 14 30·3 | 14 32·6 | 13 50·6 | 0·1 0·1 | 6·1 5·9 | 12·1 11·8 |
| 02 | 14 30·5 | 14 32·9 | 13 50·8 | 0·2 0·2 | 6·2 6·0 | 12·2 11·9 |
| 03 | 14 30·8 | 14 33·1 | 13 51·1 | 0·3 0·3 | 6·3 6·1 | 12·3 12·0 |
| 04 | 14 31·0 | 14 33·4 | 13 51·3 | 0·4 0·4 | 6·4 6·2 | 12·4 12·1 |
| 05 | 14 31·3 | 14 33·6 | 13 51·6 | 0·5 0·5 | 6·5 6·3 | 12·5 12·2 |
| 06 | 14 31·5 | 14 33·9 | 13 51·8 | 0·6 0·6 | 6·6 6·4 | 12·6 12·3 |
| 07 | 14 31·8 | 14 34·1 | 13 52·0 | 0·7 0·7 | 6·7 6·5 | 12·7 12·4 |
| 08 | 14 32·0 | 14 34·4 | 13 52·3 | 0·8 0·8 | 6·8 6·6 | 12·8 12·5 |
| 09 | 14 32·3 | 14 34·6 | 13 52·5 | 0·9 0·9 | 6·9 6·7 | 12·9 12·6 |
| 10 | 14 32·5 | 14 34·9 | 13 52·8 | 1·0 1·0 | 7·0 6·8 | 13·0 12·7 |
| 11 | 14 32·8 | 14 35·1 | 13 53·0 | 1·1 1·1 | 7·1 6·9 | 13·1 12·8 |
| 12 | 14 33·0 | 14 35·4 | 13 53·2 | 1·2 1·2 | 7·2 7·0 | 13·2 12·9 |
| 13 | 14 33·3 | 14 35·6 | 13 53·5 | 1·3 1·3 | 7·3 7·1 | 13·3 13·0 |
| 14 | 14 33·5 | 14 35·9 | 13 53·7 | 1·4 1·4 | 7·4 7·2 | 13·4 13·1 |
| 15 | 14 33·8 | 14 36·1 | 13 53·9 | 1·5 1·5 | 7·5 7·3 | 13·5 13·2 |
| 16 | 14 34·0 | 14 36·4 | 13 54·2 | 1·6 1·6 | 7·6 7·4 | 13·6 13·3 |
| 17 | 14 34·3 | 14 36·6 | 13 54·4 | 1·7 1·7 | 7·7 7·5 | 13·7 13·4 |
| 18 | 14 34·5 | 14 36·9 | 13 54·7 | 1·8 1·8 | 7·8 7·6 | 13·8 13·5 |
| 19 | 14 34·8 | 14 37·1 | 13 54·9 | 1·9 1·9 | 7·9 7·7 | 13·9 13·6 |
| 20 | 14 35·0 | 14 37·4 | 13 55·1 | 2·0 2·0 | 8·0 7·8 | 14·0 13·7 |
| 21 | 14 35·3 | 14 37·6 | 13 55·4 | 2·1 2·0 | 8·1 7·9 | 14·1 13·7 |
| 22 | 14 35·5 | 14 37·9 | 13 55·6 | 2·2 2·1 | 8·2 8·0 | 14·2 13·8 |
| 23 | 14 35·8 | 14 38·1 | 13 55·9 | 2·3 2·2 | 8·3 8·1 | 14·3 13·9 |
| 24 | 14 36·0 | 14 38·4 | 13 56·1 | 2·4 2·3 | 8·4 8·2 | 14·4 14·0 |
| 25 | 14 36·3 | 14 38·6 | 13 56·3 | 2·5 2·4 | 8·5 8·3 | 14·5 14·1 |
| 26 | 14 36·5 | 14 38·9 | 13 56·6 | 2·6 2·5 | 8·6 8·4 | 14·6 14·2 |
| 27 | 14 36·8 | 14 39·2 | 13 56·8 | 2·7 2·6 | 8·7 8·5 | 14·7 14·3 |
| 28 | 14 37·0 | 14 39·4 | 13 57·0 | 2·8 2·7 | 8·8 8·6 | 14·8 14·4 |
| 29 | 14 37·3 | 14 39·7 | 13 57·3 | 2·9 2·8 | 8·9 8·7 | 14·9 14·5 |
| 30 | 14 37·5 | 14 39·9 | 13 57·5 | 3·0 2·9 | 9·0 8·8 | 15·0 14·6 |
| 31 | 14 37·8 | 14 40·2 | 13 57·8 | 3·1 3·0 | 9·1 8·9 | 15·1 14·7 |
| 32 | 14 38·0 | 14 40·4 | 13 58·0 | 3·2 3·1 | 9·2 9·0 | 15·2 14·8 |
| 33 | 14 38·3 | 14 40·7 | 13 58·2 | 3·3 3·2 | 9·3 9·1 | 15·3 14·9 |
| 34 | 14 38·5 | 14 40·9 | 13 58·5 | 3·4 3·3 | 9·4 9·2 | 15·4 15·0 |
| 35 | 14 38·8 | 14 41·2 | 13 58·7 | 3·5 3·4 | 9·5 9·3 | 15·5 15·1 |
| 36 | 14 39·0 | 14 41·4 | 13 59·0 | 3·6 3·5 | 9·6 9·4 | 15·6 15·2 |
| 37 | 14 39·3 | 14 41·7 | 13 59·2 | 3·7 3·6 | 9·7 9·5 | 15·7 15·3 |
| 38 | 14 39·5 | 14 41·9 | 13 59·4 | 3·8 3·7 | 9·8 9·6 | 15·8 15·4 |
| 39 | 14 39·8 | 14 42·2 | 13 59·7 | 3·9 3·8 | 9·9 9·7 | 15·9 15·5 |
| 40 | 14 40·0 | 14 42·4 | 13 59·9 | 4·0 3·9 | 10·0 9·8 | 16·0 15·6 |
| 41 | 14 40·3 | 14 42·7 | 14 00·1 | 4·1 4·0 | 10·1 9·8 | 16·1 15·7 |
| 42 | 14 40·5 | 14 42·9 | 14 00·4 | 4·2 4·1 | 10·2 9·9 | 16·2 15·8 |
| 43 | 14 40·8 | 14 43·2 | 14 00·6 | 4·3 4·2 | 10·3 10·0 | 16·3 15·9 |
| 44 | 14 41·0 | 14 43·4 | 14 00·9 | 4·4 4·3 | 10·4 10·1 | 16·4 16·0 |
| 45 | 14 41·3 | 14 43·7 | 14 01·1 | 4·5 4·4 | 10·5 10·2 | 16·5 16·1 |
| 46 | 14 41·5 | 14 43·9 | 14 01·3 | 4·6 4·5 | 10·6 10·3 | 16·6 16·2 |
| 47 | 14 41·8 | 14 44·2 | 14 01·6 | 4·7 4·6 | 10·7 10·4 | 16·7 16·3 |
| 48 | 14 42·0 | 14 44·4 | 14 01·8 | 4·8 4·7 | 10·8 10·5 | 16·8 16·4 |
| 49 | 14 42·3 | 14 44·7 | 14 02·1 | 4·9 4·8 | 10·9 10·6 | 16·9 16·5 |
| 50 | 14 42·5 | 14 44·9 | 14 02·3 | 5·0 4·9 | 11·0 10·7 | 17·0 16·6 |
| 51 | 14 42·8 | 14 45·2 | 14 02·5 | 5·1 5·0 | 11·1 10·8 | 17·1 16·7 |
| 52 | 14 43·0 | 14 45·4 | 14 02·8 | 5·2 5·1 | 11·2 10·9 | 17·2 16·8 |
| 53 | 14 43·3 | 14 45·7 | 14 03·0 | 5·3 5·2 | 11·3 11·0 | 17·3 16·9 |
| 54 | 14 43·5 | 14 45·9 | 14 03·3 | 5·4 5·3 | 11·4 11·1 | 17·4 17·0 |
| 55 | 14 43·8 | 14 46·2 | 14 03·5 | 5·5 5·4 | 11·5 11·2 | 17·5 17·1 |
| 56 | 14 44·0 | 14 46·4 | 14 03·7 | 5·6 5·5 | 11·6 11·3 | 17·6 17·2 |
| 57 | 14 44·3 | 14 46·7 | 14 04·0 | 5·7 5·6 | 11·7 11·4 | 17·7 17·3 |
| 58 | 14 44·5 | 14 46·9 | 14 04·2 | 5·8 5·7 | 11·8 11·5 | 17·8 17·4 |
| 59 | 14 44·8 | 14 47·2 | 14 04·4 | 5·9 5·8 | 11·9 11·6 | 17·9 17·5 |
| 60 | 14 45·0 | 14 47·4 | 14 04·7 | 6·0 5·9 | 12·0 11·7 | 18·0 17·6 |

### 59m

| 59 | SUN PLANETS | ARIES | MOON | v or Corrn d | v or Corrn d | v or Corrn d |
|---|---|---|---|---|---|---|
| s | ° ' | ° ' | ° ' | ' ' | ' ' | ' ' |
| 00 | 14 45·0 | 14 47·4 | 14 04·7 | 0·0 0·0 | 6·0 6·0 | 12·0 11·9 |
| 01 | 14 45·3 | 14 47·7 | 14 04·9 | 0·1 0·1 | 6·1 6·0 | 12·1 12·0 |
| 02 | 14 45·5 | 14 47·9 | 14 05·2 | 0·2 0·2 | 6·2 6·1 | 12·2 12·1 |
| 03 | 14 45·8 | 14 48·2 | 14 05·4 | 0·3 0·3 | 6·3 6·2 | 12·3 12·2 |
| 04 | 14 46·0 | 14 48·4 | 14 05·6 | 0·4 0·4 | 6·4 6·3 | 12·4 12·3 |
| 05 | 14 46·3 | 14 48·7 | 14 05·9 | 0·5 0·5 | 6·5 6·4 | 12·5 12·4 |
| 06 | 14 46·5 | 14 48·9 | 14 06·1 | 0·6 0·6 | 6·6 6·5 | 12·6 12·5 |
| 07 | 14 46·8 | 14 49·2 | 14 06·4 | 0·7 0·7 | 6·7 6·6 | 12·7 12·6 |
| 08 | 14 47·0 | 14 49·4 | 14 06·6 | 0·8 0·8 | 6·8 6·7 | 12·8 12·7 |
| 09 | 14 47·3 | 14 49·7 | 14 06·8 | 0·9 0·9 | 6·9 6·8 | 12·9 12·8 |
| 10 | 14 47·5 | 14 49·9 | 14 07·1 | 1·0 1·0 | 7·0 6·9 | 13·0 12·9 |
| 11 | 14 47·8 | 14 50·2 | 14 07·3 | 1·1 1·1 | 7·1 7·0 | 13·1 13·0 |
| 12 | 14 48·0 | 14 50·4 | 14 07·5 | 1·2 1·2 | 7·2 7·1 | 13·2 13·1 |
| 13 | 14 48·3 | 14 50·7 | 14 07·8 | 1·3 1·3 | 7·3 7·2 | 13·3 13·2 |
| 14 | 14 48·5 | 14 50·9 | 14 08·0 | 1·4 1·4 | 7·4 7·3 | 13·4 13·3 |
| 15 | 14 48·8 | 14 51·2 | 14 08·3 | 1·5 1·5 | 7·5 7·4 | 13·5 13·4 |
| 16 | 14 49·0 | 14 51·4 | 14 08·5 | 1·6 1·6 | 7·6 7·5 | 13·6 13·5 |
| 17 | 14 49·3 | 14 51·7 | 14 08·7 | 1·7 1·7 | 7·7 7·6 | 13·7 13·6 |
| 18 | 14 49·5 | 14 51·9 | 14 09·0 | 1·8 1·8 | 7·8 7·7 | 13·8 13·7 |
| 19 | 14 49·8 | 14 52·2 | 14 09·2 | 1·9 1·9 | 7·9 7·8 | 13·9 13·8 |
| 20 | 14 50·0 | 14 52·4 | 14 09·5 | 2·0 2·0 | 8·0 7·9 | 14·0 13·9 |
| 21 | 14 50·3 | 14 52·7 | 14 09·7 | 2·1 2·1 | 8·1 8·0 | 14·1 14·0 |
| 22 | 14 50·5 | 14 52·9 | 14 09·9 | 2·2 2·2 | 8·2 8·1 | 14·2 14·1 |
| 23 | 14 50·8 | 14 53·2 | 14 10·2 | 2·3 2·3 | 8·3 8·2 | 14·3 14·2 |
| 24 | 14 51·0 | 14 53·4 | 14 10·4 | 2·4 2·4 | 8·4 8·3 | 14·4 14·3 |
| 25 | 14 51·3 | 14 53·7 | 14 10·6 | 2·5 2·5 | 8·5 8·4 | 14·5 14·4 |
| 26 | 14 51·5 | 14 53·9 | 14 10·9 | 2·6 2·6 | 8·6 8·5 | 14·6 14·5 |
| 27 | 14 51·8 | 14 54·2 | 14 11·1 | 2·7 2·7 | 8·7 8·6 | 14·7 14·6 |
| 28 | 14 52·0 | 14 54·4 | 14 11·4 | 2·8 2·8 | 8·8 8·7 | 14·8 14·7 |
| 29 | 14 52·3 | 14 54·7 | 14 11·6 | 2·9 2·9 | 8·9 8·8 | 14·9 14·8 |
| 30 | 14 52·5 | 14 54·9 | 14 11·8 | 3·0 3·0 | 9·0 8·9 | 15·0 14·9 |
| 31 | 14 52·8 | 14 55·2 | 14 12·1 | 3·1 3·1 | 9·1 9·0 | 15·1 15·0 |
| 32 | 14 53·0 | 14 55·4 | 14 12·3 | 3·2 3·2 | 9·2 9·1 | 15·2 15·1 |
| 33 | 14 53·3 | 14 55·7 | 14 12·6 | 3·3 3·3 | 9·3 9·2 | 15·3 15·2 |
| 34 | 14 53·5 | 14 55·9 | 14 12·8 | 3·4 3·4 | 9·4 9·3 | 15·4 15·3 |
| 35 | 14 53·8 | 14 56·2 | 14 13·0 | 3·5 3·5 | 9·5 9·4 | 15·5 15·4 |
| 36 | 14 54·0 | 14 56·4 | 14 13·3 | 3·6 3·6 | 9·6 9·5 | 15·6 15·5 |
| 37 | 14 54·3 | 14 56·7 | 14 13·5 | 3·7 3·7 | 9·7 9·6 | 15·7 15·6 |
| 38 | 14 54·5 | 14 56·9 | 14 13·8 | 3·8 3·8 | 9·8 9·7 | 15·8 15·7 |
| 39 | 14 54·8 | 14 57·2 | 14 14·0 | 3·9 3·9 | 9·9 9·8 | 15·9 15·8 |
| 40 | 14 55·0 | 14 57·5 | 14 14·2 | 4·0 4·0 | 10·0 9·9 | 16·0 15·9 |
| 41 | 14 55·3 | 14 57·7 | 14 14·5 | 4·1 4·1 | 10·1 10·0 | 16·1 16·0 |
| 42 | 14 55·5 | 14 58·0 | 14 14·7 | 4·2 4·2 | 10·2 10·1 | 16·2 16·1 |
| 43 | 14 55·8 | 14 58·2 | 14 14·9 | 4·3 4·3 | 10·3 10·2 | 16·3 16·2 |
| 44 | 14 56·0 | 14 58·5 | 14 15·2 | 4·4 4·4 | 10·4 10·3 | 16·4 16·3 |
| 45 | 14 56·3 | 14 58·7 | 14 15·4 | 4·5 4·5 | 10·5 10·4 | 16·5 16·4 |
| 46 | 14 56·5 | 14 59·0 | 14 15·7 | 4·6 4·6 | 10·6 10·5 | 16·6 16·5 |
| 47 | 14 56·8 | 14 59·2 | 14 15·9 | 4·7 4·7 | 10·7 10·6 | 16·7 16·6 |
| 48 | 14 57·0 | 14 59·5 | 14 16·1 | 4·8 4·8 | 10·8 10·7 | 16·8 16·7 |
| 49 | 14 57·3 | 14 59·7 | 14 16·4 | 4·9 4·9 | 10·9 10·8 | 16·9 16·8 |
| 50 | 14 57·5 | 15 00·0 | 14 16·6 | 5·0 5·0 | 11·0 10·9 | 17·0 16·9 |
| 51 | 14 57·8 | 15 00·2 | 14 16·9 | 5·1 5·1 | 11·1 11·0 | 17·1 17·0 |
| 52 | 14 58·0 | 15 00·5 | 14 17·1 | 5·2 5·2 | 11·2 11·1 | 17·2 17·1 |
| 53 | 14 58·3 | 15 00·7 | 14 17·3 | 5·3 5·3 | 11·3 11·2 | 17·3 17·2 |
| 54 | 14 58·5 | 15 01·0 | 14 17·6 | 5·4 5·4 | 11·4 11·3 | 17·4 17·3 |
| 55 | 14 58·8 | 15 01·2 | 14 17·8 | 5·5 5·5 | 11·5 11·4 | 17·5 17·4 |
| 56 | 14 59·0 | 15 01·5 | 14 18·0 | 5·6 5·6 | 11·6 11·5 | 17·6 17·5 |
| 57 | 14 59·3 | 15 01·7 | 14 18·3 | 5·7 5·7 | 11·7 11·6 | 17·7 17·6 |
| 58 | 14 59·5 | 15 02·0 | 14 18·5 | 5·8 5·8 | 11·8 11·7 | 17·8 17·7 |
| 59 | 14 59·8 | 15 02·2 | 14 18·8 | 5·9 5·9 | 11·9 11·8 | 17·9 17·8 |
| 60 | 15 00·0 | 15 02·5 | 14 19·0 | 6·0 6·0 | 12·0 11·9 | 18·0 17·9 |

## POLARIS (POLE STAR) TABLES, 1980
### FOR DETERMINING LATITUDE FROM SEXTANT ALTITUDE AND FOR AZIMUTH

| L.H.A. ARIES | 0°–9° | 10°–19° | 20°–29° | 30°–39° | 40°–49° | 50°–59° | 60°–69° | 70°–79° | 80°–89° | 90°–99° | 100°–109° | 110°–119° |
|---|---|---|---|---|---|---|---|---|---|---|---|---|
|   | $a_0$ | $a_0$ | $a_0$ | $a_0$ | $a_0$ | $a_0$ | $a_0$ | $a_0$ | $a_0$ | $a_0$ | $a_0$ | $a_0$ |
| ° | ° ′ | ° ′ | ° ′ | ° ′ | ° ′ | ° ′ | ° ′ | ° ′ | ° ′ | ° ′ | ° ′ | ° ′ |
| 0 | 0 17.5 | 0 13.3 | 0 10.6 | 0 09.4 | 0 09.7 | 0 11.5 | 0 14.7 | 0 19.4 | 0 25.2 | 0 32.0 | 0 39.7 | 0 48.0 |
| 1 | 17.0 | 13.0 | 10.4 | 09.3 | 09.8 | 11.7 | 15.1 | 19.9 | 25.8 | 32.8 | 40.5 | 48.8 |
| 2 | 16.5 | 12.7 | 10.2 | 09.3 | 09.9 | 12.0 | 15.5 | 20.4 | 26.5 | 33.5 | 41.3 | 49.7 |
| 3 | 16.1 | 12.4 | 10.1 | 09.3 | 10.0 | 12.3 | 16.0 | 21.0 | 27.1 | 34.3 | 42.1 | 50.5 |
| 4 | 15.7 | 12.1 | 09.9 | 09.3 | 10.2 | 12.6 | 16.4 | 21.5 | 27.8 | 35.0 | 43.0 | 51.4 |
| 5 | 0 15.2 | 0 11.8 | 0 09.8 | 0 09.3 | 0 10.4 | 0 12.9 | 0 16.9 | 0 22.1 | 0 28.5 | 0 35.8 | 0 43.8 | 0 52.2 |
| 6 | 14.8 | 11.5 | 09.7 | 09.4 | 10.6 | 13.3 | 17.3 | 22.7 | 29.2 | 36.6 | 44.6 | 53.1 |
| 7 | 14.4 | 11.3 | 09.6 | 09.4 | 10.8 | 13.6 | 17.8 | 23.3 | 29.9 | 37.3 | 45.4 | 53.9 |
| 8 | 14.1 | 11.0 | 09.5 | 09.5 | 11.0 | 14.0 | 18.3 | 23.9 | 30.6 | 38.1 | 46.3 | 54.8 |
| 9 | 13.7 | 10.8 | 09.4 | 09.6 | 11.2 | 14.3 | 18.8 | 24.6 | 31.3 | 38.9 | 47.1 | 55.7 |
| 10 | 0 13.3 | 0 10.6 | 0 09.4 | 0 09.7 | 0 11.5 | 0 14.7 | 0 19.4 | 0 25.2 | 0 32.0 | 0 39.7 | 0 48.0 | 0 56.5 |

| Lat. | $a_1$ | $a_1$ | $a_1$ | $a_1$ | $a_1$ | $a_1$ | $a_1$ | $a_1$ | $a_1$ | $a_1$ | $a_1$ | $a_1$ |
|---|---|---|---|---|---|---|---|---|---|---|---|---|
| ° | ′ | ′ | ′ | ′ | ′ | ′ | ′ | ′ | ′ | ′ | ′ | ′ |
| 0 | 0.5 | 0.6 | 0.6 | 0.6 | 0.6 | 0.5 | 0.5 | 0.4 | 0.3 | 0.3 | 0.2 | 0.2 |
| 10 | .5 | .6 | .6 | .6 | .6 | .5 | .5 | .4 | .4 | .3 | .3 | .2 |
| 20 | .5 | .6 | .6 | .6 | .6 | .6 | .5 | .5 | .4 | .4 | .3 | .3 |
| 30 | .6 | .6 | .6 | .6 | .6 | .6 | .5 | .5 | .5 | .4 | .4 | .4 |
| 40 | 0.6 | 0.6 | 0.6 | 0.6 | 0.6 | 0.6 | 0.6 | 0.5 | 0.5 | 0.5 | 0.5 | 0.5 |
| 45 | .6 | .6 | .6 | .6 | .6 | .6 | .6 | .6 | .6 | .5 | .5 | .5 |
| 50 | .6 | .6 | .6 | .6 | .6 | .6 | .6 | .6 | .6 | .6 | .6 | .6 |
| 55 | .6 | .6 | .6 | .6 | .6 | .6 | .6 | .6 | .7 | .7 | .7 | .7 |
| 60 | .6 | .6 | .6 | .6 | .6 | .6 | .7 | .7 | .7 | .7 | .8 | .8 |
| 62 | 0.7 | 0.6 | 0.6 | 0.6 | 0.6 | 0.6 | 0.7 | 0.7 | 0.8 | 0.8 | 0.8 | 0.8 |
| 64 | .7 | .6 | .6 | .6 | .6 | .6 | .7 | .7 | .8 | .8 | .9 | 0.9 |
| 66 | .7 | .6 | .6 | .6 | .6 | .7 | .7 | .8 | .8 | .9 | 0.9 | 1.0 |
| 68 | 0.7 | 0.6 | 0.6 | 0.6 | 0.6 | 0.7 | 0.7 | 0.8 | 0.9 | 1.0 | 1.0 | 1.0 |

| Month | $a_2$ | $a_2$ | $a_2$ | $a_2$ | $a_2$ | $a_2$ | $a_2$ | $a_2$ | $a_2$ | $a_2$ | $a_2$ | $a_2$ |
|---|---|---|---|---|---|---|---|---|---|---|---|---|
|   | ′ | ′ | ′ | ′ | ′ | ′ | ′ | ′ | ′ | ′ | ′ | ′ |
| Jan. | 0.7 | 0.7 | 0.7 | 0.7 | 0.7 | 0.7 | 0.7 | 0.7 | 0.7 | 0.7 | 0.6 | 0.6 |
| Feb. | .6 | .7 | .7 | .7 | .7 | .8 | .8 | .8 | .8 | .8 | .8 | .8 |
| Mar. | .5 | .5 | .6 | .6 | .7 | .7 | .8 | .8 | .8 | .9 | .9 | .9 |
| Apr. | 0.3 | 0.4 | 0.4 | 0.5 | 0.6 | 0.6 | 0.7 | 0.7 | 0.8 | 0.9 | 0.9 | 0.9 |
| May | .2 | .3 | .3 | .4 | .4 | .5 | .5 | .6 | .7 | .8 | .8 | .9 |
| June | .2 | .2 | .2 | .2 | .3 | .3 | .4 | .5 | .5 | .6 | .7 | .7 |
| July | 0.2 | 0.2 | 0.2 | 0.2 | 0.2 | 0.3 | 0.3 | 0.3 | 0.4 | 0.5 | 0.5 | 0.6 |
| Aug. | .4 | .3 | .3 | .3 | .2 | .2 | .3 | .3 | .3 | .3 | .4 | .4 |
| Sept. | .5 | .5 | .4 | .4 | .3 | .3 | .3 | .3 | .3 | .3 | .3 | .3 |
| Oct. | 0.7 | 0.7 | 0.6 | 0.5 | 0.5 | 0.4 | 0.4 | 0.3 | 0.3 | 0.3 | 0.3 | 0.3 |
| Nov. | 0.9 | 0.8 | .8 | .7 | .7 | .6 | .5 | .5 | .4 | .4 | .3 | .3 |
| Dec. | 1.0 | 1.0 | 0.9 | 0.9 | 0.8 | 0.8 | 0.7 | 0.6 | 0.6 | 0.5 | 0.4 | 0.4 |

| Lat. | | | | | | AZIMUTH | | | | | | |
|---|---|---|---|---|---|---|---|---|---|---|---|---|
| ° | ° | ° | ° | ° | ° | ° | ° | ° | ° | ° | ° | ° |
| 0 | 0.4 | 0.3 | 0.1 | 0.0 | 359.8 | 359.7 | 359.6 | 359.4 | 359.4 | 359.3 | 359.2 | 359.2 |
| 20 | 0.4 | 0.3 | 0.1 | 0.0 | 359.8 | 359.7 | 359.5 | 359.3 | 359.3 | 359.2 | 359.2 | 359.1 |
| 40 | 0.5 | 0.3 | 0.2 | 0.0 | 359.8 | 359.6 | 359.4 | 359.3 | 359.1 | 359.0 | 359.0 | 358.9 |
| 50 | 0.6 | 0.4 | 0.2 | 0.0 | 359.7 | 359.5 | 359.3 | 359.1 | 359.0 | 358.9 | 358.8 | 358.7 |
| 55 | 0.7 | 0.5 | 0.2 | 0.0 | 359.7 | 359.5 | 359.2 | 359.0 | 358.9 | 358.7 | 358.6 | 358.6 |
| 60 | 0.8 | 0.5 | 0.2 | 359.9 | 359.7 | 359.4 | 359.1 | 358.9 | 358.7 | 358.5 | 358.4 | 358.4 |
| 65 | 0.9 | 0.6 | 0.3 | 359.9 | 359.6 | 359.3 | 358.9 | 358.7 | 358.4 | 358.3 | 358.1 | 358.1 |

Latitude = Apparent altitude (corrected for refraction) − 1° + $a_0$ + $a_1$ + $a_2$

The table is entered with L.H.A. Aries to determine the column to be used; each column refers to a range of 10°. $a_0$ is taken, with mental interpolation, from the upper table with the units of L.H.A. Aries in degrees as argument; $a_1$, $a_2$ are taken, without interpolation, from the second and third tables with arguments latitude and month respectively. $a_0$, $a_1$, $a_2$ are always positive. The final table gives the azimuth of *Polaris*.

## POLARIS (POLE STAR) TABLES, 1980
### FOR DETERMINING LATITUDE FROM SEXTANT ALTITUDE AND FOR AZIMUTH

| L.H.A. ARIES | 120°–129° | 130°–139° | 140°–149° | 150°–159° | 160°–169° | 170°–179° | 180°–189° | 190°–199° | 200°–209° | 210°–219° | 220°–229° | 230°–239° |
|---|---|---|---|---|---|---|---|---|---|---|---|---|
|  | $a_0$ | $a_0$ | $a_0$ | $a_0$ | $a_0$ | $a_0$ | $a_0$ | $a_0$ | $a_0$ | $a_0$ | $a_0$ | $a_0$ |
| ° | ° ′ | ° ′ | ° ′ | ° ′ | ° ′ | ° ′ | ° ′ | ° ′ | ° ′ | ° ′ | ° ′ | ° ′ |
| 0 | 0 56·5 | 1 05·1 | 1 13·6 | 1 21·5 | 1 28·8 | 1 35·1 | 1 40·4 | 1 44·4 | 1 47·0 | 1 48·2 | 1 48·0 | 1 46·2 |
| 1 | 57·4 | 06·0 | 14·4 | 22·3 | 29·4 | 35·7 | 40·8 | 44·7 | 47·2 | 48·3 | 47·8 | 46·0 |
| 2 | 58·2 | 06·8 | 15·2 | 23·0 | 30·1 | 36·3 | 41·3 | 45·0 | 47·4 | 48·3 | 47·7 | 45·7 |
| 3 | 0 59·1 | 07·7 | 16·0 | 23·8 | 30·8 | 36·8 | 41·7 | 45·3 | 47·5 | 48·3 | 47·6 | 45·4 |
| 4 | 1 00·0 | 08·5 | 16·8 | 24·5 | 31·4 | 37·4 | 42·1 | 45·6 | 47·7 | 48·3 | 47·4 | 45·1 |
| 5 | 1 00·8 | 1 09·4 | 1 17·6 | 1 25·2 | 1 32·1 | 1 37·9 | 1 42·5 | 1 45·9 | 1 47·8 | 1 48·3 | 1 47·3 | 1 44·8 |
| 6 | 01·7 | 10·2 | 18·4 | 26·0 | 32·7 | 38·4 | 42·9 | 46·1 | 47·9 | 48·2 | 47·1 | 44·5 |
| 7 | 02·6 | 11·1 | 19·2 | 26·7 | 33·3 | 38·9 | 43·3 | 46·4 | 48·0 | 48·2 | 46·9 | 44·1 |
| 8 | 03·4 | 11·9 | 20·0 | 27·4 | 33·9 | 39·4 | 43·7 | 46·6 | 48·1 | 48·1 | 46·7 | 43·8 |
| 9 | 04·3 | 12·7 | 20·7 | 28·1 | 34·5 | 39·9 | 44·0 | 46·8 | 48·2 | 48·0 | 46·4 | 43·4 |
| 10 | 1 05·1 | 1 13·6 | 1 21·5 | 1 28·8 | 1 35·1 | 1 40·4 | 1 44·4 | 1 47·0 | 1 48·2 | 1 48·0 | 1 46·2 | 1 43·0 |

| Lat. | $a_1$ | $a_1$ | $a_1$ | $a_1$ | $a_1$ | $a_1$ | $a_1$ | $a_1$ | $a_1$ | $a_1$ | $a_1$ | $a_1$ |
|---|---|---|---|---|---|---|---|---|---|---|---|---|
| ° | ′ | ′ | ′ | ′ | ′ | ′ | ′ | ′ | ′ | ′ | ′ | ′ |
| 0 | 0·2 | 0·2 | 0·2 | 0·3 | 0·4 | 0·4 | 0·5 | 0·6 | 0·6 | 0·6 | 0·6 | 0·5 |
| 10 | ·2 | ·3 | ·3 | ·3 | ·4 | ·5 | ·5 | ·6 | ·6 | ·6 | ·6 | ·5 |
| 20 | ·3 | ·3 | ·3 | ·4 | ·4 | ·5 | ·5 | ·6 | ·6 | ·6 | ·6 | ·6 |
| 30 | ·4 | ·4 | ·4 | ·4 | ·5 | ·5 | ·6 | ·6 | ·6 | ·6 | ·6 | ·6 |
| 40 | 0·5 | 0·5 | 0·5 | 0·5 | 0·5 | 0·6 | 0·6 | 0·6 | 0·6 | 0·6 | 0·6 | 0·6 |
| 45 | ·5 | ·5 | ·5 | ·6 | ·6 | ·6 | ·6 | ·6 | ·6 | ·6 | ·6 | ·6 |
| 50 | ·6 | ·6 | ·6 | ·6 | ·6 | ·6 | ·6 | ·6 | ·6 | ·6 | ·6 | ·6 |
| 55 | ·7 | ·7 | ·7 | ·7 | ·6 | ·6 | ·6 | ·6 | ·6 | ·6 | ·6 | ·6 |
| 60 | ·8 | ·8 | ·8 | ·7 | ·7 | ·7 | ·6 | ·6 | ·6 | ·6 | ·6 | ·6 |
| 62 | 0·8 | 0·8 | 0·8 | 0·8 | 0·7 | 0·7 | 0·7 | 0·6 | 0·6 | 0·6 | 0·6 | 0·6 |
| 64 | 0·9 | 0·9 | ·9 | ·8 | ·8 | ·7 | ·7 | ·6 | ·6 | ·6 | ·6 | ·6 |
| 66 | 1·0 | 1·0 | 0·9 | ·9 | ·8 | ·7 | ·7 | ·6 | ·6 | ·6 | ·6 | ·7 |
| 68 | 1·1 | 1·0 | 1·0 | 0·9 | 0·9 | 0·8 | 0·7 | 0·6 | 0·6 | 0·6 | 0·6 | 0·7 |

| Month | $a_2$ | $a_2$ | $a_2$ | $a_2$ | $a_2$ | $a_2$ | $a_2$ | $a_2$ | $a_2$ | $a_2$ | $a_2$ | $a_2$ |
|---|---|---|---|---|---|---|---|---|---|---|---|---|
|  | ′ | ′ | ′ | ′ | ′ | ′ | ′ | ′ | ′ | ′ | ′ | ′ |
| Jan. | 0·6 | 0·6 | 0·6 | 0·5 | 0·5 | 0·5 | 0·5 | 0·5 | 0·5 | 0·5 | 0·5 | 0·5 |
| Feb. | ·8 | ·7 | ·7 | ·7 | ·6 | ·6 | ·6 | ·5 | ·5 | ·5 | ·5 | ·4 |
| Mar. | 0·9 | 0·9 | 0·9 | 0·8 | ·8 | ·8 | ·7 | ·7 | ·6 | ·6 | ·5 | ·5 |
| Apr. | 1·0 | 1·0 | 1·0 | 1·0 | 0·9 | 0·9 | 0·9 | 0·8 | 0·8 | 0·7 | 0·6 | 0·6 |
| May | 0·9 | 1·0 | 1·0 | 1·0 | 1·0 | 1·0 | 1·0 | 0·9 | 0·9 | 0·8 | ·8 | ·7 |
| June | ·8 | 0·9 | 0·9 | 1·0 | 1·0 | 1·0 | 1·0 | 1·0 | 1·0 | 1·0 | 0·9 | ·8 |
| July | 0·7 | 0·7 | 0·8 | 0·8 | 0·9 | 0·9 | 1·0 | 1·0 | 1·0 | 1·0 | 1·0 | 0·9 |
| Aug. | ·5 | ·6 | ·6 | ·7 | ·7 | ·8 | 0·8 | 0·9 | 0·9 | 0·9 | 1·0 | 1·0 |
| Sept. | ·4 | ·4 | ·4 | ·5 | ·6 | ·6 | ·7 | ·7 | ·8 | ·8 | 0·9 | 0·9 |
| Oct. | 0·3 | 0·3 | 0·3 | 0·3 | 0·4 | 0·4 | 0·5 | 0·5 | 0·6 | 0·7 | 0·7 | 0·8 |
| Nov. | ·2 | ·2 | ·2 | ·2 | ·2 | ·3 | ·3 | ·4 | ·4 | ·5 | ·5 | ·6 |
| Dec. | 0·3 | 0·3 | 0·2 | 0·2 | 0·2 | 0·2 | 0·2 | 0·2 | 0·3 | 0·3 | 0·4 | 0·4 |

| Lat. | AZIMUTH |||||||||||||
|---|---|---|---|---|---|---|---|---|---|---|---|---|
| ° | ° | ° | ° | ° | ° | ° | ° | ° | ° | ° | ° | ° |
| 0 | 359·2 | 359·2 | 359·2 | 359·3 | 359·4 | 359·5 | 359·6 | 359·7 | 359·9 | 0·0 | 0·2 | 0·3 |
| 20 | 359·1 | 359·1 | 359·2 | 359·3 | 359·3 | 359·5 | 359·6 | 359·7 | 359·9 | 0·0 | 0·2 | 0·3 |
| 40 | 358·9 | 358·9 | 359·0 | 359·1 | 359·2 | 359·3 | 359·5 | 359·7 | 359·8 | 0·0 | 0·2 | 0·4 |
| 50 | 358·7 | 358·7 | 358·8 | 358·9 | 359·1 | 359·2 | 359·4 | 359·6 | 359·8 | 0·0 | 0·3 | 0·5 |
| 55 | 358·6 | 358·6 | 358·7 | 358·8 | 358·9 | 359·1 | 359·3 | 359·6 | 359·8 | 0·0 | 0·3 | 0·5 |
| 60 | 358·4 | 358·4 | 358·5 | 358·6 | 358·8 | 359·0 | 359·2 | 359·5 | 359·8 | 0·1 | 0·3 | 0·6 |
| 65 | 358·1 | 358·1 | 358·2 | 358·4 | 358·6 | 358·8 | 359·1 | 359·4 | 359·7 | 0·1 | 0·4 | 0·7 |

ILLUSTRATION
On 1980 April 21 at G.M.T.
$23^h$ $18^m$ $56^s$ in longitude
W. 37° 14′ the apparent altitude
(corrected for refraction), Ho, of
Polaris was 49° 31′·6.

From the daily pages:
G.H.A. Aries ($23^h$)   195° 09·8′
Increment ($18^m$ $56^s$)   4 44·8
Longitude (west)   −37 14
L.H.A. Aries   162 41

Ho   49° 31′·6
$a_0$ (argument 162° 41′)   1 30·6
$a_1$ (lat. 50° approx.)   0·6
$a_2$ (April)   0·9
Sum −1° = Lat. =   50° 03′·7

## POLARIS (POLE STAR) TABLES, 1980
### FOR DETERMINING LATITUDE FROM SEXTANT ALTITUDE AND FOR AZIMUTH

| L.H.A. ARIES | 240°–249° | 250°–259° | 260°–269° | 270°–279° | 280°–289° | 290°–299° | 300°–309° | 310°–319° | 320°–329° | 330°–339° | 340°–349° | 350°–359° |
|---|---|---|---|---|---|---|---|---|---|---|---|---|
|  | $a_0$ | $a_0$ | $a_0$ | $a_0$ | $a_0$ | $a_0$ | $a_0$ | $a_0$ | $a_0$ | $a_0$ | $a_0$ | $a_0$ |
| ° | ° ′ | ° ′ | ° ′ | ° ′ | ° ′ | ° ′ | ° ′ | ° ′ | ° ′ | ° ′ | ° ′ | ° ′ |
| 0 | 1 43·0 | 1 38·6 | 1 32·9 | 1 26·2 | 1 18·6 | 1 10·5 | 1 01·9 | 0 53·3 | 0 44·8 | 0 36·8 | 0 29·4 | 0 22·9 |
| 1 | 42·7 | 38·0 | 32·2 | 25·4 | 17·8 | 09·6 | 01·1 | 52·4 | 44·0 | 36·0 | 28·7 | 22·3 |
| 2 | 42·2 | 37·5 | 31·6 | 24·7 | 17·0 | 08·8 | 1 00·2 | 51·6 | 43·2 | 35·2 | 28·0 | 21·7 |
| 3 | 41·8 | 37·0 | 31·0 | 24·0 | 16·2 | 07·9 | 0 59·3 | 50·7 | 42·4 | 34·5 | 27·3 | 21·1 |
| 4 | 41·4 | 36·4 | 30·3 | 23·2 | 15·4 | 07·1 | 58·5 | 49·9 | 41·5 | 33·7 | 26·7 | 20·6 |
| 5 | 1 41·0 | 1 35·9 | 1 29·6 | 1 22·5 | 1 14·6 | 1 06·2 | 0 57·6 | 0 49·0 | 0 40·7 | 0 33·0 | 0 26·0 | 0 20·0 |
| 6 | 40·5 | 35·3 | 29·0 | 21·7 | 13·8 | 05·4 | 56·7 | 48·2 | 39·9 | 32·2 | 25·4 | 19·5 |
| 7 | 40·0 | 34·7 | 28·3 | 20·9 | 12·9 | 04·5 | 55·9 | 47·3 | 39·1 | 31·5 | 24·7 | 19·0 |
| 8 | 39·6 | 34·1 | 27·6 | 20·2 | 12·1 | 03·7 | 55·0 | 46·5 | 38·3 | 30·8 | 24·1 | 18·5 |
| 9 | 39·1 | 33·5 | 26·9 | 19·4 | 11·3 | 02·8 | 54·2 | 45·7 | 37·5 | 30·1 | 23·5 | 18·0 |
| 10 | 1 38·6 | 1 32·9 | 1 26·2 | 1 18·6 | 1 10·5 | 1 01·9 | 0 53·3 | 0 44·8 | 0 36·8 | 0 29·4 | 0 22·9 | 0 17·5 |

| Lat. | $a_1$ | $a_1$ | $a_1$ | $a_1$ | $a_1$ | $a_1$ | $a_1$ | $a_1$ | $a_1$ | $a_1$ | $a_1$ | $a_1$ |
|---|---|---|---|---|---|---|---|---|---|---|---|---|
| ° | ′ | ′ | ′ | ′ | ′ | ′ | ′ | ′ | ′ | ′ | ′ | ′ |
| 0 | 0·5 | 0·4 | 0·3 | 0·3 | 0·2 | 0·2 | 0·2 | 0·2 | 0·2 | 0·3 | 0·4 | 0·4 |
| 10 | ·5 | ·4 | ·4 | ·3 | ·3 | ·2 | ·2 | ·3 | ·3 | ·3 | ·4 | ·5 |
| 20 | ·5 | ·5 | ·4 | ·4 | ·3 | ·3 | ·3 | ·3 | ·3 | ·4 | ·4 | ·5 |
| 30 | ·5 | ·5 | ·5 | ·4 | ·4 | ·4 | ·4 | ·4 | ·4 | ·4 | ·5 | ·5 |
| 40 | 0·6 | 0·5 | 0·5 | 0·5 | 0·5 | 0·5 | 0·5 | 0·5 | 0·5 | 0·5 | 0·5 | 0·6 |
| 45 | ·6 | ·6 | ·6 | ·5 | ·5 | ·5 | ·5 | ·5 | ·5 | ·6 | ·6 | ·6 |
| 50 | ·6 | ·6 | ·6 | ·6 | ·6 | ·6 | ·6 | ·6 | ·6 | ·6 | ·6 | ·6 |
| 55 | ·6 | ·6 | ·7 | ·7 | ·7 | ·7 | ·7 | ·7 | ·7 | ·7 | ·6 | ·6 |
| 60 | ·7 | ·7 | ·7 | ·7 | ·8 | ·8 | ·8 | ·8 | ·8 | ·7 | ·7 | ·7 |
| 62 | 0·7 | 0·7 | 0·8 | 0·8 | 0·8 | 0·8 | 0·8 | 0·8 | 0·8 | 0·8 | 0·7 | 0·7 |
| 64 | ·7 | ·7 | ·8 | ·8 | ·9 | 0·9 | 0·9 | 0·9 | 0·9 | ·8 | ·8 | ·7 |
| 66 | ·7 | ·8 | ·8 | 0·9 | 0·9 | 1·0 | 1·0 | 1·0 | 0·9 | ·9 | ·8 | ·7 |
| 68 | 0·7 | 0·8 | 0·9 | 1·0 | 1·0 | 1·0 | 1·1 | 1·0 | 1·0 | 0·9 | 0·9 | 0·8 |

| Month | $a_2$ | $a_2$ | $a_2$ | $a_2$ | $a_2$ | $a_2$ | $a_2$ | $a_2$ | $a_2$ | $a_2$ | $a_2$ | $a_2$ |
|---|---|---|---|---|---|---|---|---|---|---|---|---|
|  | ′ | ′ | ′ | ′ | ′ | ′ | ′ | ′ | ′ | ′ | ′ | ′ |
| Jan. | 0·5 | 0·5 | 0·5 | 0·5 | 0·6 | 0·6 | 0·6 | 0·6 | 0·6 | 0·7 | 0·7 | 0·7 |
| Feb. | ·4 | ·4 | ·4 | ·4 | ·4 | ·4 | ·4 | ·5 | ·5 | ·5 | ·6 | ·6 |
| Mar. | ·4 | ·4 | ·4 | ·3 | ·3 | ·3 | ·3 | ·3 | ·3 | ·4 | ·4 | ·4 |
| Apr. | 0·5 | 0·5 | 0·4 | 0·3 | 0·3 | 0·3 | 0·2 | 0·2 | 0·2 | 0·2 | 0·3 | 0·3 |
| May | ·7 | ·6 | ·5 | ·4 | ·4 | ·3 | ·3 | ·2 | ·2 | ·2 | ·2 | ·2 |
| June | ·8 | ·7 | ·7 | ·6 | ·5 | ·5 | ·4 | ·3 | ·3 | ·2 | ·2 | ·2 |
| July | 0·9 | 0·9 | 0·8 | 0·7 | 0·7 | 0·6 | 0·5 | 0·5 | 0·4 | 0·4 | 0·3 | 0·3 |
| Aug. | ·9 | ·9 | ·9 | ·9 | ·8 | ·8 | ·7 | ·6 | ·6 | ·5 | ·5 | ·4 |
| Sept. | ·9 | ·9 | ·9 | ·9 | ·9 | ·9 | ·8 | ·8 | ·8 | ·7 | ·6 | ·6 |
| Oct. | 0·8 | 0·9 | 0·9 | 0·9 | 0·9 | 0·9 | 0·9 | 0·9 | 0·9 | 0·9 | 0·8 | 0·8 |
| Nov. | ·7 | ·7 | ·8 | ·8 | ·9 | ·9 | 1·0 | 1·0 | 1·0 | 1·0 | 1·0 | 0·9 |
| Dec. | 0·5 | 0·6 | 0·6 | 0·7 | 0·8 | 0·8 | 0·9 | 0·9 | 1·0 | 1·0 | 1·0 | 1·0 |

| Lat. | AZIMUTH |||||||||||||
|---|---|---|---|---|---|---|---|---|---|---|---|---|
| ° | ° | ° | ° | ° | ° | ° | ° | ° | ° | ° | ° | ° |
| 0 | 0·4 | 0·6 | 0·6 | 0·7 | 0·8 | 0·8 | 0·8 | 0·8 | 0·8 | 0·8 | 0·6 | 0·5 |
| 20 | 0·5 | 0·6 | 0·7 | 0·8 | 0·8 | 0·9 | 0·9 | 0·9 | 0·8 | 0·7 | 0·7 | 0·5 |
| 40 | 0·6 | 0·7 | 0·8 | 0·9 | 1·0 | 1·1 | 1·1 | 1·1 | 1·0 | 0·9 | 0·8 | 0·7 |
| 50 | 0·7 | 0·8 | 1·0 | 1·1 | 1·2 | 1·3 | 1·3 | 1·3 | 1·2 | 1·1 | 1·0 | 0·8 |
| 55 | 0·7 | 0·9 | 1·1 | 1·3 | 1·4 | 1·4 | 1·4 | 1·4 | 1·3 | 1·2 | 1·1 | 0·9 |
| 60 | 0·9 | 1·1 | 1·3 | 1·4 | 1·6 | 1·6 | 1·7 | 1·6 | 1·5 | 1·4 | 1·2 | 1·0 |
| 65 | 1·0 | 1·3 | 1·5 | 1·7 | 1·8 | 1·9 | 2·0 | 1·9 | 1·8 | 1·7 | 1·5 | 1·2 |

Latitude = Apparent altitude (corrected for refraction) − 1° + $a_0$ + $a_1$ + $a_2$

The table is entered with L.H.A. Aries to determine the column to be used; each column refers to a range of 10°. $a_0$ is taken, with mental interpolation, from the upper table with the units of L.H.A. Aries in degrees as argument; $a_1$, $a_2$ are taken, without interpolation, from the second and third tables with arguments latitude and month respectively. $a_0$, $a_1$, $a_2$ are always positive. The final table gives the azimuth of *Polaris*.

# TABLES FOR INTERPOLATING SUNRISE, MOONRISE, ETC.
## TABLE I—FOR LATITUDE

| Tabular Interval | | | Difference between the times for consecutive latitudes | | | | | | | | | | | | | | | |
|---|---|---|---|---|---|---|---|---|---|---|---|---|---|---|---|---|---|---|
| 10° | 5° | 2° | 5ᵐ | 10ᵐ | 15ᵐ | 20ᵐ | 25ᵐ | 30ᵐ | 35ᵐ | 40ᵐ | 45ᵐ | 50ᵐ | 55ᵐ | 60ᵐ | 1ʰ 05ᵐ | 1ʰ 10ᵐ | 1ʰ 15ᵐ | 1ʰ 20ᵐ |
| ° ′ | ° ′ | ° ′ | m | m | m | m | m | m | m | m | m | m | m | m | h m | h m | h m | h m |
| 0 30 | 0 15 | 0 06 | 0 | 0 | 1 | 1 | 1 | 1 | 1 | 2 | 2 | 2 | 2 | 2 | 0 02 | 0 02 | 0 02 | 0 02 |
| 1 00 | 0 30 | 0 12 | 0 | 1 | 1 | 2 | 2 | 3 | 3 | 3 | 4 | 4 | 4 | 5 | 05 | 05 | 05 | 05 |
| 1 30 | 0 45 | 0 18 | 1 | 1 | 2 | 3 | 3 | 4 | 4 | 5 | 5 | 6 | 7 | 7 | 07 | 07 | 07 | 07 |
| 2 00 | 1 00 | 0 24 | 1 | 2 | 3 | 4 | 5 | 5 | 6 | 7 | 7 | 8 | 9 | 10 | 10 | 10 | 10 | 10 |
| 2 30 | 1 15 | 0 30 | 1 | 2 | 4 | 5 | 6 | 7 | 8 | 9 | 9 | 10 | 11 | 12 | 12 | 13 | 13 | 13 |
| 3 00 | 1 30 | 0 36 | 1 | 3 | 4 | 6 | 7 | 8 | 9 | 10 | 11 | 12 | 13 | 14 | 0 15 | 0 15 | 0 16 | 0 16 |
| 3 30 | 1 45 | 0 42 | 2 | 3 | 5 | 7 | 8 | 10 | 11 | 12 | 13 | 14 | 16 | 17 | 18 | 18 | 19 | 19 |
| 4 00 | 2 00 | 0 48 | 2 | 4 | 6 | 8 | 9 | 11 | 13 | 14 | 15 | 16 | 18 | 19 | 20 | 21 | 22 | 22 |
| 4 30 | 2 15 | 0 54 | 2 | 4 | 7 | 9 | 11 | 13 | 15 | 16 | 18 | 19 | 21 | 22 | 23 | 24 | 25 | 26 |
| 5 00 | 2 30 | 1 00 | 2 | 5 | 7 | 10 | 12 | 14 | 16 | 18 | 20 | 22 | 23 | 25 | 26 | 27 | 28 | 29 |
| 5 30 | 2 45 | 1 06 | 3 | 5 | 8 | 11 | 13 | 16 | 18 | 20 | 22 | 24 | 26 | 28 | 0 29 | 0 30 | 0 31 | 0 32 |
| 6 00 | 3 00 | 1 12 | 3 | 6 | 9 | 12 | 14 | 17 | 20 | 22 | 24 | 26 | 29 | 31 | 32 | 33 | 34 | 36 |
| 6 30 | 3 15 | 1 18 | 3 | 6 | 10 | 13 | 16 | 19 | 22 | 24 | 26 | 29 | 31 | 34 | 36 | 37 | 38 | 40 |
| 7 00 | 3 30 | 1 24 | 3 | 7 | 10 | 14 | 17 | 20 | 23 | 26 | 29 | 31 | 34 | 37 | 39 | 41 | 42 | 44 |
| 7 30 | 3 45 | 1 30 | 4 | 7 | 11 | 15 | 18 | 22 | 25 | 28 | 31 | 34 | 37 | 40 | 43 | 44 | 46 | 48 |
| 8 00 | 4 00 | 1 36 | 4 | 8 | 12 | 16 | 20 | 23 | 27 | 30 | 34 | 37 | 41 | 44 | 0 47 | 0 48 | 0 51 | 0 53 |
| 8 30 | 4 15 | 1 42 | 4 | 8 | 13 | 17 | 21 | 25 | 29 | 33 | 36 | 40 | 44 | 48 | 0 51 | 0 53 | 0 56 | 0 58 |
| 9 00 | 4 30 | 1 48 | 4 | 9 | 13 | 18 | 22 | 27 | 31 | 35 | 39 | 43 | 47 | 52 | 0 55 | 0 58 | 1 01 | 1 04 |
| 9 30 | 4 45 | 1 54 | 5 | 9 | 14 | 19 | 24 | 28 | 33 | 38 | 42 | 47 | 51 | 56 | 1 00 | 1 04 | 1 08 | 1 12 |
| 10 00 | 5 00 | 2 00 | 5 | 10 | 15 | 20 | 25 | 30 | 35 | 40 | 45 | 50 | 55 | 60 | 1 05 | 1 10 | 1 15 | 1 20 |

Table I is for interpolating the L.M.T. of sunrise, twilight, moonrise, etc., for latitude. It is to be entered, in the appropriate column on the left, with the difference between true latitude and the nearest tabular latitude which is *less* than the true latitude; and with the argument at the top which is the nearest value of the difference between the times for the tabular latitude and the next higher one; the correction so obtained is applied to the time for the tabular latitude; the sign of the correction can be seen by inspection. It is to be noted that the interpolation is not linear, so that when using this table it is essential to take out the tabular phenomenon for the latitude *less* than the true latitude.

## TABLE II—FOR LONGITUDE

| Long. East or West | Difference between the times for given date and preceding date (for east longitude) or for given date and following date (for west longitude) | | | | | | | | | | | | | | | | | |
|---|---|---|---|---|---|---|---|---|---|---|---|---|---|---|---|---|---|---|
| | 10ᵐ | 20ᵐ | 30ᵐ | 40ᵐ | 50ᵐ | 60ᵐ | 1ʰ+10ᵐ | 1ʰ+20ᵐ | 1ʰ+30ᵐ | 40ᵐ | 50ᵐ | 60ᵐ | 2ʰ 10ᵐ | 2ʰ 20ᵐ | 2ʰ 30ᵐ | 2ʰ 40ᵐ | 2ʰ 50ᵐ | 3ʰ 00ᵐ |
| ° | m | m | m | m | m | m | m | m | m | m | m | m | h m | h m | h m | h m | h m | h m |
| 0 | 0 | 0 | 0 | 0 | 0 | 0 | 0 | 0 | 0 | 0 | 0 | 0 | 0 00 | 0 00 | 0 00 | 0 00 | 0 00 | 0 00 |
| 10 | 0 | 1 | 1 | 1 | 1 | 2 | 2 | 2 | 2 | 3 | 3 | 3 | 04 | 04 | 04 | 04 | 05 | 05 |
| 20 | 1 | 1 | 2 | 2 | 3 | 3 | 4 | 4 | 5 | 6 | 6 | 7 | 07 | 08 | 08 | 09 | 09 | 10 |
| 30 | 1 | 2 | 2 | 3 | 4 | 5 | 6 | 7 | 7 | 8 | 9 | 10 | 11 | 12 | 12 | 13 | 14 | 15 |
| 40 | 1 | 2 | 3 | 4 | 6 | 7 | 8 | 9 | 10 | 11 | 12 | 13 | 14 | 16 | 17 | 18 | 19 | 20 |
| 50 | 1 | 3 | 4 | 6 | 7 | 8 | 10 | 11 | 12 | 14 | 15 | 17 | 0 18 | 0 19 | 0 21 | 0 22 | 0 24 | 0 25 |
| 60 | 2 | 3 | 5 | 7 | 8 | 10 | 12 | 13 | 15 | 17 | 18 | 20 | 22 | 23 | 25 | 27 | 28 | 30 |
| 70 | 2 | 4 | 6 | 8 | 10 | 12 | 14 | 16 | 17 | 19 | 21 | 23 | 25 | 27 | 29 | 31 | 33 | 35 |
| 80 | 2 | 4 | 7 | 9 | 11 | 13 | 16 | 18 | 20 | 22 | 24 | 27 | 29 | 31 | 33 | 36 | 38 | 40 |
| 90 | 2 | 5 | 7 | 10 | 12 | 15 | 17 | 20 | 22 | 25 | 27 | 30 | 32 | 35 | 37 | 40 | 42 | 45 |
| 100 | 3 | 6 | 8 | 11 | 14 | 17 | 19 | 22 | 25 | 28 | 31 | 33 | 0 36 | 0 39 | 0 42 | 0 44 | 0 47 | 0 50 |
| 110 | 3 | 6 | 9 | 12 | 15 | 18 | 21 | 24 | 27 | 31 | 34 | 37 | 40 | 43 | 46 | 49 | 0 52 | 0 55 |
| 120 | 3 | 7 | 10 | 13 | 17 | 20 | 23 | 27 | 30 | 33 | 37 | 40 | 43 | 47 | 50 | 53 | 0 57 | 1 00 |
| 130 | 4 | 7 | 11 | 14 | 18 | 22 | 25 | 29 | 32 | 36 | 40 | 43 | 47 | 51 | 54 | 0 58 | 1 01 | 1 05 |
| 140 | 4 | 8 | 12 | 16 | 19 | 23 | 27 | 31 | 35 | 39 | 43 | 47 | 51 | 54 | 0 58 | 1 02 | 1 06 | 1 10 |
| 150 | 4 | 8 | 13 | 17 | 21 | 25 | 29 | 33 | 38 | 42 | 46 | 50 | 0 54 | 0 58 | 1 03 | 1 07 | 1 11 | 1 15 |
| 160 | 4 | 9 | 13 | 18 | 22 | 27 | 31 | 36 | 40 | 44 | 49 | 53 | 0 58 | 1 02 | 1 07 | 1 11 | 1 16 | 1 20 |
| 170 | 5 | 9 | 14 | 19 | 24 | 28 | 33 | 38 | 42 | 47 | 52 | 57 | 1 01 | 1 06 | 1 11 | 1 16 | 1 20 | 1 25 |
| 180 | 5 | 10 | 15 | 20 | 25 | 30 | 35 | 40 | 45 | 50 | 55 | 60 | 1 05 | 1 10 | 1 15 | 1 20 | 1 25 | 1 30 |

Table II is for interpolating the L.M.T. of moonrise, moonset and the Moon's meridian passage for longitude. It is entered with longitude and with the difference between the times for the given date and for the preceding date (in east longitudes) or following date (in west longitudes). The correction is normally *added* for west longitudes and *subtracted* for east longitudes, but if, as occasionally happens, the times become earlier each day instead of later, the signs of the corrections must be reversed.

## ALTITUDE CORRECTION TABLES 0°–35°—MOON

| App. Alt. | 0°–4° Corrⁿ | 5°–9° Corrⁿ | 10°–14° Corrⁿ | 15°–19° Corrⁿ | 20°–24° Corrⁿ | 25°–29° Corrⁿ | 30°–34° Corrⁿ | App. Alt. |
|---|---|---|---|---|---|---|---|---|
| 00 | 0 33.8 | 5 58.2 | 10 62.1 | 15 62.8 | 20 62.2 | 25 60.8 | 30 58.9 | 00 |
| 10 | 35.9 | 58.5 | 62.2 | 62.8 | 62.1 | 60.8 | 58.8 | 10 |
| 20 | 37.8 | 58.7 | 62.2 | 62.8 | 62.1 | 60.7 | 58.8 | 20 |
| 30 | 39.6 | 58.9 | 62.3 | 62.8 | 62.1 | 60.7 | 58.7 | 30 |
| 40 | 41.2 | 59.1 | 62.3 | 62.8 | 62.0 | 60.6 | 58.6 | 40 |
| 50 | 42.6 | 59.3 | 62.4 | 62.7 | 62.0 | 60.6 | 58.5 | 50 |
| 00 | 1 44.0 | 6 59.5 | 11 62.4 | 16 62.7 | 21 62.0 | 26 60.5 | 31 58.5 | 00 |
| 10 | 45.2 | 59.7 | 62.4 | 62.7 | 61.9 | 60.4 | 58.4 | 10 |
| 20 | 46.3 | 59.9 | 62.5 | 62.7 | 61.9 | 60.4 | 58.3 | 20 |
| 30 | 47.3 | 60.0 | 62.5 | 62.7 | 61.9 | 60.3 | 58.2 | 30 |
| 40 | 48.3 | 60.2 | 62.5 | 62.7 | 61.8 | 60.3 | 58.2 | 40 |
| 50 | 49.2 | 60.3 | 62.6 | 62.7 | 61.8 | 60.2 | 58.1 | 50 |
| 00 | 2 50.0 | 7 60.5 | 12 62.6 | 17 62.7 | 22 61.7 | 27 60.1 | 32 58.0 | 00 |
| 10 | 50.8 | 60.6 | 62.6 | 62.6 | 61.7 | 60.1 | 57.9 | 10 |
| 20 | 51.4 | 60.7 | 62.6 | 62.6 | 61.6 | 60.0 | 57.8 | 20 |
| 30 | 52.1 | 60.9 | 62.7 | 62.6 | 61.6 | 59.9 | 57.8 | 30 |
| 40 | 52.7 | 61.0 | 62.7 | 62.6 | 61.5 | 59.9 | 57.7 | 40 |
| 50 | 53.3 | 61.1 | 62.7 | 62.6 | 61.5 | 59.8 | 57.6 | 50 |
| 00 | 3 53.8 | 8 61.2 | 13 62.7 | 18 62.5 | 23 61.5 | 28 59.7 | 33 57.5 | 00 |
| 10 | 54.3 | 61.3 | 62.7 | 62.5 | 61.4 | 59.7 | 57.4 | 10 |
| 20 | 54.8 | 61.4 | 62.7 | 62.5 | 61.4 | 59.6 | 57.4 | 20 |
| 30 | 55.2 | 61.5 | 62.8 | 62.5 | 61.3 | 59.6 | 57.3 | 30 |
| 40 | 55.6 | 61.6 | 62.8 | 62.4 | 61.3 | 59.5 | 57.2 | 40 |
| 50 | 56.0 | 61.6 | 62.8 | 62.4 | 61.2 | 59.4 | 57.1 | 50 |
| 00 | 4 56.4 | 9 61.7 | 14 62.8 | 19 62.4 | 24 61.2 | 29 59.3 | 34 57.0 | 00 |
| 10 | 56.7 | 61.8 | 62.8 | 62.3 | 61.1 | 59.3 | 56.9 | 10 |
| 20 | 57.1 | 61.9 | 62.8 | 62.3 | 61.1 | 59.2 | 56.9 | 20 |
| 30 | 57.4 | 61.9 | 62.8 | 62.3 | 61.0 | 59.1 | 56.8 | 30 |
| 40 | 57.7 | 62.0 | 62.8 | 62.2 | 60.9 | 59.1 | 56.7 | 40 |
| 50 | 57.9 | 62.1 | 62.8 | 62.2 | 60.9 | 59.0 | 56.6 | 50 |
| H.P. | L U | L U | L U | L U | L U | L U | L U | H.P. |
| 54.0 | 0.3 0.9 | 0.3 0.9 | 0.4 1.0 | 0.5 1.1 | 0.6 1.2 | 0.7 1.3 | 0.9 1.5 | 54.0 |
| 54.3 | 0.7 1.1 | 0.7 1.2 | 0.7 1.2 | 0.8 1.3 | 0.9 1.4 | 1.1 1.5 | 1.2 1.7 | 54.3 |
| 54.6 | 1.1 1.4 | 1.1 1.4 | 1.1 1.4 | 1.2 1.5 | 1.3 1.6 | 1.4 1.7 | 1.5 1.8 | 54.6 |
| 54.9 | 1.4 1.6 | 1.5 1.6 | 1.5 1.6 | 1.6 1.7 | 1.6 1.8 | 1.8 1.9 | 1.9 2.0 | 54.9 |
| 55.2 | 1.8 1.8 | 1.8 1.8 | 1.9 1.9 | 1.9 1.9 | 2.0 2.0 | 2.1 2.1 | 2.2 2.2 | 55.2 |
| 55.5 | 2.2 2.0 | 2.2 2.0 | 2.3 2.1 | 2.3 2.1 | 2.4 2.2 | 2.4 2.3 | 2.5 2.4 | 55.5 |
| 55.8 | 2.6 2.2 | 2.6 2.2 | 2.6 2.3 | 2.7 2.3 | 2.7 2.4 | 2.8 2.4 | 2.9 2.5 | 55.8 |
| 56.1 | 3.0 2.4 | 3.0 2.5 | 3.0 2.5 | 3.0 2.5 | 3.1 2.6 | 3.1 2.6 | 3.2 2.7 | 56.1 |
| 56.4 | 3.4 2.7 | 3.4 2.7 | 3.4 2.7 | 3.4 2.7 | 3.4 2.8 | 3.5 2.8 | 3.5 2.9 | 56.4 |
| 56.7 | 3.7 2.9 | 3.7 2.9 | 3.8 2.9 | 3.8 2.9 | 3.8 3.0 | 3.8 3.0 | 3.9 3.0 | 56.7 |
| 57.0 | 4.1 3.1 | 4.1 3.1 | 4.1 3.1 | 4.1 3.1 | 4.2 3.1 | 4.2 3.2 | 4.2 3.2 | 57.0 |
| 57.3 | 4.5 3.3 | 4.5 3.3 | 4.5 3.3 | 4.5 3.3 | 4.5 3.3 | 4.5 3.4 | 4.6 3.4 | 57.3 |
| 57.6 | 4.9 3.5 | 4.9 3.5 | 4.9 3.5 | 4.9 3.5 | 4.9 3.5 | 4.9 3.5 | 4.9 3.6 | 57.6 |
| 57.9 | 5.3 3.8 | 5.3 3.8 | 5.3 3.8 | 5.2 3.8 | 5.2 3.7 | 5.2 3.7 | 5.2 3.7 | 57.9 |
| 58.2 | 5.6 4.0 | 5.6 4.0 | 5.6 4.0 | 5.6 4.0 | 5.6 3.9 | 5.6 3.9 | 5.6 3.9 | 58.2 |
| 58.5 | 6.0 4.2 | 6.0 4.2 | 6.0 4.2 | 6.0 4.2 | 6.0 4.1 | 5.9 4.1 | 5.9 4.1 | 58.5 |
| 58.8 | 6.4 4.4 | 6.4 4.4 | 6.4 4.4 | 6.3 4.4 | 6.3 4.3 | 6.3 4.3 | 6.2 4.2 | 58.8 |
| 59.1 | 6.8 4.6 | 6.8 4.6 | 6.7 4.6 | 6.7 4.6 | 6.7 4.5 | 6.6 4.5 | 6.6 4.4 | 59.1 |
| 59.4 | 7.2 4.8 | 7.1 4.8 | 7.1 4.8 | 7.1 4.8 | 7.0 4.7 | 7.0 4.7 | 6.9 4.6 | 59.4 |
| 59.7 | 7.5 5.1 | 7.5 5.0 | 7.5 5.0 | 7.5 5.0 | 7.4 4.9 | 7.3 4.8 | 7.2 4.7 | 59.7 |
| 60.0 | 7.9 5.3 | 7.9 5.3 | 7.9 5.2 | 7.8 5.2 | 7.8 5.1 | 7.7 5.0 | 7.6 4.9 | 60.0 |
| 60.3 | 8.3 5.5 | 8.3 5.5 | 8.2 5.4 | 8.2 5.4 | 8.1 5.3 | 8.0 5.2 | 7.9 5.1 | 60.3 |
| 60.6 | 8.7 5.7 | 8.7 5.7 | 8.6 5.7 | 8.6 5.6 | 8.5 5.5 | 8.4 5.4 | 8.2 5.3 | 60.6 |
| 60.9 | 9.1 5.9 | 9.0 5.9 | 9.0 5.9 | 8.9 5.8 | 8.8 5.7 | 8.7 5.6 | 8.6 5.4 | 60.9 |
| 61.2 | 9.5 6.2 | 9.4 6.1 | 9.4 6.1 | 9.3 6.0 | 9.2 5.9 | 9.1 5.8 | 8.9 5.6 | 61.2 |
| 61.5 | 9.8 6.4 | 9.8 6.3 | 9.7 6.3 | 9.7 6.2 | 9.5 6.1 | 9.4 5.9 | 9.2 5.8 | 61.5 |

### DIP

| Ht. of Eye | Corrⁿ | Ht. of Eye | Ht. of Eye | Corrⁿ | Ht. of Eye |
|---|---|---|---|---|---|
| m | | ft. | m | | ft. |
| 2.4 | −2.8 | 8.0 | 9.5 | −5.5 | 31.5 |
| 2.6 | −2.9 | 8.6 | 9.9 | −5.6 | 32.7 |
| 2.8 | −3.0 | 9.2 | 10.3 | −5.7 | 33.9 |
| 3.0 | −3.1 | 9.8 | 10.6 | −5.8 | 35.1 |
| 3.2 | −3.2 | 10.5 | 11.0 | −5.9 | 36.3 |
| 3.4 | −3.3 | 11.2 | 11.4 | −6.0 | 37.6 |
| 3.6 | −3.4 | 11.9 | 11.8 | −6.1 | 38.9 |
| 3.8 | −3.5 | 12.6 | 12.2 | −6.2 | 40.1 |
| 4.0 | −3.6 | 13.3 | 12.6 | −6.3 | 41.5 |
| 4.3 | −3.7 | 14.1 | 13.0 | −6.4 | 42.8 |
| 4.5 | −3.8 | 14.9 | 13.4 | −6.5 | 44.2 |
| 4.7 | −3.9 | 15.7 | 13.8 | −6.6 | 45.5 |
| 5.0 | −4.0 | 16.5 | 14.2 | −6.7 | 46.9 |
| 5.2 | −4.1 | 17.4 | 14.7 | −6.8 | 48.4 |
| 5.5 | −4.2 | 18.3 | 15.1 | −6.9 | 49.8 |
| 5.8 | −4.3 | 19.1 | 15.5 | −7.0 | 51.3 |
| 6.1 | −4.4 | 20.1 | 16.0 | −7.1 | 52.8 |
| 6.3 | −4.5 | 21.0 | 16.5 | −7.2 | 54.3 |
| 6.6 | −4.6 | 22.0 | 16.9 | −7.3 | 55.8 |
| 6.9 | −4.7 | 22.9 | 17.4 | −7.4 | 57.4 |
| 7.2 | −4.8 | 23.9 | 17.9 | −7.5 | 58.9 |
| 7.5 | −4.9 | 24.9 | 18.4 | −7.6 | 60.5 |
| 7.9 | −5.0 | 26.0 | 18.8 | −7.7 | 62.1 |
| 8.2 | −5.1 | 27.1 | 19.3 | −7.8 | 63.8 |
| 8.5 | −5.2 | 28.1 | 19.8 | −7.9 | 65.4 |
| 8.8 | −5.3 | 29.2 | 20.4 | −8.0 | 67.1 |
| 9.2 | −5.4 | 30.4 | 20.9 | −8.1 | 68.8 |
| 9.5 | | 31.5 | 21.4 | | 70.5 |

### MOON CORRECTION TABLE

The correction is in two parts; the first correction is taken from the upper part of the table with argument apparent altitude, and the second from the lower part, with argument H.P., in the same column as that from which the first correction was taken. Separate corrections are given in the lower part for lower (L) and upper (U) limbs. All corrections are to be **added** to apparent altitude, *but 30′ is to be subtracted from the altitude of the upper limb.*

For corrections for pressure and temperature see page A4.

For bubble sextant observations ignore dip, take the mean of upper and lower limb corrections and subtract 15′ from the altitude.

App. Alt. = Apparent altitude = Sextant altitude corrected for index error and dip.

## ALTITUDE CORRECTION TABLES 35°–90°—MOON

| App. Alt. | 35°–39° Corrn | 40°–44° Corrn | 45°–49° Corrn | 50°–54° Corrn | 55°–59° Corrn | 60°–64° Corrn | 65°–69° Corrn | 70°–74° Corrn | 75°–79° Corrn | 80°–84° Corrn | 85°–89° Corrn | App. Alt. |
|---|---|---|---|---|---|---|---|---|---|---|---|---|
| 00 | 35 56.5 | 40 53.7 | 45 50.5 | 50 46.9 | 55 43.1 | 60 38.9 | 65 34.6 | 70 30.1 | 75 25.3 | 80 20.5 | 85 15.6 | 00 |
| 10 | 56.4 | 53.6 | 50.4 | 46.8 | 42.9 | 38.8 | 34.4 | 29.9 | 25.2 | 20.4 | 15.5 | 10 |
| 20 | 56.3 | 53.5 | 50.2 | 46.7 | 42.8 | 38.7 | 34.3 | 29.7 | 25.0 | 20.2 | 15.3 | 20 |
| 30 | 56.2 | 53.4 | 50.1 | 46.5 | 42.7 | 38.5 | 34.1 | 29.6 | 24.9 | 20.0 | 15.1 | 30 |
| 40 | 56.2 | 53.3 | 50.0 | 46.4 | 42.5 | 38.4 | 34.0 | 29.4 | 24.7 | 19.9 | 15.0 | 40 |
| 50 | 56.1 | 53.2 | 49.9 | 46.3 | 42.4 | 38.2 | 33.8 | 29.3 | 24.5 | 19.7 | 14.8 | 50 |
| 00 | 36 56.0 | 41 53.1 | 46 49.8 | 51 46.2 | 56 42.3 | 61 38.1 | 66 33.7 | 71 29.1 | 76 24.4 | 81 19.6 | 86 14.6 | 00 |
| 10 | 55.9 | 53.0 | 49.7 | 46.0 | 42.1 | 37.9 | 33.5 | 29.0 | 24.2 | 19.4 | 14.5 | 10 |
| 20 | 55.8 | 52.8 | 49.5 | 45.9 | 42.0 | 37.8 | 33.4 | 28.8 | 24.1 | 19.2 | 14.3 | 20 |
| 30 | 55.7 | 52.7 | 49.4 | 45.8 | 41.8 | 37.7 | 33.2 | 28.7 | 23.9 | 19.1 | 14.1 | 30 |
| 40 | 55.6 | 52.6 | 49.3 | 45.7 | 41.7 | 37.5 | 33.1 | 28.5 | 23.8 | 18.9 | 14.0 | 40 |
| 50 | 55.5 | 52.5 | 49.2 | 45.5 | 41.6 | 37.4 | 32.9 | 28.3 | 23.6 | 18.7 | 13.8 | 50 |
| 00 | 37 55.4 | 42 52.4 | 47 49.1 | 52 45.4 | 57 41.4 | 62 37.2 | 67 32.8 | 72 28.2 | 77 23.4 | 82 18.6 | 87 13.7 | 00 |
| 10 | 55.3 | 52.3 | 49.0 | 45.3 | 41.3 | 37.1 | 32.6 | 28.0 | 23.3 | 18.4 | 13.5 | 10 |
| 20 | 55.2 | 52.2 | 48.8 | 45.2 | 41.2 | 36.9 | 32.5 | 27.9 | 23.1 | 18.2 | 13.3 | 20 |
| 30 | 55.1 | 52.1 | 48.7 | 45.0 | 41.0 | 36.8 | 32.3 | 27.7 | 22.9 | 18.1 | 13.2 | 30 |
| 40 | 55.0 | 52.0 | 48.6 | 44.9 | 40.9 | 36.6 | 32.2 | 27.6 | 22.8 | 17.9 | 13.0 | 40 |
| 50 | 55.0 | 51.9 | 48.5 | 44.8 | 40.8 | 36.5 | 32.0 | 27.4 | 22.6 | 17.8 | 12.8 | 50 |
| 00 | 38 54.9 | 43 51.8 | 48 48.4 | 53 44.6 | 58 40.6 | 63 36.4 | 68 31.9 | 73 27.2 | 78 22.5 | 83 17.6 | 88 12.7 | 00 |
| 10 | 54.8 | 51.7 | 48.2 | 44.5 | 40.5 | 36.2 | 31.7 | 27.1 | 22.3 | 17.4 | 12.5 | 10 |
| 20 | 54.7 | 51.6 | 48.1 | 44.4 | 40.3 | 36.1 | 31.6 | 26.9 | 22.1 | 17.3 | 12.3 | 20 |
| 30 | 54.6 | 51.5 | 48.0 | 44.2 | 40.2 | 35.9 | 31.4 | 26.8 | 22.0 | 17.1 | 12.2 | 30 |
| 40 | 54.5 | 51.4 | 47.9 | 44.1 | 40.1 | 35.8 | 31.3 | 26.6 | 21.8 | 16.9 | 12.0 | 40 |
| 50 | 54.4 | 51.2 | 47.8 | 44.0 | 39.9 | 35.6 | 31.1 | 26.5 | 21.7 | 16.8 | 11.8 | 50 |
| 00 | 39 54.3 | 44 51.1 | 49 47.6 | 54 43.9 | 59 39.8 | 64 35.5 | 69 31.0 | 74 26.3 | 79 21.5 | 84 16.6 | 89 11.7 | 00 |
| 10 | 54.2 | 51.0 | 47.5 | 43.7 | 39.6 | 35.3 | 30.8 | 26.1 | 21.3 | 16.5 | 11.5 | 10 |
| 20 | 54.1 | 50.9 | 47.4 | 43.6 | 39.5 | 35.2 | 30.7 | 26.0 | 21.2 | 16.3 | 11.4 | 20 |
| 30 | 54.0 | 50.8 | 47.3 | 43.5 | 39.4 | 35.0 | 30.5 | 25.8 | 21.0 | 16.1 | 11.2 | 30 |
| 40 | 53.9 | 50.7 | 47.2 | 43.3 | 39.2 | 34.9 | 30.4 | 25.7 | 20.9 | 16.0 | 11.0 | 40 |
| 50 | 53.8 | 50.6 | 47.0 | 43.2 | 39.1 | 34.7 | 30.2 | 25.5 | 20.7 | 15.8 | 10.9 | 50 |

| H.P. | L U | L U | L U | L U | L U | L U | L U | L U | L U | L U | L U | H.P. |
|---|---|---|---|---|---|---|---|---|---|---|---|---|
| 54.0 | 1.1 1.7 | 1.3 1.9 | 1.5 2.1 | 1.7 2.4 | 2.0 2.6 | 2.3 2.9 | 2.6 3.2 | 2.9 3.5 | 3.2 3.8 | 3.5 4.1 | 3.8 4.5 | 54.0 |
| 54.3 | 1.4 1.8 | 1.6 2.0 | 1.8 2.2 | 2.0 2.5 | 2.3 2.7 | 2.5 3.0 | 2.8 3.2 | 3.0 3.5 | 3.3 3.8 | 3.6 4.1 | 3.9 4.4 | 54.3 |
| 54.6 | 1.7 2.0 | 1.9 2.2 | 2.1 2.4 | 2.3 2.6 | 2.5 2.8 | 2.7 3.0 | 3.0 3.3 | 3.2 3.5 | 3.5 3.8 | 3.7 4.1 | 4.0 4.3 | 54.6 |
| 54.9 | 2.0 2.2 | 2.2 2.3 | 2.3 2.5 | 2.5 2.7 | 2.7 2.9 | 2.9 3.1 | 3.2 3.3 | 3.4 3.5 | 3.6 3.8 | 3.9 4.0 | 4.1 4.3 | 54.9 |
| 55.2 | 2.3 2.3 | 2.5 2.4 | 2.6 2.6 | 2.8 2.8 | 3.0 2.9 | 3.2 3.1 | 3.4 3.3 | 3.6 3.5 | 3.8 3.7 | 4.0 4.0 | 4.2 4.2 | 55.2 |
| 55.5 | 2.7 2.5 | 2.8 2.6 | 2.9 2.7 | 3.1 2.9 | 3.2 3.0 | 3.4 3.2 | 3.6 3.4 | 3.7 3.5 | 3.9 3.7 | 4.1 3.9 | 4.3 4.1 | 55.5 |
| 55.8 | 3.0 2.6 | 3.1 2.7 | 3.2 2.8 | 3.3 3.0 | 3.5 3.1 | 3.6 3.3 | 3.8 3.4 | 3.9 3.6 | 4.1 3.7 | 4.2 3.9 | 4.4 4.0 | 55.8 |
| 56.1 | 3.3 2.8 | 3.4 2.9 | 3.5 3.0 | 3.6 3.1 | 3.7 3.2 | 3.8 3.3 | 4.0 3.4 | 4.1 3.6 | 4.2 3.7 | 4.4 3.8 | 4.5 4.0 | 56.1 |
| 56.4 | 3.6 2.9 | 3.7 3.0 | 3.8 3.1 | 3.9 3.2 | 3.9 3.3 | 4.0 3.4 | 4.1 3.5 | 4.3 3.6 | 4.4 3.7 | 4.5 3.8 | 4.6 3.9 | 56.4 |
| 56.7 | 3.9 3.1 | 4.0 3.1 | 4.1 3.2 | 4.1 3.3 | 4.2 3.3 | 4.3 3.4 | 4.4 3.5 | 4.4 3.6 | 4.5 3.7 | 4.6 3.8 | 4.7 3.8 | 56.7 |
| 57.0 | 4.3 3.2 | 4.3 3.3 | 4.3 3.3 | 4.4 3.4 | 4.4 3.4 | 4.5 3.5 | 4.5 3.5 | 4.6 3.6 | 4.7 3.6 | 4.7 3.7 | 4.8 3.8 | 57.0 |
| 57.3 | 4.6 3.4 | 4.6 3.4 | 4.6 3.4 | 4.6 3.5 | 4.7 3.5 | 4.7 3.5 | 4.7 3.5 | 4.8 3.6 | 4.8 3.6 | 4.8 3.7 | 4.9 3.7 | 57.3 |
| 57.6 | 4.9 3.6 | 4.9 3.6 | 4.9 3.6 | 4.9 3.6 | 4.9 3.6 | 4.9 3.6 | 4.9 3.6 | 4.9 3.6 | 5.0 3.6 | 5.0 3.6 | 5.1 3.6 | 57.6 |
| 57.9 | 5.2 3.7 | 5.2 3.7 | 5.2 3.7 | 5.2 3.7 | 5.2 3.7 | 5.1 3.6 | 5.1 3.6 | 5.1 3.6 | 5.1 3.6 | 5.1 3.6 | 5.1 3.6 | 57.9 |
| 58.2 | 5.5 3.9 | 5.5 3.8 | 5.5 3.8 | 5.4 3.8 | 5.4 3.7 | 5.4 3.7 | 5.3 3.7 | 5.3 3.6 | 5.2 3.6 | 5.2 3.5 | 5.2 3.5 | 58.2 |
| 58.5 | 5.9 4.0 | 5.8 4.0 | 5.8 3.9 | 5.7 3.9 | 5.6 3.8 | 5.6 3.8 | 5.5 3.7 | 5.5 3.6 | 5.4 3.6 | 5.3 3.5 | 5.3 3.4 | 58.5 |
| 58.8 | 6.2 4.2 | 6.1 4.1 | 6.0 4.1 | 6.0 4.0 | 5.9 3.9 | 5.8 3.9 | 5.7 3.8 | 5.6 3.7 | 5.5 3.5 | 5.4 3.5 | 5.3 3.4 | 58.8 |
| 59.1 | 6.5 4.3 | 6.4 4.3 | 6.3 4.2 | 6.2 4.1 | 6.1 4.0 | 6.0 3.9 | 5.9 3.8 | 5.8 3.7 | 5.7 3.5 | 5.6 3.4 | 5.4 3.3 | 59.1 |
| 59.4 | 6.8 4.5 | 6.7 4.4 | 6.6 4.3 | 6.5 4.2 | 6.4 4.1 | 6.2 3.9 | 6.1 3.8 | 6.0 3.7 | 5.8 3.5 | 5.7 3.4 | 5.5 3.2 | 59.4 |
| 59.7 | 7.1 4.6 | 7.0 4.5 | 6.9 4.4 | 6.8 4.3 | 6.6 4.1 | 6.5 4.0 | 6.3 3.8 | 6.2 3.7 | 6.0 3.5 | 5.8 3.3 | 5.6 3.2 | 59.7 |
| 60.0 | 7.5 4.8 | 7.3 4.7 | 7.2 4.5 | 7.0 4.4 | 6.9 4.2 | 6.7 4.0 | 6.5 3.9 | 6.3 3.7 | 6.1 3.5 | 5.9 3.3 | 5.7 3.1 | 60.0 |
| 60.3 | 7.8 5.0 | 7.6 4.8 | 7.5 4.7 | 7.3 4.5 | 7.1 4.3 | 6.9 4.1 | 6.7 3.9 | 6.5 3.7 | 6.3 3.5 | 6.0 3.2 | 5.8 3.0 | 60.3 |
| 60.6 | 8.1 5.1 | 7.9 5.0 | 7.7 4.8 | 7.6 4.6 | 7.3 4.4 | 7.1 4.2 | 6.9 3.9 | 6.7 3.7 | 6.4 3.4 | 6.2 3.2 | 5.9 2.9 | 60.6 |
| 60.9 | 8.4 5.3 | 8.2 5.1 | 8.0 4.9 | 7.8 4.7 | 7.6 4.5 | 7.3 4.2 | 7.1 4.0 | 6.8 3.7 | 6.6 3.4 | 6.3 3.2 | 6.0 2.9 | 60.9 |
| 61.2 | 8.7 5.4 | 8.5 5.2 | 8.3 5.0 | 8.1 4.8 | 7.8 4.5 | 7.6 4.3 | 7.3 4.0 | 7.0 3.7 | 6.7 3.4 | 6.4 3.1 | 6.1 2.8 | 61.2 |
| 61.5 | 9.1 5.6 | 8.8 5.4 | 8.6 5.1 | 8.3 4.9 | 8.1 4.6 | 7.8 4.3 | 7.5 4.0 | 7.2 3.7 | 6.9 3.4 | 6.5 3.1 | 6.2 2.7 | 61.5 |

# EXTRACTS FROM THE ADMIRALTY TIDE TABLES VOL. I 1980

Reproduced from British Admiralty Tide Tables with the sanction of the Controller, H.M. Stationery Office and the Hydrographer of the Navy.

# AVONMOUTH
## MEAN SPRING AND NEAP CURVES
For instructions see page xiv

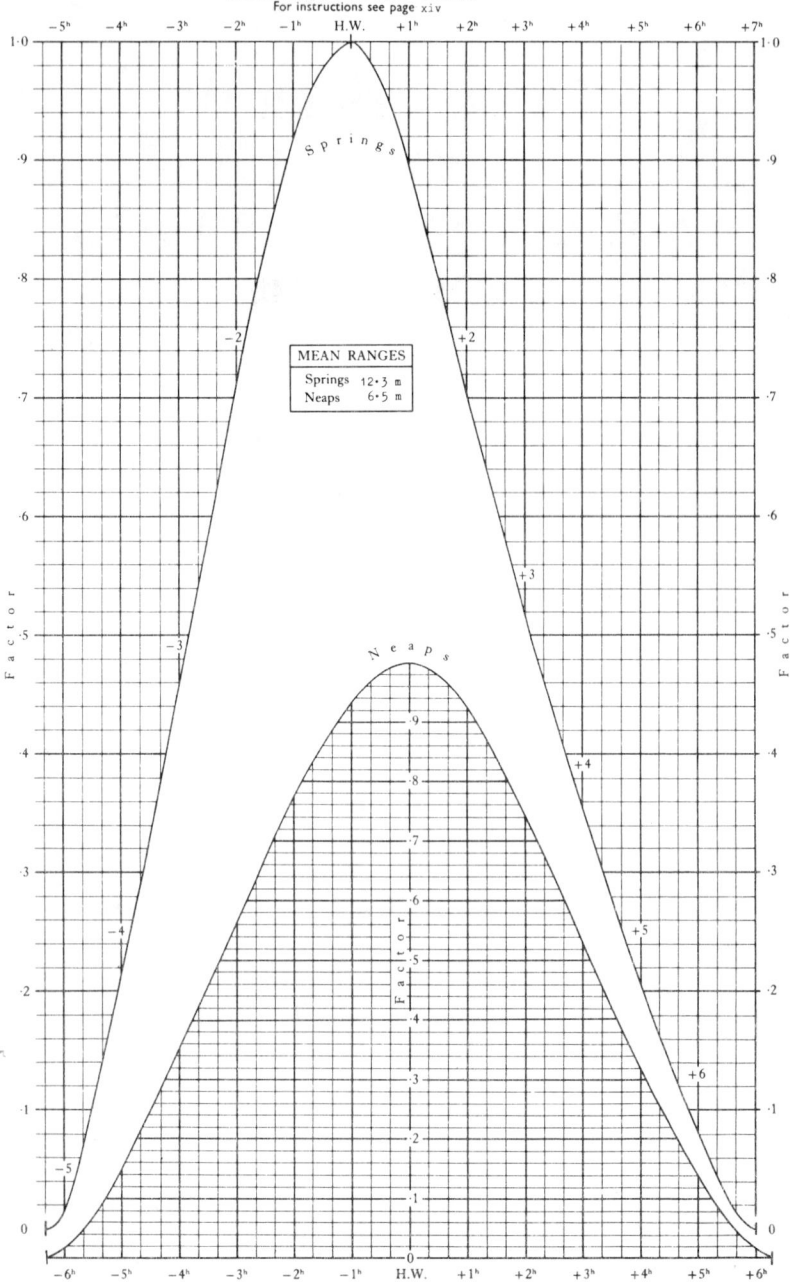

**MEAN RANGES**
Springs 12·3 m
Neaps 6·5 m

# ENGLAND, WEST COAST - PORT OF BRISTOL (AVONMOUTH)

LAT 51°30'N  LONG 2°43'W

TIME ZONE GMT         TIMES AND HEIGHTS OF HIGH AND LOW WATERS         YEAR 1980

|  | JANUARY | | | FEBRUARY | | | MARCH | | | APRIL | |
|---|---|---|---|---|---|---|---|---|---|---|---|
|  | TIME M | TIME M |  | TIME M | TIME M |  | TIME M | TIME M |  | TIME M | TIME M |
| **1** TU | 0038 1.3<br>0617 12.6<br>1307 1.3<br>1845 12.8 | **16** 0548 12.2<br>1230 1.6<br>W 1811 12.4 | **1** F | 0203 1.0<br>0726 12.9<br>1422 1.1<br>1949 12.9 | **16** 0154 0.8<br>0712 13.7<br>SA 1428 0.5<br>1934 13.8 | **1** SA | 0149 1.1<br>0708 12.7<br>1409 1.1<br>1930 12.7 | **16** 0141 0.7<br>0654 13.9<br>SU 1417 0.4<br>1920 13.9 | **1** TU | 0232 1.0<br>0748 12.8<br>1448 1.1<br>2009 12.9 | **16** 0304 0.1<br>0809 14.4<br>W 1529 0.0<br>2025 14.2 |
| **2** W | 0132 1.1<br>0703 13.0<br>1356 1.1<br>1929 13.0 | **17** 0102 1.3<br>0639 13.0<br>TH 1334 1.1<br>1901 13.1 | **2** SA | 0242 1.0<br>0801 13.0<br>1457 1.2<br>2024 12.9 | **17** 0251 0.4<br>0757 14.2<br>SU 1519 0.2<br>2020 14.1 | **2** SU | 0227 1.1<br>0742 12.9<br>1442 1.1<br>2003 12.9 | **17** 0239 0.3<br>0742 14.4<br>M 1510 0.1<br>2005 14.2 | **2** W | 0302 1.0<br>0821 12.9<br>1518 1.1<br>2037 13.0 | **17** 0343 0.1<br>0851 14.3<br>TH 1604 0.1<br>2102 14.0 |
| **3** TH | 0217 0.9<br>0743 13.2<br>1435 1.1<br>2008 13.0 | **18** 0201 0.9<br>0725 13.5<br>F 1430 0.7<br>1947 13.5 | **3** SU | 0314 1.1<br>0836 13.0<br>1529 1.3<br>2059 12.9 | **18** 0340 0.2<br>0841 14.4<br>M 1606 0.1<br>2104 14.1 | **3** M | 0258 1.0<br>0814 13.0<br>1513 1.1<br>2036 13.0 | **18** 0329 0.1<br>0827 14.6<br>TU 1554 0.0<br>2047 14.3 | **3** TH | 0332 1.0<br>0848 13.0<br>1544 1.2<br>2103 13.0 | **18** 0416 0.3<br>0929 13.8<br>F 1630 0.5<br>2138 13.3 |
| **4** F | 0254 1.0<br>0819 13.2<br>1509 1.2<br>2043 13.0 | **19** 0253 0.7<br>0810 13.9<br>SA 1520 0.6<br>2033 13.7 | **4** M | 0348 1.3<br>0908 13.0<br>1602 1.5<br>2131 12.8 | **19** 0423 0.2<br>0926 14.3<br>TU 1648 0.2<br>2146 13.8 | **4** TU | 0330 1.1<br>0845 13.1<br>1545 1.2<br>2105 13.0 | **19** 0409 0.0<br>0909 14.5<br>W 1632 0.0<br>2126 14.0 | **4** F | 0358 1.1<br>0914 12.9<br>1605 1.4<br>2126 12.7 | **19** 0440 0.7<br>1006 12.9<br>SA 1650 1.0<br>2214 12.4 |
| **5** SA | 0327 1.2<br>0853 13.0<br>1540 1.5<br>2118 12.7 | **20** 0340 0.6<br>0854 13.9<br>SU 1608 0.5<br>2118 13.7 | **5** TU | 0419 1.5<br>0937 12.7<br>1631 1.7<br>2157 12.5 | **20** 0459 0.4<br>1009 13.9<br>W 1721 0.5<br>2225 13.2 | **5** W | 0400 1.2<br>0914 13.0<br>1612 1.4<br>2130 12.8 | **20** 0443 0.2<br>0949 14.0<br>TH 1700 0.4<br>2202 13.4 | **5** SA | 0416 1.3<br>0937 12.5<br>1618 1.6<br>2150 12.2 | **20** 0500 1.3<br>1043 11.9<br>SU 1708 1.7<br>2255 11.5 |
| **6** SU | 0358 1.5<br>0928 12.7<br>1612 1.8<br>2152 12.4 | **21** 0424 0.7<br>0940 13.8<br>M 1653 0.6<br>2202 13.4 | **6** W | 0445 1.7<br>1005 12.4<br>1654 2.0<br>2221 12.0 | **21** 0526 0.8<br>1049 13.1<br>TH 1746 1.0<br>2304 12.3 | **6** TH | 0425 1.3<br>0937 12.8<br>1632 1.6<br>2151 12.5 | **21** 0506 0.6<br>1026 13.1<br>F 1719 1.0<br>2238 12.4 | **6** SU | 0429 1.7<br>1003 11.9<br>M 1629 2.1<br>2220 11.5 | **21** 0520 2.1<br>1127 10.9<br>TU 1735 2.4<br>2346 10.6 |
| **7** M | 0429 1.8<br>1001 12.3<br>1643 2.1<br>2222 11.9 | **22** 0504 0.9<br>1025 13.4<br>TU 1732 0.9<br>2247 12.9 | **7** TH | 0502 2.1<br>1029 11.8<br>1709 2.3<br>2244 11.4 | **22** 0548 1.4<br>1131 12.0<br>F 1808 1.8<br>2346 11.3 | **7** F | 0439 1.6<br>1000 12.3<br>1642 1.9<br>2212 11.8 | **22** 0522 1.3<br>1106 12.0<br>SA 1735 1.7<br>2319 11.4 | **7** M | 0443 2.2<br>1037 11.1<br>1650 2.6<br>2302 10.8 | **22** 0556 2.7<br>1223 10.1<br>TU 1823 3.0 |
| **8** TU | 0457 2.2<br>1033 11.8<br>1710 2.4<br>2253 11.4 | **23** 0538 1.2<br>1111 12.8<br>W 1806 1.3<br>2330 12.1 | **8** F | 0515 2.6<br>1056 11.1<br>1722 2.9<br>2315 10.6 | **23** 0611 2.2<br>1219 11.0<br>SA 1837 2.5 | **8** SA | 0447 2.1<br>1024 11.6<br>1651 2.4<br>2239 11.1 | **23** 0542 2.1<br>1151 10.9<br>SU 1801 2.5 | **8** TU | 0516 2.8<br>1128 10.3<br>1736 3.2<br>1940 3.2 | **23** 0050 10.0<br>0657 3.1<br>W 1337 9.8<br>1940 3.2 |
| **9** W | 0523 2.6<br>1105 11.2<br>1736 2.8<br>2325 10.7 | **24** 0608 1.8<br>1158 12.0<br>TH 1837 1.9 | **9** SA | 0531 3.1<br>1133 10.4<br>1746 3.4 | **24** 0040 10.4<br>0653 2.9<br>SU 1324 10.2<br>1932 3.1 | **9** SU | 0458 2.6<br>1056 10.8<br>1707 3.0<br>2321 10.3 | **24** 0010 10.4<br>0618 2.8<br>M 1251 10.0<br>1851 3.1 | **9** W | 0005 10.2<br>0620 3.2<br>TH 1245 10.0<br>1901 3.5 | **24** 0212 9.9<br>0831 3.1<br>F 1504 10.0<br>2118 3.0 |
| **10** TH | 0549 3.1<br>1143 10.6<br>1806 3.3 | **25** 0019 11.2<br>0644 2.4<br>F 1253 11.2<br>1921 2.5 | **10** SU | 0000 9.9<br>0607 3.7<br>1228 9.8<br>1837 3.9 | **25** 0155 9.9<br>0808 3.3<br>M 1450 10.0<br>2058 3.1 | **10** M | 0526 3.3<br>1147 10.0<br>1751 3.6 | **25** 0120 9.8<br>0727 3.3<br>TU 1416 9.7<br>2018 3.2 | **10** TH | 0137 10.1<br>0806 3.2<br>F 1426 10.2<br>2056 2.9 | **25** 0337 10.5<br>1007 2.6<br>SA 1618 10.7<br>2245 2.5 |
| **11** F | 0008 10.1<br>0624 3.6<br>1230 10.0<br>1849 3.7 | **26** 0119 10.6<br>0739 2.9<br>SA 1404 10.6<br>2022 2.8 | **11** M | 0110 9.5<br>0719 3.9<br>1349 9.7<br>2006 3.8 | **26** 0327 10.1<br>0947 3.0<br>TU 1617 10.5<br>2233 2.5 | **11** TU | 0026 9.7<br>0633 3.6<br>1307 9.7<br>1920 3.8 | **26** 0253 9.8<br>0912 3.1<br>W 1548 10.1<br>2201 2.8 | **11** F | 0317 10.8<br>0956 2.5<br>SA 1557 11.2<br>2240 2.3 | **26** 0441 11.1<br>1122 2.0<br>SA 1712 11.4<br>2352 1.8 |
| **12** SA | 0105 9.6<br>0718 3.9<br>1334 9.8<br>1953 3.9 | **27** 0236 10.3<br>0856 3.0<br>SU 1523 10.6<br>2141 2.7 | **12** TU | 0242 9.8<br>0859 3.5<br>1526 10.3<br>2148 3.2 | **27** 0447 10.9<br>1119 2.3<br>W 1725 11.3<br>2355 1.9 | **12** W | 0201 9.8<br>0823 3.5<br>1450 10.0<br>2117 3.3 | **27** 0421 10.5<br>1051 2.4<br>TH 1700 11.0<br>2330 2.1 | **12** SA | 0435 11.7<br>1131 1.6<br>SU 1708 12.2<br>2354 1.2 | **27** 0528 11.7<br>1219 1.6<br>SU 1754 12.0 |
| **13** SU | 0220 9.7<br>0832 3.7<br>1452 10.0<br>2115 3.5 | **28** 0355 10.6<br>1021 2.6<br>M 1638 11.1<br>2301 2.2 | **13** W | 0410 10.8<br>1042 2.6<br>1715 11.3<br>2325 2.2 | **28** 0541 11.7<br>1231 1.6<br>TH 1815 12.0 | **13** TH | 0336 10.6<br>1018 2.7<br>1620 11.1<br>2303 2.3 | **28** 0523 11.4<br>1207 1.8<br>F 1751 11.8 | **13** SU | 0536 13.0<br>1249 0.9<br>M 1806 13.1 | **28** 0043 1.5<br>0607 12.0<br>M 1303 1.4<br>1830 12.3 |
| **14** M | 0338 10.3<br>0955 3.1<br>1607 10.8<br>2236 2.8 | **29** 0505 11.4<br>1140 2.0<br>TU 1741 12.0 | **14** TH | 0522 12.0<br>1211 1.6<br>1751 12.3 | **29** 0101 1.3<br>0633 12.3<br>F 1327 1.2<br>1855 12.5 | **14** F | 0500 11.9<br>1154 1.6<br>1731 12.3 | **29** 0036 1.5<br>0607 12.0<br>SA 1304 1.3<br>1831 12.3 | **14** M | 0117 0.8<br>0632 13.5<br>TU 1353 0.4<br>1857 13.8 | **29** 0121 1.3<br>0642 12.3<br>TU 1339 1.3<br>1904 12.5 |
| **15** TU | 0448 11.3<br>1116 2.3<br>1714 11.6<br>2353 2.0 | **30** 0013 1.7<br>0602 12.1<br>W 1249 1.5<br>1831 12.5 | **15** F | 0046 1.4<br>0621 13.0<br>1326 1.0<br>1846 13.2 | | **15** SA | 0030 1.4<br>0601 13.0<br>1313 0.9<br>1830 13.2 | **30** 0125 1.2<br>0643 12.4<br>SU 1345 1.1<br>1904 12.6 | **15** TU | 0214 0.2<br>0722 14.2<br>W 1445 0.2<br>1944 14.1 | **30** 0155 1.2<br>0718 12.5<br>W 1412 1.2<br>1938 12.8 |
| | | **31** 0115 1.2<br>0648 12.6<br>TH 1341 1.2<br>1913 12.7 | | | | | | **31** 0201 1.1<br>0716 12.6<br>M 1416 1.1<br>1938 12.8 | | | |

HEIGHTS IN METRES

ENGLAND, WEST COAST

| No. | PLACE | Lat. N. | Long. W. | TIME DIFFERENCES High Water (Zone G.M.T.) | | Low Water | | HEIGHT DIFFERENCES (IN METRES) | | | | M.L. Z₀ m. | M g° |
|---|---|---|---|---|---|---|---|---|---|---|---|---|---|
| | | | | | | | | MHWS | MHWN | MLWN | MLWS | | |
| 523 | PORT OF BRISTOL (AVONMOUTH) | . (see page 122) | | 0000 and 1200 | 0600 and 1800 | 0000 and 1200 | 0700 and 1900 | 13·2 | 10·0 | 3·5 | 0·9 | | |
| | *River Severn* | | | | | | | | | | | | |
| 517 | Sudbrook | 51 35 | 2 43 | +0010 | +0010 | +0025 | +0015 | +0·2 | +0·1 | −0·1 | +0·1 | ⊗ | ⊗ |
| 518 | Beachley (Aust) | 51 36 | 2 38 | +0010 | +0015 | +0040 | +0025 | −0·2 | −0·2 | −0·5 | −0·3 | 6·64 | 211 |
| 519 | Inward Rocks | 51 39 | 2 37 | +0020 | +0020 | +0105 | +0045 | −1·0 | −1·1 | −1·4 | −0·6 | ⊗ | ⊗ |
| 520 | Narlwood Rocks | 51 39 | 2 36 | +0025 | +0025 | +0120 | +0100 | −1·9 | −2·0 | −2·3 | −0·8 | ⊗ | ⊗ |
| 521 | White House | 51 40 | 2 33 | +0025 | +0025 | +0145 | +0120 | −3·0 | −3·1 | −3·6 | −1·0 | ⊗ | ⊗ |
| 522 | Berkeley | 51 42 | 2 30 | +0030 | +0045 | +0245 | +0220 | −3·8 | −3·9 | −3·4 | −0·5 | ⊗ | ⊗ |
| 522a | Sharpness Dock | 51 43 | 2 29 | +0035 | +0050 | +0305 | +0245 | −3·9 | −4·2 | −3·3 | −0·4 | ⊗ | ⊗ |
| 522b | Wellhouse Rock | 51 44 | 2 29 | +0040 | +0055 | +0320 | +0305 | −4·1 | −4·4 | −3·1 | −0·2 | ⊗ | ⊗ |
| 523 | PORT OF BRISTOL (AVONMOUTH) | . (see page 122) | | 0200 and 1400 | 0800 and 2000 | 0300 and 1500 | 0800 and 2000 | 13·2 | 10·0 | 3·5 | 0·9 | 6·96 | 202 |
| | *River Avon* | | | | | | | | | | | | |
| 523a | Shirehampton | 51 29 | 2 41 | 0000 | 0000 | +0035 | +0010 | −0·7 | −0·7 | −0·8 | 0·0 | ⊗ | ⊗ |
| 523b | Sea Mills | 51 29 | 2 39 | +0005 | +0005 | +0105 | +0030 | −1·4 | −1·5 | −1·7 | −0·1 | ⊗ | ⊗ |
| 524 | Bristol (Cumberland Basin) | 51 27 | 2 37 | +0010 | +0010 | § | § | −2·9 | −3·0 | § | § | ⊗ | ⊗ |
| 524a | Portishead | 51 29 | 2 45 | −0002 | 0000 | ⊗ | ⊗ | −0·1 | −0·1 | ⊗ | ⊗ | ⊗ | ⊗ |
| 525 | Clevedon | 51 27 | 2 52 | −0010 | −0020 | −0025 | −0015 | −0·4 | −0·2 | +0·2 | 0·0 | ⊗ | ⊗ |
| 527 | Weston-super-Mare | 51 21 | 2 59 | −0020 | −0030 | −0130 | −0030 | −1·2 | −1·0 | −0·8 | −0·2 | 6·14 | 195 |
| | *River Parrett* | | | | | | | | | | | | |
| 528 | Burnham | 51 14 | 3 00 | −0020 | −0025 | −0030 | 0000 | −2·3 | −1·9 | −1·4 | −1·1 | ⊗ | ⊗ |
| 529 | Bridgwater | 51 08 | 3 00 | −0015 | −0030 | +0305 | +0455 | −8·6 | −8·1 | § | § | ⊗ | ⊗ |
| 531 | Watchet | 51 11 | 3 20 | −0035 | −0050 | −0145 | −0040 | −1·9 | −1·5 | +0·1 | +0·1 | 5·88 | 179 |
| 532 | Minehead | 51 13 | 3 28 | −0035 | −0045 | ⊗ | ⊗ | −2·6 | −1·9 | −0·1 | −0·1 | ⊗ | ⊗ |
| 533 | Porlock Bay | 51 13 | 3 38 | −0045 | −0055 | −0205 | −0050 | −3·0 | −2·2 | −0·1 | −0·1 | 5·62 | 189 |
| 534 | Lynmouth | 51 14 | 3 49 | −0055 | −0115 | ⊗ | ⊗ | −3·6 | −2·7 | ⊗ | ⊗ | ⊗ | ⊗ |

# INDEX

## A

A.B.C. tables   112, 183
Admiralty list of lights   45
Almanac   105
Altitude, correction of   157
   correction tables   104, 166
Amplitude problem   112, 129
   tables   130
Angle on the bow, doubling the   82
Aries, first point of   102
Augmentation   163
Azimuth   112
   conversion to bearing   112
   problem   112

## B

Bearing, 3 figure notation   8
   calculation of   112
   compass   11
   magnetic   11
   position from   33
   relative   16
   transit   41

## C

'C' correction   199
Celestial equator   101
   meridian   102
   poles   101, 211
   sphere   101
Chart datum   87, 88
Co-lat   182
Compass   8
   bearing   11
   errors   8, 10, 43
     calculation of   37, 41, 122
   gyro   8
   north   9, 10
Composite great circle   222

Convergency   33, 34
Correction of altitudes   157
Course   8, 10
Current   84

## D

Danger angles   47
'd' correction   106
Declination   102
Departure   50
Deviation   9, 12
Difference of latitude   2
Difference of longitude   2, 50
Difference of meridional parts   60
Dip, formula for   166
   table   166
Direction, abeam   24
   measurement of   8
Distance, of sea horizon   45
   measurement of   6
Doubling the angle on the bow   82
Drift   21, 27, 70
Drying height   87, 97

## E

Earth, the shape of   6
Ecliptic   102
   obliquity of   102
Equator   1
   celestial   101
Ex-meridian problem   202
   tables   205

## F

First point of Aries   102

## INDEX

### G
Geographical mile 7
 poles 1, 8, 9
 position 101
 range 45
Great circle 33
 composite 222
 sailing 215
Greenwich hour angle 104
 rate of change of 108
Gyro compass 8

### H
Height, drying 97
 of tide 87
Highest astronomical tide 88
Horizon, distance of 45
 rational 130, 157
 sensible 157
 visible 157
Horizontal parallax 163
Horizontal sextant angle 37
Hour angle 110, 182
Hour circles 102
Hyperbolic position line 33, 49

### I
Increment tables 104
Intercept 142, 183
 terminal point 143
International nautical mile 6

### K
Knot 7

### L
Latitude 2
 by ex-meridian 202
 by meridian altitude 148
 by pole star 211
 difference of 2
 mean 55
 middle 59
 parallel of 2
Leading marks 43
Leeway 29

Lights, height of 36, 98
 range of 45
Local hour angle 110, 182
Longitude 2
 by chronometer 146, 192
 correction 113
 difference of 2
Lower meridian passage 118, 180
Lowest astronomical tide 88
Luminous range 46
 diagram 46, 47
Lunar tide 84

### M
Magnetic compass 9
 meridian 9
 north 9, 10
 variation 9
Marcq St. Hilaire method 142, 183, 205
Mean high water neaps 87
Mean high water springs 87, 98
Mean latitude 59
 sailing 60
Mean low water neaps 87
Mean low water springs 87
Mercator sailing 60
Meridian 1
 celestial 102
 prime 1
Meridian altitude, latitude by 148
Meridian passage 111
 lower 118
 times of 111
Meridional parts 60
Middle latitude 59
 sailing 60
Moon 103
 correction of altitude 163
 GHA and declination of 107
 rising and setting 119
 SHA of 103
 times of meridian passage of 113
Moonrise and moonset 119

### N
Nautical almanac 104
Nautical mile 6, 141
Neap tides 84

# INDEX

Nominal range 45
Noon position 198

## O

Obliquity of the ecliptic 102

## P

Parallax 159
Parallel of latitude 2
Parallel sailing 50
  formula for 51, 65
Pelorus 16
Plane sailing 55, 65
Planet, correction of altitude of 162
  GHA and declination of 107
  meridian passage of 115
Polar distance 182
Poles, celestial 211
  geographical 1, 8, 9
Pole star problem 211
  tables 104, 212
Position circle 36, 37, 43, 80, 141
  hyperbola 49
  line 33, 141
  measurement of 1
  transferred 65, 72, 76, 80
Prime meridian 1
PZX triangle 182

## R

Radio direction finding 16, 33
Ranges of lights 45
Rational horizon 129, 157
Refraction 45, 158
Relative bearing 16
Rhumb line bearing 34
Rising and dipping distance 43
Rising and setting 129
Running fix 76, 78, 82

## S

Sailings, the 50
  great circle 215
Sea horizon, distance of 45
Sea mile 6
Secondary port 89, 92
Selected stars 104
Semi diameter 159
Sensible horizon 157

Set and drift 70
Sextant angles 34, 47
Sidereal hour angle 102
Small circle 1, 33
Solar tide 84
Speed 7
  made good 25
  through the water 25
Spring tides 84
Standard port 89
Star, correction of altitude 162
  GHA and declination of 108
  meridian passage of 116
Streams, tidal 84
  tidal atlases 84, 85
Sun, change of declination of 102
  change of SHA of 103
  correction of altitude of 160
  GHA and declination of 106
  meridian passage of 112
Sunrise and sunset, times of 119

## T

Three figure notation 8
Tidal calculations 89
  information 85
  streams 84
Tide 20, 84
  counteracting the 21
  height of 87
  rate of 21, 27
Tide tables 85
Time of meridian passage 113
Time zones 89
Total correction tables 116
Transferred position line 65, 72, 76, 80
Transit bearing 41
Transit, lower 118
Traverse table 65, 80
True bearing, calculation of 112

## V

Variation 9, 12
'v' correction 105
Vertex 216, 222
Vertical sextant angle 34, 47
  tables 37
Visible horizon 157

## Z

Zenith distance 141, 182